RENEWALS: 691-4574
DATE DUE

Ecological Studies, Vol. 75

Analysis and Synthesis

Edited by

W. D. Billings, Durham, USA
F. Golley, Athens, USA
O. L. Lange, Würzburg, FRG
J. S. Olson, Oak Ridge, USA
H. Remmert, Marburg, FRG

Ecological Studies

Volume 59
Ecology of Biological Invasions of North America and Hawaii (1986)
Edited by H.A. Mooney and J.A. Drake

Volume 60
Amazonian Rain Forests Ecosystem Disturbance and Recovery (1987)
Edited by C.F. Jordan

Volume 61
Potentials and Limitations of Ecosystem Analysis (1987)
Edited by E.-D. Schulze and H. Zwölfer

Volume 62
Frost Survival of Plants (1987)
By A. Sakai and W. Larcher

Volume 63
Long-Term Forest Dynamics of the Temperate Zone (1987)
By P.A. Delcourt and H.R. Delcourt

Volume 64
Landscape Heterogeneity and Disturbance (1987)
Edited by M. Goigel Turner

Volume 65
Community Ecology of Sea Otters (1987)
Edited by G.R. van Blaricom and J.A. Estes

Volume 66
Forest Hydrology and Ecology at Coweeta (1987)
Edited by W.T. Swank and D.A. Crossley, Jr.

Volume 67
Concepts of Ecosystem Ecology A Comparative View (1988)
Edited by L.R. Pomeroy and J.J. Alberts

Volume 68
Stable Isotopes in Ecological Research (1989)
Edited by P.W. Rundel, J.R. Ehleringer, and K.A. Nagy

Volume 69
Vertebrates in Complex Tropical Systems (1989)
Edited by M.L. Harmelin-Vivien and F. Bourlière

Volume 70
The Northern Forest Border in Canada and Alaska (1989)
By J.A. Larsen

Volume 71
Tidal Flat Estuaries: Simulation and Analysis of the Ems Estuary (1988)
Edited by J. Baretta and P. Ruardij

Volume 72
Acidic Deposition and Forest Soils (1989)
By D. Binkley, C.T. Driscoll, H.L. Allen, P. Schoeneberger, and D. McAvoy

Volume 73
Toxic Organic Chemicals in Porous Media (1989)
Edited by Z. Gerstl, Y. Chen, U. Mingelgrin, and B. Yaron

Volume 74
Inorganic Contaminants in the Vadose Zone (1989)
Edited by B. Bar-Yosef, N.J. Barrow, and J. Goldshmid

Volume 75
The Grazing Land Ecosystems of the African Sahel (1989)
By H.N. Le Houérou

Henry Noël Le Houérou

The Grazing Land Ecosystems of the African Sahel

With 114 Figures

Springer-Verlag Berlin Heidelberg New York
London Paris Tokyo Hong Kong

Dr. HENRY NOËL LE HOUÉROU
Villa 67
327, rue A. L. de Jussieu
34090 Montpellier, France

ISBN 3-540-50791-4 Springer-Verlag Berlin Heidelberg New York
ISBN 0-387-50791-4 Springer-Verlag New York Berlin Heidelberg

Library of Congress Cataloging-in-Publication Data. Le Houérou, H. N. (Henry Noël) The grazing land ecosystems of the African Sahel / Henry Noël Le Houérou. p. cm. – (Ecological studies: vol. 75) Includes index. 1. Rangelands – Sahel. 2. Range ecology – Sahel. 3. Grazing – Sahel. I. Title. II. Series: Ecological studies; v. 75. SF85.4.S15L4 1989 636.08'4'096 – dc 20 89-11481

This work is subject to copyright. All rights are reserved, whether the whole or part of the material is concerned, specifically the rights of translation, reprinting, reuse of illustrations, recitation, broadcasting, reproduction on microfilms or in other ways, and storage in data banks. Duplication of this publication or parts thereof is only permitted under the provisions of the German Copyright Law of September 9, 1965, in its version of June 24, 1985, and a copyright fee must always be paid. Violations fall under the prosecution act of the German Copyright Law.

© Springer-Verlag Berlin Heidelberg 1989
Printed in the United States of America

The use of general descriptive names, registered names, trademarks, etc. in this publication does not imply, even in the absence of a specific statement, that such names are exempt from the relevant protective laws and regulations and therefore free for general use.

Typesetting: International Typesetters Inc., Makati, Philippines
2131/3145-543210 – Printed on acid-free paper

Library
University of Texas
at San Antonio

Contents

Abbreviations and Acronyms

AVHRR	Advanced Very High Resolution Radiometer
CILSS	Comité Inter-Etats de Lutte contre la Sècheresse au Sahel
CIPEA	Centre International pour l'Elevage en Afrique
CNRS	Centre National de la Recherche Scientifique
CP	Crude Protein (N×6.25)
CTFT	Centre Technique Forestier Tropical
DGRST	Délégation Générale à la Recherche Scientifique et Technique
DM	Dry Matter
DSPA	Direction de la Santé et de la Production Animales (Senegal)
EMASAR	Ecological Management of Arid and Semi-Arid Rangelands (FAO)
ERTS	Earth Resources Terrestrial Satellite
FAO	Food and Agriculture Organization of the United Nations
FAPIS	Formation en Aménagement Pastoral Intégré au Sahel
FIT	Front Inter-Tropical
FU	Feed Units (Scandinavian)
GAC	Global Area Coverage (NOAA/AVHRR)
GEMS	Global Ecological Monitoring System (UNEP)
GERDAT	Groupement d'Etudes et de Recherches pour le Développement de l'Agriculture Tropicale (later CIRAD = Centre de Coopération Internationale en Recherche Agronomique pour le Développement)
GIS	Geographic Information System (UNEP)
GREP	Global Radiative Evaporation Potential
GRIZA	Groupe de Recherches Interdisciplinaires en Zones Arides
GVI	Grazing Value Index
IBP	International Biological Programme
IEMVT	Institut d'Elevage et de Médecine Vétérinaire des Pays Tropicaux
IFAN	Institut Fondamental d'Afrique Noire (Dakar)
ILCA	International Livestock Center for Africa (= CIPEA)
IMRES	Inventory and Monitoring of the Rangelands Ecosystems of the Sahel (FAO/UNEP/Senegal Research Project)
INDVI	Integrated Normalized Difference of Vegetation Index
IPAL	Integrated Project on Arid Lands (UNESCO)
IR	Infra-Red
ISRA	Institut Sénégalais de Recherches Agricoles
ITCZ	Intertropical Convergence Zone
KREMU	Kenya Rangelands Ecological Monitoring Unit
LAC	Local Area Coverage (NOAA/AVHRR)
LAI	Leaf Area Index
LAT	Lutte Contre l'Aridité en Milieu tropical (DGRST programme)
LWt	Live Weight
LNERV	Laboratoire National d'Elevage et de Recherches Vétérinaires (Dakar)
LTP	Laboratory of Terrestrial Physics (NASA)
MAB	Man and Biosphere programme (UNESCO)
MNDVI	Maximum Normalized Difference of Vegetation Index

MPW	Mean Population Weight (Livestock)
$MPW^{0.75}$	Mean Population Metabolic Weight
MSC	Maximum Standing Crop
MSS	Multi-Spectral Scanner
NASA	National Aeronautics and Space Administration (USA)
NDVI	Normalized Difference of Vegetation Index
NOAA	National Oceanic and Atmospheric Administration (USA)
NIR	Near-Infra-Red
NPP	Net Primary Production
ORSTOM	Office de la Recherche Scientifique et Technique d'Outre-Mer (Paris)
PET	Potential Evapotranspiration
R	Red
RBV	Return Beam Vidicon
RUE	Rain-Use Efficiency
RUSA	Annual Rainfall Erosivity Index (Wischmeir)
SEM	Standard Error of the Mean ($\sigma/\sqrt{n}$)
SFU	Scandinavian Feed Unit (Net energy of 1 kg barley grain)
SISCOMA	Société Industrielle Sénégalaise de Construction Mécanique et Agricole
SODESP	Société de Développement de l'Elevage dans la zone Sylvo-Pastorale (Ferlo, Senegal)
SRF	Low Altitude Systematic Reconnaissance Flights
TLU	Tropical Livestock Unit
TM	Thematic Mapper
UNEP	United Nations Environment Programme (Nairobi)
UMT/UTM	Universal Mercator Transverse
$Wt^{0.75}$	Metabolic weight
$\bar{X}$	Mean of a given variable
σ	Standard deviation from the mean
ν	Coefficient of variation
λ	Wave Length

1 Definition, Geographical Limits, Contacts with Other Ecoclimatic Zones

The word "Sahel", pertaining to the ecoclimatic and biogeographic region bordering the Sahara to the south, was first used by the French botanist and explorer, Auguste Chevalier (1900, p. 206), who then defined a "zone sahélienne" around Timbuctoo, Mali. The word later became part of the scientific vocabulary in African phytogeography (Chevalier 1933; Murat 1937; Zolotarevsky and Murat 1938; Monod 1937, 1957; Trochain 1940, 1952, 1957, 1969; Lebrun 1947; Troupin 1960; Aubreville 1949; AETFAT 1959; White 1983, etc.).

The noun "Sahel" has two possible etymologies from the Arabic words ساحل (sahel) meaning shore, coast; which implies that the region is the southern shore of the Sahara and سهل (s'hel) meaning plain or flat land which would refer to this vast monotonous flat or gently rolling country. The word, and the region, became familiar to the common folk, worldwide, during the 1969/73 great drought which was largely publicized in the world media (Figs. 1–4).

Most ecologists and phytogeographers now agree on the geographical definition of the Sahel Zone or Sahelian Phytogeographical Domain (part of the Sudano-Zambesian or Sudano-Angolan or Sudano-Deccanian Region belonging to the Paleo-Tropical Floristic Empire), although there may be some slight differences between authors on the upper and lower limits. The proposed upper limit in latitude is sometimes the 50-mm or the 150/200-mm isohyets; but most specialists now agree on the 100-mm isohyet (Le Houérou 1976b; Le Houérou and Popov 1981; Boudet 1984b; Davy et al. 1976; Keay 1959; Rodier 1964; Le Houérou and Gillet 1985) as the approximate limit between the Sahara and the Sahel. This corresponds roughly to the borderline between contracted and scattered vegetation types as defined by Monod (1954). Contracted denotes that perennial vegetation is concentrated in depressions and along the hydrological network, while the scattered ("diffus", in French) refers to vegetation more or less uniformly, albeit sparsely, distributed on the interfluves and pediments. The limit between contracted and scattered types actually depends on the nature of the substratum (Le Houérou 1959, 1969); on both sides of the Sahara vegetation becomes contracted towards the 50-mm isohyet on sands and sandy soils and around 100–150 mm on the shallow substrata and soils of the "reg" and "tenere" or on the finely textured soils and the pebble plains. Both on its northern and southern edge the limits of the Sahara can thus be realistically defined as the 100 ± 50 mm on isohyet (Le Houérou 1987c).

The 100-mm isohyet is also the approximate southern limit of a large number of typically Saharan species such as:

Calligonum spp.	*Moltkiopsis ciliata*
Cornulaca monacantha	*Monsonia nivea*
Ephedra spp.	*Stipagrostis acutiflora*
Erodium glaucophyllum	*Stipagrostis plumosa*
Fagonia spp.	*Stipagrostis pungens*
Farsetia spp.	*Traganum nudatum*
Malcolmia aegyptiaca	*Zilla spinosa*

and several hundred others (Gillet 1968; Boudet 1975a; Le Houérou 1959, 1969, 1976b).

Tucker and Dregne (1988) attempted to define the border between the Sahara and the Sahel from quantified orbital data on the green herbage biomass produced. The attempt integrated the Normalized Difference of Vegetation Index (NDVI) for each 6–8 day period between 1981 and 1987. The discriminating value corresponded to a green herbage biomass of 50 g m^{-2} which is the minimum dry matter (DM) accumulation that can be dependably estimated from NOAA/AVHRR data. This boundary of 500 kg DM ha^{-1} moved 100–300 km southward from 1981 to 1984, then it moved 50–200 km northward in 1985–86, and southward again in 1987.

The authors concluded that the limit of 500 kg ha^{-1} of above-ground green herbage production showed such great annual variations that only a decade-long study would be able to determine whether or not this limit tends to move southward and therefore whether or not the Sahara itself is expanding to the south.

These findings are based on a more or less arbitrary limit corresponding to the sensitivity of the device used; given the Rain-Use Efficiencies (RUE) measured in these areas, the 500 kg ha^{-1} DM line would roughly correspond to the 200–250 mm isohyets; whereas most specialists in Sahelian ecology set that limit around 100 mm (Le Houérou and Gillet 1985). The Sahara-Sahel border would rather correspond to 15–20 g m^{-2}. Moreover, there is a consensus among ecologists that the best ecoclimatic limits are derived from the spatial distribution of species and communities, since such distributions are relatively stable over time spans of several decades, whereas biomass depends essentially on the vagaries of seasonal rains.

The southern limit of the Sahel Zone corresponds to the isohyets of 500 to 700 mm, depending on the author; but most specialists of the region now accept the mean value of 600 mm as the delineation between the Sahel and the Sudan ecological zones, which corresponds to the northern limit of a number of typically Sudanian species and vegetation[1] types belonging to the Sudanian and Guinean domains, e.g. for trees and shrubs:

[1]The adjective "Sudanian" refers to the so-called ecological zone, while "Sudanese" refers to the Republic of Sudan. Similarly "Sudan" is the ecological zone, whereas "Republic of Sudan" is the country.

Albizia chevalieri
Anogeissus leiocarpus
Annona senegalensis
Bombax costatum
Borassus ethiopum
Butyrospermum parkii
Combretum nigricans
Dalbergia melanoxylon
Entada africana
Feretia apodanthera
Guiera senegalensis
Khaya senegalensis
Lannea acida
Mitragyna inermis
Parkia biglobosa
Piliostigma reticulata
Piliostigma thonningii
Sterculia setigera
Tamarindus indica
Terminalia avicennoides
Terminalia macroptera
Ximenia americana

to quote only the most common species. Some of these may be found in the Sahel around ponds and watercourses, but are not zonally Sahelian. Among the Sudanian grasses found along the southern limit of the Sahel and further southwards, we have:

Perennial grasses

Andropogon ascinodis
Andropogon gayanus
Andropogon tectorum
Ctenium newtonii
Cymbopogon giganteus
Diheteropogon amplectens
Hyparrhenia rufa
Hyparrhenia smithiana
Hyparrhenia subplumosa
Hyperthelia dissoluta
Loudetia simplex

Annual grasses

Andropogon pseudapricus
Ctenium elegans
Diheteropogon hagerupii
Eragrostis gangetica
Loudetia togoensis
Pennisetum subangustum

Thus defined, the Sahel stretches across the African continent from west to east over a strip 400 to 600 km wide and nearly 6000 km long, i.e. over an area of approximately 3 million km^2 (over 42% of the conterminous United States) (Figs. 1–4). The Sahel bears many similarities with the East-African arid zone, in vegetation, soils and fauna; but there are two major differences: (1) the rainfall regime in East Africa is equatorial, i.e. bimodal, and thus there are two rainy seasons annually; (2) as a consequence, the grass layer is dominated by perennial species and not by annuals; the dry season is therefore much shorter and nutritional stress on herbivores much less acute in East Africa than in the Sahel.

The eastern limit of the Sahel corresponds to the water divide between the Red Sea and the Nile-Atbara Basin. The western Red Sea shores are bordered by a 50–75-km-wide coastal plain dominated by a continuous range of hills and mountains running parallel to the coast (NW-SE) at an elevation of 700–2500 m. These reliefs continue south-eastwards, paralell to the coast, right to the Horn of

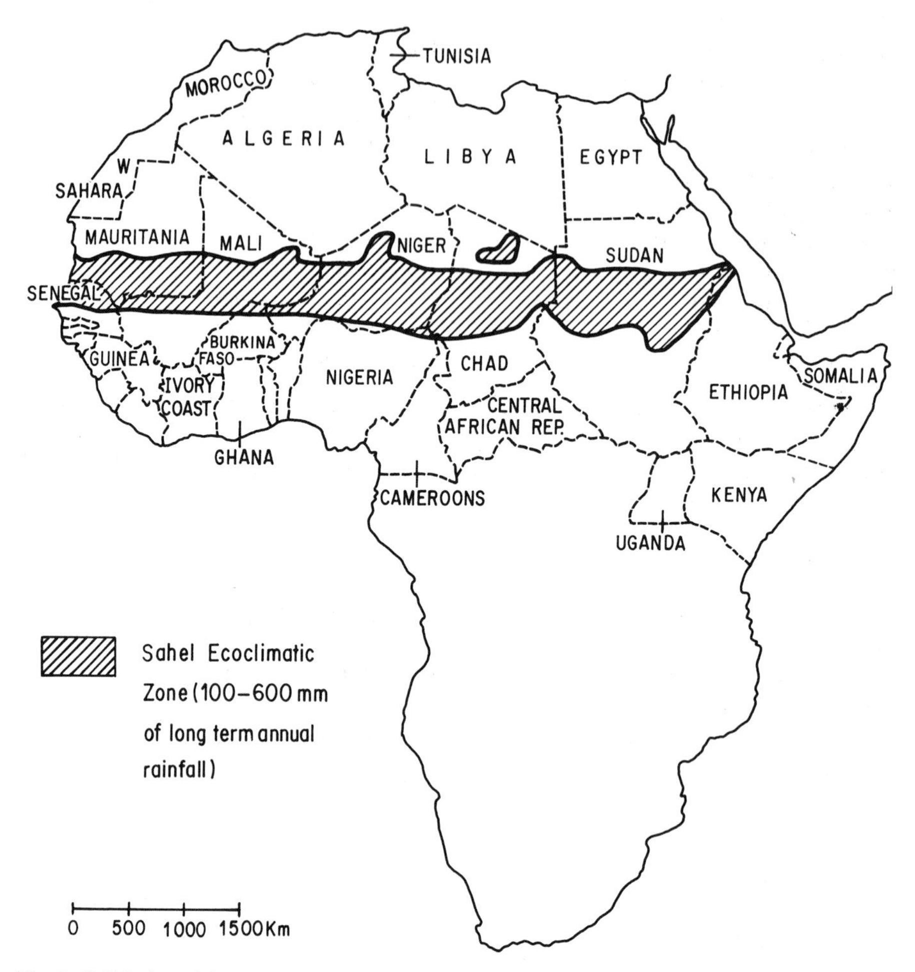

Fig. 1. Political partition of the Sahel Zone as from 1988

Africa (Cape Gaddafui) to the NE tip of Somalia. To the north of the 13° latitude the western sides of these reliefs are submitted to a typical Sahelian climate with a narrowly monomodal annual rainfall pattern (see the ombrothermal diagrams of Sinkat, Gebeit and Derudeb in the Rep. of Sudan and of Keren, Adigrat, Asmara, Barentu, Agordat, Adigrat and Mekele in Ethiopia; Figs. 29 to 32.

The coastal plain, by contrast, has a typical Mediterranean rainfall regime with precipitations occurring from October to February, and a very weak peak of ca. 10 mm m^{-1} in July-August. The coastal plain thus does not belong to the Sahel (see Fig. 29: ombrothermal diagrams of Masawa, Tokar, Suakin and Port-Sudan).

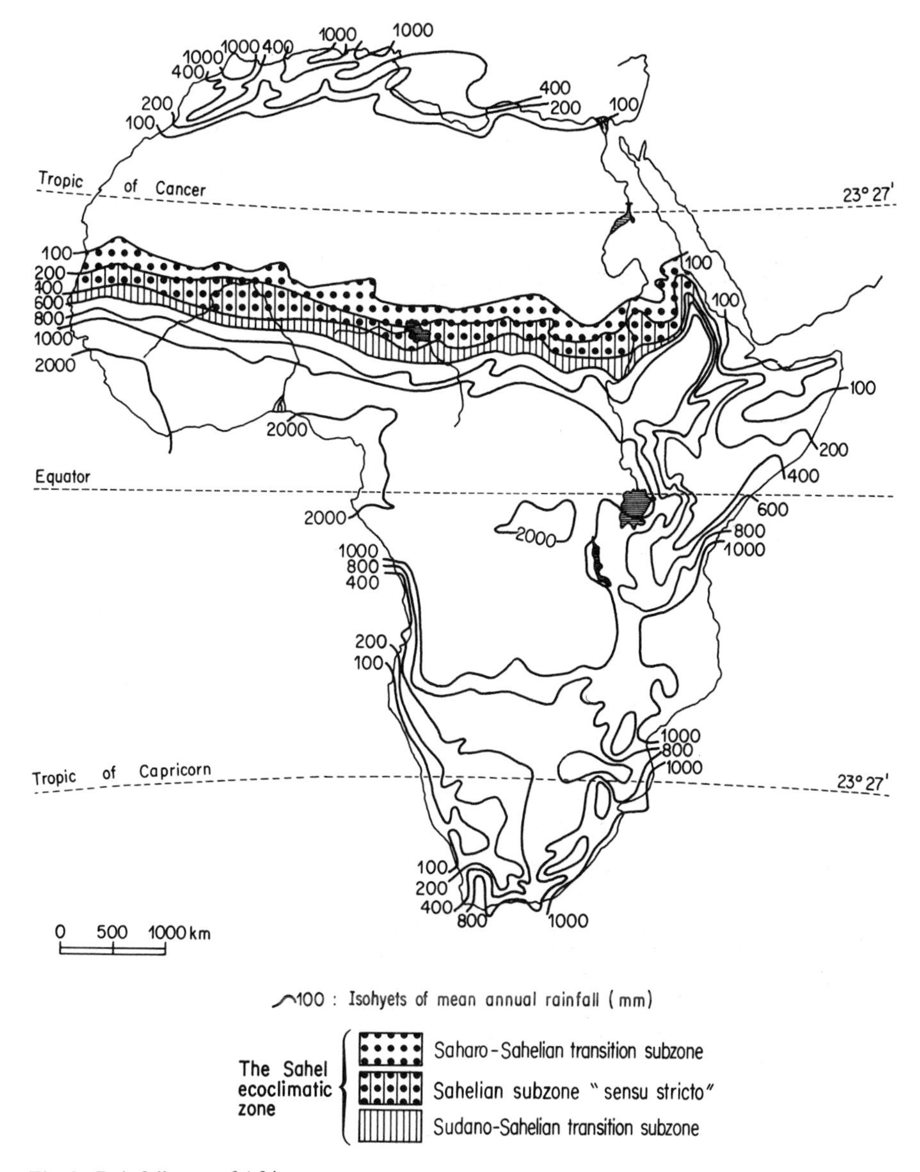

Fig. 2. Rainfall map of Africa

The subcoastal mountain range is submitted on its eastern side to a bimodal rainfall regime with winter and summer precipitations very similar to that of East Africa. Vegetation therefore bears many strong affinities with that of East Africa (Le Houérou, 1988c) and Jebel Marra [Wickens, 1976; see the ombrothermal diagrams of Erkwit (Sudan) and Faghena (Ethiopia) Fig. 30a and b]. This narrow

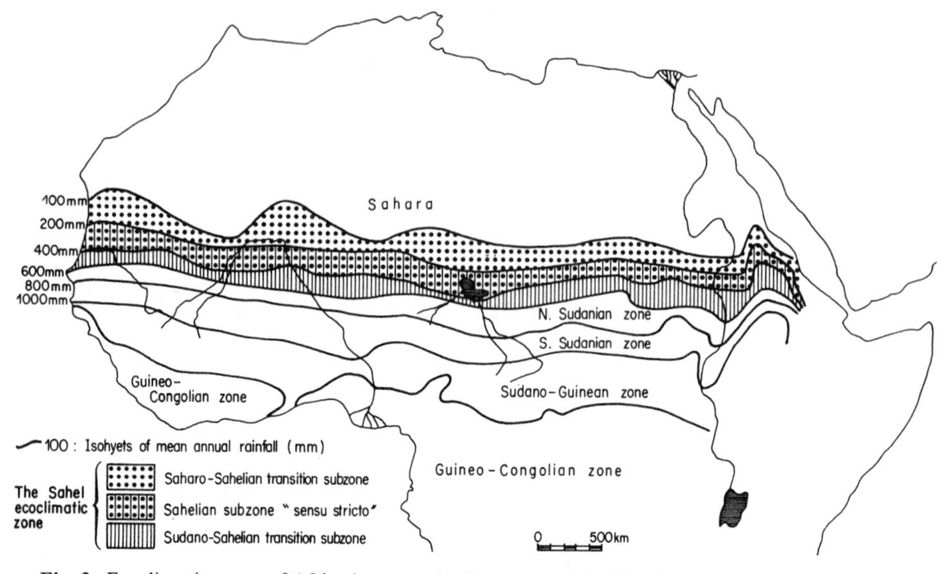

Fig. 3. Ecoclimatic zones of Africa between the Equator and the Tropic Cancer

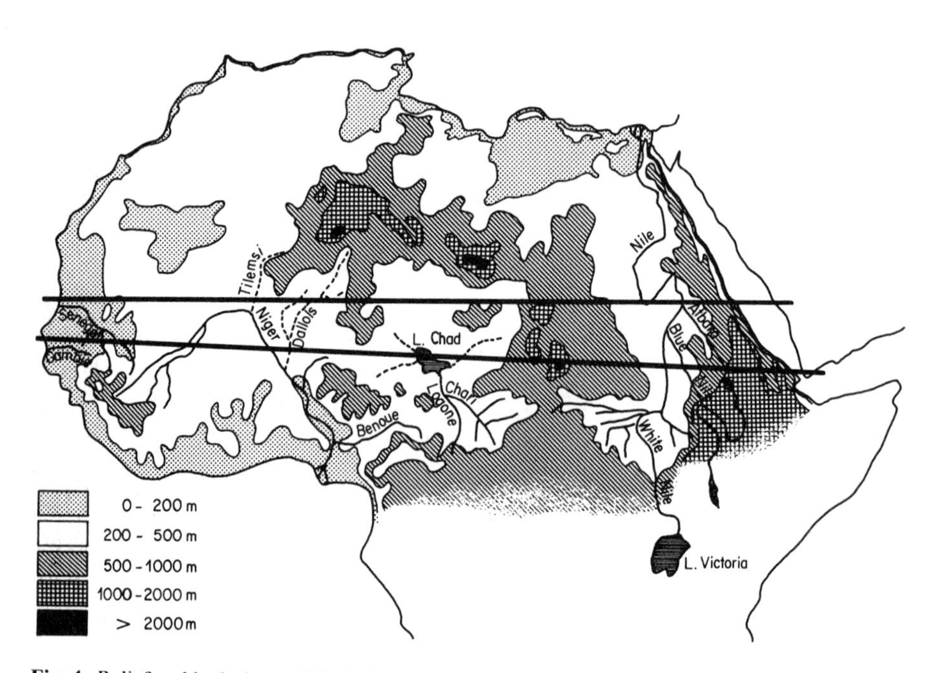

Fig. 4. Relief and hydrology of the Sahel and neighbouring regions

belt on the eastern side of the subcoastal chains actually bears an almost typical East-African montane vegetation (Kassas 1956a; Hassan 1974):

Euphorbia abyssinica
E. thi
E. cuneata
Lannea schimperi
Maytenus senegalensis
Coleus spp.
Kalanchoe spp.
Nepeta spp.
Asparagus racemosus
Dracaena ombet
Euclea schimperi
Carissa edulis
Olea africana
Clematis spp.
Aloe spp.
Caralluma spp.
Micromeria spp.
Acacia etbaica
Dodonaea viscosa
Rhus abyssinica
Rhus flexicaulis
Jasminum floribundum
Cissus spp.
Ximenia americana
Lavandula coronopifolia
etc.

Moreover, on the eastern side of the subcoastal range between the elevations of 700 and 1500 m exists a narrow belt on frequent mist (hence the name of the "mist oasis" of Erkwit). This mist belt extends all the way to Cape Gaddafui; further south, in Somalia, this mist belt is the exclusive habitat of Frankincense trees (*Boswellia carteri* and *B. freeri*).

At its western end the harshness of the Sahel climate is mitigated by the vicinity of the Atlantic Ocean and the upwellings tied to the cold Canary Current, which brings about cool air and moisture; this is particularly felt in the Cape Verde Peninsula where Dakar was established towards the end of the 19th century. The oceanic influence on the Sahel climate is felt some 30 to 50 km inland (Giffard 1974).

The Cape Verde Islands, some 600 km to the west of Dakar, are submitted to an Insulo-Atlantic climate substantially different from that of the Sahel because of the strong influence of the trade winds which bring moisture and mist in the windward exposures facing the NE.

Mist and hidden precipitations occur above 300–400 m on NE slopes for a long period of the dry season; conversely, leeward conditions are much drier due to the foehn effect. In contrast to the Sahelian climate, the Cape-Verdian climate is submitted to extreme rainfall variability. The coefficient of variation of annual rainfall at Praia ($\bar{x} = 250$ mm) is almost 70%, so that a large number of typically Saharan plant species occur in areas having a mean annual rainfall of 200–300 mm in the leeward lowlands of Maio and Santiago (Le Houérou 1980e):

Aerva persica
Suaeda volkensi
Salvia aegyptiaca
Launaea arborescens
Trichodesma africanum
Caylusea hexagyna
Eremopogon foveolatus
Pulicaria crispa
Zygophyllum simplex
Z. fontanesii
Fagonia latifolia
Heliotropium undulatum
Sclerocephalus arabicus
Forskhalea procidifolia
etc.

The Sahelian zone, broadly speaking, is usually broken down into three subzones, as shown in Table 1:

1. Saharo-Sahelian transition subzone, between 100 and 200 mm;
2. Sahel zone proper between the 200- and 400-mm isohyets;
3. Sudano-Sahelian transition subzone, between 400 and 600 mm (Chevalier 1933; Trochain 1940; Aubreville 1949; Monod 1957; Keay 1959; Rodier 1964; Gillet 1968; Boudet 1974a,b, 1984b; Le Houérou 1976a,b, 1977, 1979a,b, 1980b; Davy et al. 1976; Le Houérou and Popov 1981; Le Houérou and Gillet 1985).

These subdivisions are based on floristic and vegetation distribution grounds, wildlife and livestock distribution and land-use patterns as summarily shown in Tables 1, 2 and 3 and discussed in some detail later.

Table 1. Distribution of some dominant or common trees shrubs and perennial grasses in the various ecoclimatic zones between the Sahara and the Equator (ecoclimatic zones as defined by Le Houérou and Popov 1981)

	1	2	3	4	5	6	7	8
Trees and shrubs	Saharan	Saharo-Sahelian	Sahelian proper	Sudano-Sahelian	Northern Sudanian	Southern Sudanian	Guinea	Montane
Balanites aegyptiaca	(+)	+	+	+	+	–	–	+
Calotropis procera	+	+	+	+	+	+	(+)	+
Acacia tortilis	+	+	+	(+)	–	–	–	+
Capparis decidua	+	+	(+)	–	–	–	–	(+)
Leptadenia pyrotechnica	+	+	(+)	–	–	–	–	(+)
Ochradenus baccatus	+	+	(+)	–	–	–	–	(+)
Grewia tenax	(+)	+	(+)	–	–	–	–	(+)
Hyphaene thebaica	+	+	+	(+)	–	–	–	+
Maerua crassifolia	+	+	+	(+)	–	–	–	+
Acacia ehrenbergiana	+	+	+	(+)	–	–	–	+
Acacia laeta	–	+	+	(+)	–	–	–	+
Acacia senegal	–	(+)	+	(+)	–	–	–	+
Salvadora persica	+	+	+	(+)	–	–	–	+
Cordia sinensis	(+)	+	+	(+)	–	–	–	+
Boscia senegalensis	(+)	+	+	+	–	–	–	+
Cadaba farinosa	–	(+)	+	+	–	–	–	+
Cadaba glandulosa	–	(+)	+	+	–	–	–	+
Dobera glabra	–	(+)	+	+	–	–	–	+
Euphorbia balsamifera	(+)	+	+	+	(+)	–	–	+
Combretum aculeatum	–	(+)	+	+	(+)	–	–	+
Commiphora africana	–	(+)	+	+	(+)	–	–	+
Adenium obesum	–	(+)	+	+	(+)	–	–	+
Ziziphus mauritiana	–	(+)	+	+	+	–	–	+
Ziziphus mucronata	–	(+)	+	+	+	–	–	+
Feretia apodanthera	–	(+)	+	+	+	–	–	+
Acacia mellifera	–	(+)	+	+	+	(+)	–	+
Faidherbia albida	–	(+)	+	+	+	+	–	+
Acacia nilotica	–	(+)	+	+	–	+	–	+

Table 1. *Continued*

Trees and shrubs	1	2	3	4	5	6	7	8
Bauhinia rufescens	–	(+)	(+)	+	+	(+)	–	+
Acacia nubica	–	(+)	+	+	+	–	–	+
Grewia bicolor	–	(+)	+	+	+	+	–	+
Boscia salicifolia	–	–	(+)	+	+	+	–	+
Maerua angustifolia	–	–	(+)	+	+	(+)	–	+
Crataeva adansoni	–	–	+	+	+	(+)	–	+
Combretum cordofanianum	–	–	(+)	+	+	+	–	+
Combretum glutinosum	–	–	(+)	+	+	(+)	–	(+)
Combretum micranthum	–	–	(+)	+	+	(+)	–	(+)
Guiera senegalensis	–	–	(+)	+	+	(+)	–	+
Mitragyna inermis	–	–	(+)	+	+	+	–	+
Anogeissus leiocarpus	–	–	(+)	+	+	+	+	+
Celtis integrifolia	–	–	(+)	+	+	+	+	+
Maytenus senegalensis	–	–	(+)	+	+	+	–	+
Stereospermum kunthianum	–	–	(+)	+	+	+	–	+
Diospyros mespiliformis	–	–	(+)	+	+	+	(+)	+
Adansonia digitata	–	–	(+)	+	+	+	(+)	+
Bombax costatum	–	–	(+)	+	+	+	(+)	+
Hymenocardia acida	–	–	(+)	+	+	+	(+)	+
Sclerocarya birrea	–	–	(+)	+	+	+	(+)	+
Sterculia setigera	–	–	(+)	+	+	+	–	+
Pterocarpus lucens	–	–	(+)	+	+	(+)	–	–
Lannea acida	–	–	(+)	+	+	+	–	–
Leptadenia hastata	–	–	(+)	+	+	(+)	–	+
Terminalia avicennoides	–	–	(+)	+	+	+	(+)	+
Boswellia papyrifera	–	–	(+)	+	+	(+)	–	+
Terminalia brownii	–	–	(+)	+	+	(+)	–	+
Acacia seyal	–	–	(+)	+	+	(+)	–	+
Acacia ataxacantha	–	–	(+)	+	+	+	(+)	+
Dalbergia melanoxylon	–	–	(+)	+	+	+	(+)	(+)
Grewia flavescens	–		(+)	+	+	(+)	–	+
Grewia mollis	–	–	(+)	+	+	(+)	–	+
Grewia villosa	–	–	(+)	+	+	(+)	–	+
Maerua oblongifolia	–	–	(+)	+	+	(+)	–	+
Maerua angustifolia	–	–	(+)	+	+	(+)	–	+
Capparis corymbosa	–	–	–	+	+	+	(+)	+
Capparis tomentosa	–	–	–	+	+	+	(+)	+
Combretum racemosum	–	–	–	(+)	+	+	(+)	–
Combretum nigricans	–	–	–	+	+	+	–	+
Combretum geitonophyllum	–	–	–	(+)	+	+	(+)	–
Combretum molle	–	–	–	+	+	+	(+)	+
Combretum ghazalense	–	–	–	+	+	+	–	–
Borassus ethiopum	–	–	–	+	+	+	+	(+)
Khaya senegalensis	–	–	–	+	+	+	(+)	–
Parkia biglobosa	–	–	–	+	+	+	(+)	(+)
Butyrospermum paradoxum	–	–	–	+	+	+	(+)	–
Ceiba pentandra	–	–	–	(+)	+	+	+	–
Xeroderris stuhlmanni	–	–	–	(+)	+	+	–	–
Piliostigma thonningii	–	–	–	(+)	+	+	(+)	+
Cassia sieberiana	–	–	–	(+)	+	+	(+)	+
Pterocarpus erinaceus	–	–	–	(+)	+	+	+	–
Cordyla pinnata	–	–	–	(+)	(+)	+	+	–

Table 1. *Continued*

Trees and shrubs	1 Saharan	2 Saharo-Sahelian	3 Sahelian proper	4 Sudano-Sahelian	5 Northern Sudanian	6 Southern Sudanian	7 Guinea	8 Montane
Entada africana	–	–	–	(+)	+	+	+	+
Ozoroa insignis	–	–	–	(+)	+	+	+	+
Albizia chevalieri	–	–	–	(+)	+	+	+	–
Albizia amara	–	–	–	(+)	+	+	–	+
Ximenia americana	–	–	–	(+)	+	+	+	+
Strychnos spinosa	–	–	–	(+)	+	+	+	+
Tamarindus indica	–	–	–	(+)	+	+	(+)	+
Parinari macrophylla	–	–	–	(+)	+	+	+	–
Crossopterix febrifuga	–	–	–	(+)	+	+	+	–
Daniellia oliveri	–	–	–	–	(+)	+	+	+
Isoberlinia doka	–	–	–	–	(+)	+	+	–
Lonchocarpus laxiflorus	–	–	–	–	(+)	+	+	–
Oxytenanthera abyssinica	–	–	–	–	(+)	+	+	–
Cussonia kirkii	–	–	–	–	(+)	+	+	–
Cussonia barteri	–	–	–	–	(+)	+	+	(+)
Gardenia erubescens	–	–	–	–	(+)	+	+	+
Gardenia ternifolia	–	–	–	–	(+)	+	+	+
Acacia dudgeoni	–	–	–	–	(+)	+	+	+
Moringa oleifera	–	–	–	–	(+)	+	+	+
Vitex doniana	–	–	–	–	–	+	+	+
Vitex madiensis	–	–	–	–	–	+	+	+
Lophira lanceolata	–	–	–	–	–	(+)	+	+
Trema guineensis	–	–	–	–	–	(+)	+	–
Ficus capensis	–	–	–	–	–	+	+	+
Ficus glumosa	–	–	–	–	–	+	+	+
Ficus thonningii	–	–	–	–	–	+	+	+
Cola cordifolia	–	–	–	–	–	+	+	–
Bridelia ferruginea	–	–	–	–	–	+	+	+
Burkea africana	–	–	–	–	–	+	+	+
Ficus asperofilia	–	–	–	–	–	+	+	–
Albizia zygia	–	–	–	–	–	+	+	(+)
Annona senegalensis	–	–	–	–	–	+	+	–
Terminalia macroptera	–	–	–	–	–	+	+	(+)
Newbouldia laevis	–	–	–	–	–	–	+	–
Aspilia latifolia	–	–	–	–	–	–	+	–
Spondias mombin	–	–	–	–	–	–	+	–
Cyclodicus gabonensis	–	–	–	–	–	–	+	–
Perocopsis laxiflora	–	–	–	–	–	–	+	–
Uapaca togoensis	–	–	–	–	–	–	+	(+)
Harungana madascariensis	–	–	–	–	–	–	+	(+)
Syzygium guineense	–	–	–	–	–	–	+	(+)
Perennial grasses								
Stipagrostis pungens	+	(+)	–	–	–	–	–	–
Stipagrostis ciliata	+	(+)	–	–	–	–	–	+
Stipagrostis acutiflora	+	(+)	–	–	–	–	–	+

Table 1. *Continued*

Perennial grasses	1	2	3	4	5	6	7	8
Stipagrostis plumosa	+	(+)	–	–	–	–	–	+
Panicum turgidum	+	+	–	–	–	–	–	+
Lasiurus hirsutus	+	+	–	–	–	–	–	+
Cymbopogon schoenanthus	+	+	–	–	–	–	–	+
Aristida pallida	+	+	(+)	–	–	–	–	+
Aristida papposa	–	+	(+)	+	–	–	–	+
Cymbopogon proximus	–	–	(+)	+	–	–	–	–
Aristida longiflora	–	–	(+)	+	–	–	–	+
Andropogon gayanus	–	–	–	(+)	+	+	(+)	+
Hyperthelia dissoluta	–	–	–	(+)	+	+	+	+
Cymbopogon giganteus	–	–	–	(+)	+	+	–	+
Hyparrhenia rufa	–	–	–	–	+	+	+	(+)
Hyparrhenia smithiana	–	–	–	–	–	+	+	–
Andropogon ascinodis	–	–	–	–	–	+	+	–
Andropogon tectorum	–	–	–	–	–	+	(+)	–
Diheteropogon amplectens	–	–	–	–	–	+	+	–
Loudetia simplex	–	–	–	–	–	+	+	–
Loudetia arundinacea	–	–	–	–	–	(+)	+	–
Panicum phragmitoides	–	–	–	–	–	(+)	+	+
Panicum maximum	–	–	–	–	–	–	+	–
Pennisetum purpureum	–	–	–	–	–	–	+	–
Hyparrhenia diplandra	–	–	–	–	–	–	+	–
Hyparrhenia filipendula	–	–	–	–	–	–	+	–
Imperata cylindrica	–	–	–	–	–	–	+	–
Andropogon macrophyllus	–	–	–	–	–	–	+	–
Themeda triandra	–	–	–	–	–	–	–	+
Heteropogon contortus	–	–	–	–	(+)	(+)	–	+
Hyparrhenia cymbaria	–	–	–	–	–	–	–	+
Chloris gayana	–	–	–	–	–	–	–	+
Cenchrus ciliaris	(+)	(+)	–	–	–	–	–	+
Botriochloa pertusa	–	–	–	–	–	–	–	+
Chrysopogon plumulosus	–	(+)	–	–	–	–	–	+
Antephora hochstetteri	–	–	–	–	–	–	–	+
Dichanthium annulatum	–	–	–	–	–	–	–	+
Sporobolus ioclados	–	–	–	–	–	–	–	+
Eremopogon foveolatus	(+)	(+)	–	–	–	–	–	+
Enneapogon brachystachyus	–	–	–	–	–	–	–	+
Sehima ischaemoides	–	–	–	–	–	–	–	+
Schmidtia pappophoroides	–	–	–	–	–	–	–	+
Hyparrhenia hirta	–	–	–	–	–	–	–	+

Table 2. Zonal distribution of mammals between the Tropic of Cancer and the Equator[a]

Species	Ecoclimatic Zones [b]						
	Saharan	Saharo-Sahelian	Sahelian	Sudano-Sahelian	Northern Sudanian	Southern Sudanian	Guinean/Congolian
Fernecus zerda	+	–	–	–	–	–	–
Addax nasomaculatus	(+)	(+)	–	–	–	–	–
Ammotragus lervia	+	+	+	–	–	–	–
Vulpes ruppelii	+	+	+	–	–	–	–
Vulpes pallida	(+)	+	+	(+)	–	–	–
Gazella dama	(+)	(+)	(+)	–	–	–	–
Felis margarita	+	+	+	–	–	–	–
Gazella leptoceros	(+)	(+)	(+)	–	–	–	–
Gazella dorcas	+	+	+	(+)	–	–	–
Lepus capensis	+	+	+	+	+	–	–
Pelictis libyca	+	+	+	+	+	–	–
Canis aureus	+	+	+	+	+	+	–
Hyaena hyaena	+	+	+	+	+	+	–
Felis libyca	(+)	+	+	+	+	+	–
Acinonyx jubatus	(+)	+	+	+	+	+	–
Procavia capensis	(+)	+	+	+	+	+	–
Gazella rufifrons	–	+	+	(+)	–	–	–
Oryx dammah	–	(+)	(+)	(+)	–	–	–
Zorilla striatus	–	(+)	+	+	+	–	–
Lepus crawshayi	–	(+)	+	+	+	+	–
Felis caracal	–	+	+	+	+	+	–
Lycaon pictus	–	(+)	+	+	+	+	–
Panthera leo	–	(+)	+	+	+	(+)	–
Galago senegalensis	–	–	+	+	–	–	–
Erythrocebus patas	–	–	+	+	–	–	–
Cercopithecus aethiops	–	–	+	+	–	–	–
Crocuta crocuta	–	–	+	+	+	+	–
Canis adustus	–	–	+	+	+	(+)	–
Ichneumia albicauda	–	–	(+)	+	+	+	–
Herpestes ichneumon	–	–	+	+	+	+	–
Xerus erythropus	–	–	+	+	+	+	–
Genetta genetta	–	–	+	+	+	+	–
Giraffa camelopardalis	–	–	+	+	+	(+)	–
Felis serval	–	–	+	+	+	+	–
Hippotragus equinus	–	–	+	+	+	+	–
Redunca redunca	–	–	+	+	+	+	–
Alcephalus caama	–	–	+	+	+	+	–
Damaliscus korrigum	–	–	(+)	+	+	+	–
Kobus defassa	–	–	(+)	+	+	+	–
Papio anubis	–	–	(+)	+	+	+	–
Orycteropus afer	–	–	+	+	+	+	+
Phacochoerus aethiopicus	–	(+)	+	+	+	+	+
Kobus kob	–	–	–	(+)	+	+	+
Alcephalus buselaphus	–	–	–	(+)	(+)	+	+
Sylvicapra grimmia	–	–	–	(+)	(+)	+	+
Panthera pardus	–	–	–	(+)	+	+	+

Table 2. *Continued*

Species	Ecoclimatic Zones[b]						
	Saharan	Saharo-Sahelian	Sahelian	Sudano-Sahelian	Northern Sudanian	Southern Sudanian	Guinean/Congolian
Loxodonta africana	–	–	–	+	+	+	+
Hyppopotamus amphibius	–	–	–	(+)	+	+	+
Syncerus caffer	–	–	–	(+)	+	+	+
Ourebia ourebi	–	–	–	(+)	+	+	+
Taurotragus derbyanus	–	–	–	(+)	+	+	+
Tragelaphus strepticeros	–	–	–	(+)	+	+	+
Tragelaphus scriptus	–	–	–	–	(+)	+	+
Cephalophus rufilatus	–	–	–	–	(+)	+	+
Potamochaerus porcus	–	–	–	–	–	+	+
Naudinia binotata	–	–	–	–	–	+	+
Genetta tigrina	–	–	–	–	–	+	+
Genetta servalina	–	–	–	–	–	+	+
Potamogale velox	–	–	–	–	–	–	+
Manis gigantea	–	–	–	–	–	–	+
Periodictius potto	–	–	–	–	–	–	+
Galagoides demidovii	–	–	–	–	–	–	+
Cercocebus spp.[c]	–	–	–	–	–	–	+
Cercopithecus spp.[d]	–	–	–	–	–	–	+
Gorilla gorilla	–	–	–	–	–	–	+
Papio sphinx	–	–	–	–	–	–	+
Pan troglodytes	–	–	–	–	–	–	+
Colobus spp.[e]	–	–	–	–	–	–	+
Aonyx congica	–	–	–	–	–		+
Polana richardsonii	–	–	–	–	–	–	+
Bdeogale nigripes	–	–	–	–	–	–	+
Herpestes naso	–	–	–	–	–	–	+
Crossacrus obscurus	–	–	–	–	–	–	+
Dendrohyrax arboreus	–	–	–	–	–	–	+
Hylochaerus meinetzageni	–	–	–	–	–	–	+
Okapia johnstoni	–	–	–	–	–	–	+
Boocercus euryceros	–	–	–	–	–	–	+
Tragelaphus spekei	–	–	–	–	–	–	+
Cephalophus spp.[f]	–	–	–	–	–	–	+
Rodents in Senegal and Mauritania[g]							
Gerbillus nanus	+	(+)	–	–	–	–	–
Gerbillus gerbillus	+	(+)	–	–	–	–	–
Gerbillus pyramidum	+	(+)	–	–	–	–	–
Jaculus jaculus	+	(+)	–	–	–	–	–
Meriones libycus	+	(+)	–	–	–	–	–
Pachyuromys duprasi	+	(+)	–	–	–	–	–
Psammomys obesus	+	(+)	–	–	–	–	–

Table 2. *Continued*

Species	Ecoclimatic Zones [b]						
	Saharan	Saharo-Sahelian	Sahelian	Sudano-Sahelian	Northern Sudanian	Southern Sudanian	Guinean/Congolian
Taterillus arenairus	+	+	+	–	–	–	–
Desmodilliscus braueri	(+)	+	+	+	(+)	–	–
Taterillus pygargus	–	(+)	+	+	+	–	–
Arvicanthis niloticus	–	(+)	+	+	+	–	–
Mus haussa	–	(+)	+	+	+	–	–
Mastomys huberti	–	–	+	+	+	–	–
Euxerus erythropus	–	–	+	+	+	+	–
Hystrix cristata	–	–	+	+	+	+	–
Taterillus gracilis	–	–	(+)	+	+	+	–
Mastomys erythroleucus	–	–	+	+	+	–	–
Tatera guineae	–	–	–	–	+	+	+
Heliosciurus gambianus	–	–	–	–	+	+	+
Cricetemys gambianus	–	–	–	–	+	+	+
Steatomys caurinus	–	–	–	–	+	+	+
Mus mattheyi	–	–	–	–	+	+	+
Mus musculoides	–	–	–	–	+	+	+
Rattus rattus	–	–	–	–	+	+	+
Lemniscomys barbarus	–	–	–	–	+	+	+
Myomys daltoni	–	–	–	–	+	+	+
Mastomys erythroleucus	–	–	–	–	+	+	+
Graphiurus murinus	–	–	–	–	+	+	+
Tatera gambiana	–	–	–	–	+	+	+

[a] An interpretation of the indications given by Dekeyser (1955); Dorst and Dandelot (1970); Happold (1973); Delany and Happold (1979); Poulet (1972); Haltenorth et al. (1985).
[b] Ecoclimatic zones as defined by Le Houérou and Popov (1981).
[c] *Cercocebus aterrimus, C. galeritus, C. torguatus, C. albigena.*
[d] *Cercopithecus ascanius, C. cephus, C. neglectus, C. nictitans, C. diana, C. erythrogaster, C. mona, C. mitis, C. neglectus.*
[e] *Colobus angolensis, C. badius, C. guereza, C. pennanti, C. satanas.*
[f] *Cephalophus callipygus, C. dorsalis, C. leucogaster, C. niger, C. nigrifons, C. monticola, C. silvicultor.*
[g] From Poulet (1982).

Table 3. Ecoclimatic zones in Africa between 10° and 24° N latitude (Le Houérou 1976b, 1977, 1980f; Davy et al. 1976; Le Houérou and Popov 1981; Boudet 1984b)

Ecoclimatic zone (No.)	Tropical climates	Biogeographical zones	Mean annual rainfall (mm)	Length of growing season (days)	Vegetation types	Dominant trees and shrubs	Dominant grasses	Livestock	Land-use patterns
7	Desert	Southern Sahara			Contracted along hydrological network	*Acacia raddiana, A. ehrenbergiana*	*Stipagrostis pungens* (perennial)	Camels, goats (sheep)	Nomadic pastoralism, no rain-fed cropping
			100	15					
6a	Very arid	Saharo-Sahelian			Contracted and diffuse steppes	*A. raddiana, A. ehrenbergiana*	*Panicum turgidum* (perennial)	Camels goats, sheep (zebu cattle)	Nomadic and transhumant pastoralism, no rain-fed cropping
			200	30					
6b	Arid	Sahelian *stricto sensu*			Mimosaceae Savanna with annual grass layer	*A. senegal, A. raddiana, Balanites aegyptiaca, Commiphora africana*	*Aristida mutabilis, Schoenefeldia gracilis* (annuals)	Camels sheep, goats zebus	Transhumant and nomadic pastoralism, some flood cropping (millet)
			400	75					
5	Semi-arid	Sudano-Sahelian			Combretaceae Savanna with annual grass layer	*Combretum micranthum, Combr.* spp., *A. seyal, Sclerocrya birrea*	*Cenchrus biflorus, Eragrostis tremula* (annuals)	Zebus Sheep Goats Camels	Settled and nomadic pastoralism Millet cropping
			600	120					
4	Dry subhumid	Sahelo-Sudanian			Combretaceae Savanna and woodland perennial grass layer	*Khaya senegalensis, Parkia biglobosa, Butyrospermum paradoxum*	*Andropogon gayanus* (perennial) *Diheteropogon hagerupii* (annual)	Zebus Sheep Goats (Camels)	Settled and nomadic pastoralism Settled farming: millet, groundnut, sorghum, cowpea
			900	180					
3	Subhumid	Sudanian *stricto sensu*			Mixed savanna and woodland perennial grass layer	*Anogeissus leiocarpus Combr. nigricans*	*Andropogon gayanus, Andr. pseudapricus*	Taurine and zebu cattle Trypanosomic infestation	Settled farming: cotton, corn, cassava groundnut, mango
			1200	210					

Table

Country	Ecoclimatic Zones (in 10^3 Km^2 and %)									
	Saharan		Sahelian		Sudanian		Guinean		Total	
	area	%	area	%	area	%	area	%	area	%
Burkina-Faso	0.0	–	25	9	249	91	0.0	–	274	3
Chad	630	49	366	29	288	22	0.0	–	1284	15
Gambia	0.0	–	0.0	–	11	100	0.0	–	11	0.1
Mali	550	44	300	24	390	32	0.0	–	1241	14
Mauritania	808	78	223	22	0.0	–	0.0	–	1031	12
Niger	540	43	723	57	0.4	0.1	0.0	–	1267	14
Nigeria	0.0	–	50	5	674	73	200	22	924	10
Senegal	0.0	–	90	44	111	56	0.0	–	201	2
Sudan (Rep. of)	800	32	800	32	480	19	420	17	2500	29
Various: Cameroons, Ethiopia, Cape Verde Islands			67						67	0.9
Total	3328	38	2644	30	2203	25	620	7	8800	100

Source: Le Houérou and Popov, 1981.

2 Environmental Characteristics

2.1 Climatic Factors

2.1.1 General

The climate of the Sahel is typically tropical with a monomodal precipitation pattern and rainfall occurring in a short summer season, i.e. during the long-day period. The rainy season follows the apparent course of the sun at the zenith; the rains are therefore sometimes called "zenithal rains". The rains are provoked by the monsoon of the Gulf of Guinea and the northward movement of the Intertropical Convergence Zone (ITCZ) also called Intertropical Front (FIT, in French). The northward progress and southward retreat of the ITCZ roughly follow the apparent march of the sun at the zenith which is at the equator at the equinoxes, on the Tropic of Cancer at the June solstice and on the Tropic of Capricorn at the December solstice, (Figs. 6–7).

2.1.2 Rainfall (Figs 2, 3, 5-39)

The rainy season starts in June and ends in September; the period from September to June is virtually rainless. The rainy season and the growing season from the agronomical and ecological viewpoints are defined as the period in which monthly precipitations in millimeters are equal to or greater than twice the mean temperature in degrees Celsius: $P > 2t$ (Bagnouls and Gaussen 1953; Walter and Lieth 1967). This empirical rule was shown to correspond closely to a more rational and scientific basis: a statistical study over some 800 African rainfall stations found that the threshold $P > 2t$ corresponded in 92% of the cases with the threshold $P > 0.35$ PET, where PET is the Potential Evapotranspiration either measured or computed via Penman's formula. On the other hand, 0.35 PET corresponds to the "Crop Coefficient" used by irrigation agronomists for the period from sowing to emergence (Doorenbos and Kassam 1979, t. 18, p 25). In other words, the amount of water evaporated from a bare, non-satured, soil surface is usually close to 0.35 PET, and therefore the amount of water available in the soil in excess of 0.35 PET is also available for plant growth (Le Houérou and Popov 1981). As the two above mentioned criteria are virtually equivalent, it follows that $P > 2t$ is a good substitute for 0.35 PET, whenever PET is unknown due to lack of adequate basic data. It also explains why this empirical finding of

Bagnouls and Gaussen has been so widely and successfully used by many ecologists and agronomists throughout the world in distinguishing dry season from growing season (Figs. 8–32).

As the mean daily temperature is about 30 °C in the Sahel at the beginning of the rainy season (Cochemé and Franquin 1967) and as PET is around 6 mm day^{-1} at that time (Davy et al. 1976), it follows that the rainy/growing season starts, on average, when precipitation reaches:

$$30\,^{\circ}\text{C} \times 2/30 \text{ days} = 2.0 \text{ mm day}^{-1} \text{ or } 60 \text{ mm month}^{-1};$$
$$6 \text{ mm day}^{-1} \times 0.35 = 2.10 \text{ mm day}^{-1} \text{ or } 63\text{–}65 \text{ mm month}^{-1}.$$

The opposite, naturally, is true for the end of the rainy season after due consideration of the amount of water that may have accumulated in the soil.

Thus defined, the growing season lasts 15 to 30 days in the Saharo-Sahelian transition subzone, 1–3 months in the Sahel stricto sensu and 3–4 months in the Sudano-Sahelian transition subzone (Figs. 6–7).

The beginning and the end of the rainy season occur at fairly regular periods every year.

The N-S mean annual precipitation gradient is about 1 mm km^{-1} in latitude (Le Houérou 1976a).

The variability in annual precipitation increases with aridity, i.e. from S to N. The coefficient of variation of annual rainfall increases from 12–15% in the Guinean rain forest to over 80% at the southern fringe of the Sahara; it varies from about 25% in the Sudano-Sahelian transition subzone to 40–45% in the Saharo-Sahelian transition subzone (Tables 4–6; Figs. 33–38).

Annual rainfall is about 135–150% of the long-term mean 1 year in 10 and some 50–65% of the mean or less 1 year in 10 (Tables 7–8).

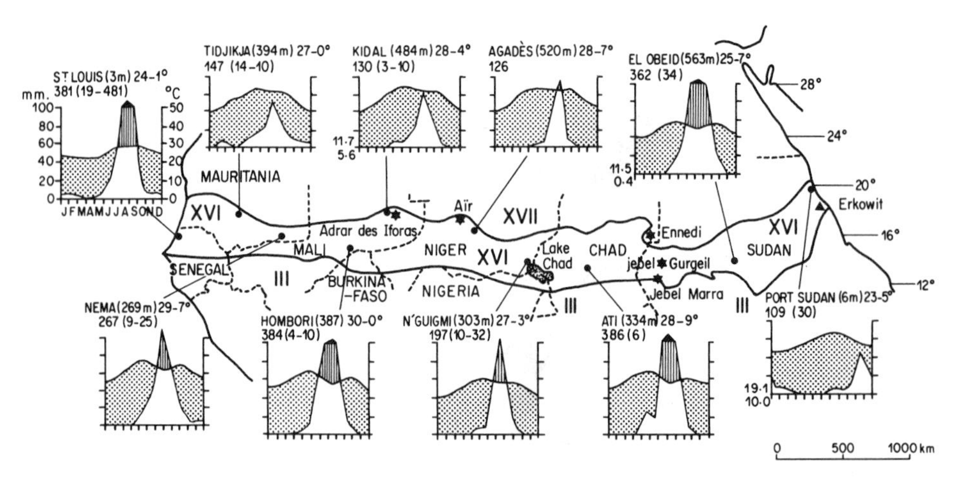

Fig. 5. Zonal distribution of ombrothermic diagrams and water balance in the Sahel (After White 1983)

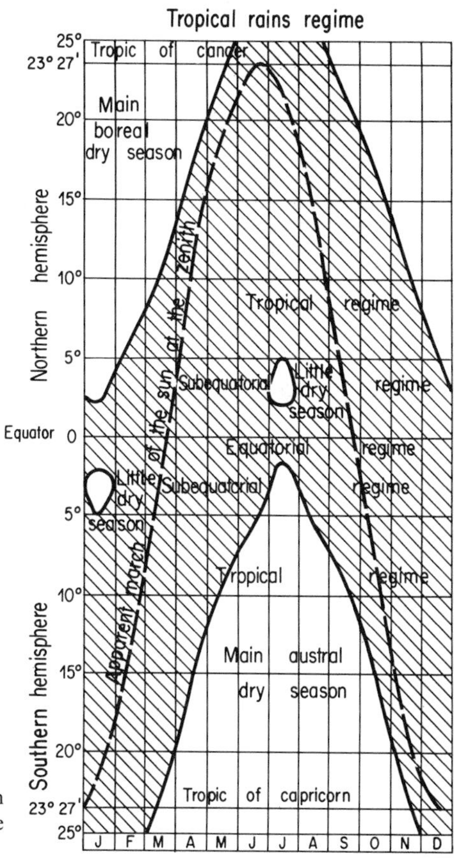

Fig. 6. Distribution of rainy and dry seasons in the intertropical African zone (After Aubreville 1949)

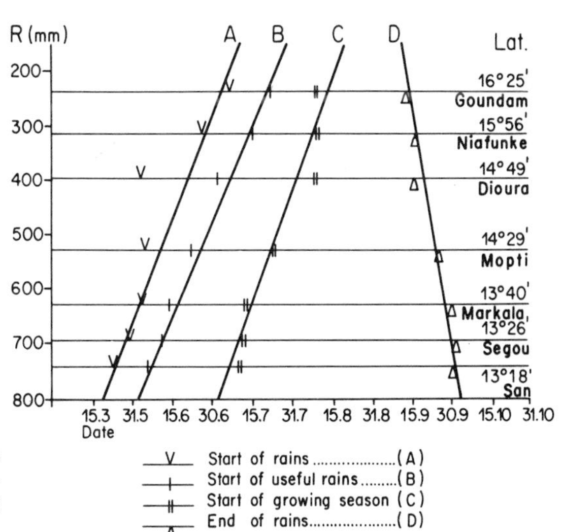

Fig. 7. Relationship between mean annual rainfall, latitude and characteristics of the rainy season in Central Mali (After Wilson 1982b)

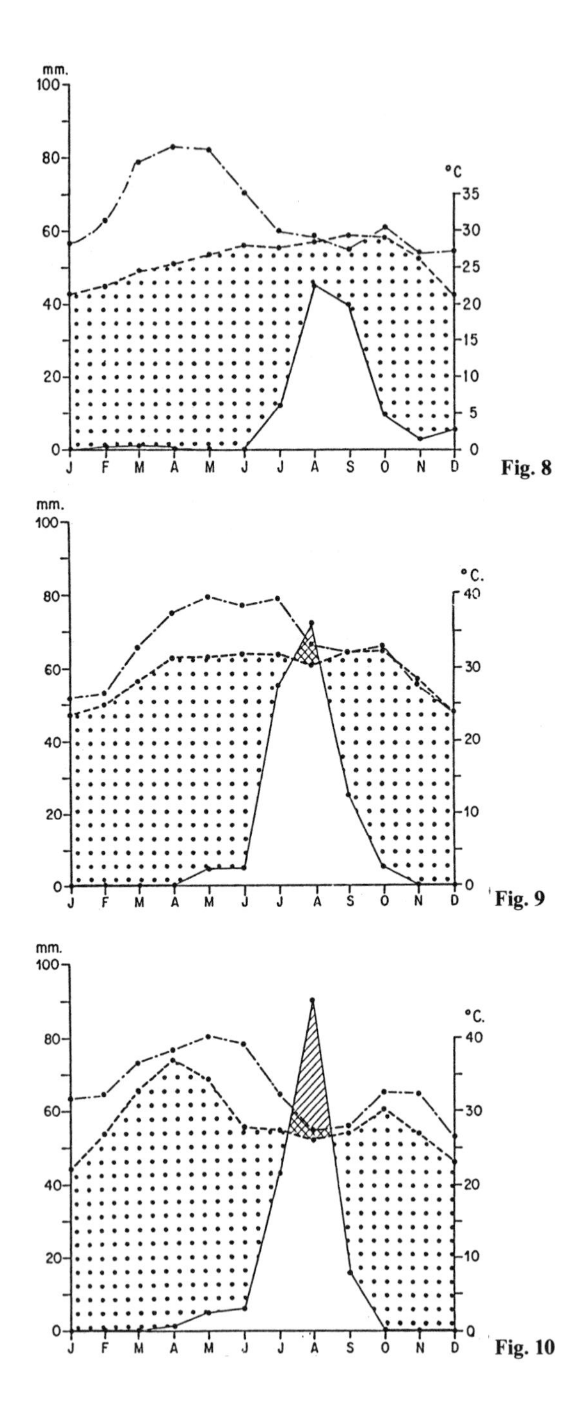

Fig. 8

Fig. 9

Fig. 10

Fig. 8. *Nouakchott (Mauritania)* tropical Saharan ecoclimatic zone, ecotone with the Saharo-Sahelian ecoclimatic zone:

P = 117 mm; t = 26.0°C; Alt. = 3 m asl
2 t = 26 × 2 × 12 = 624 mm; PET (68.6 t) = 1784 mm
0.35 PET (Penman) = 790 mm; PET (Penman) = 2257 mm
Water deficit (2 t) = 624 – 117 = 507 mm
(0.35 PET) = 790 – 117 = 673 mm

Growing (rainy) season (P > 2 t) = 0 day
(P > 0.35 PET) = 0 day
Dry season (P < 2 t) = 365 days
(P < 0.35 PET) = 365 days

Fig. 9. *Khartoum (Rep. of Sudan)* Saharo-Sahelian ecoclimatic zone:

P = 164 mm; t = 29.8°C; Alt. = 380 m
2 t = 29.8 × 2 × 12 = 715 mm; PET (68.6 t) = 2044 mm
0.35 PET (Penman) = 788 mm; PET (Penman) = 2254 mm
Water deficit (2 t) = 715 – 164 = 551 mm
(0.35 PET) = 788 – 164 = 624 mm

Growing (rainy) season (P > 2 t) = 24 days
(P > 0.35 PET) = 12 days
Dry season (P > 2 t) = 341 days
(P < 0.35 PET) = 353 days

Fig. 10. *Agadès (Niger)* Saharo-Sahelian ecoclimatic zone:

P = 164 mm; t = 28.2°C; Alt. = 503 m
2 t = 28.2 × 2 × 12 = 677 mm; PET (68.6 t) = 1935 mm
0.35 PET (Penman) = 800 mm; PET (Penman) = 2288 mm
Water deficit (2 t) = 677 – 164 = 513 mm
(0.35 PET) = 800 – 164 = 636 mm

Growing (rainy) season (P > 2 t) = 40 days
(P > 0.35 PET) = 35 days
Dry season (P > 2 t) = 325 days
(P < 0.35 PET) = 332 days

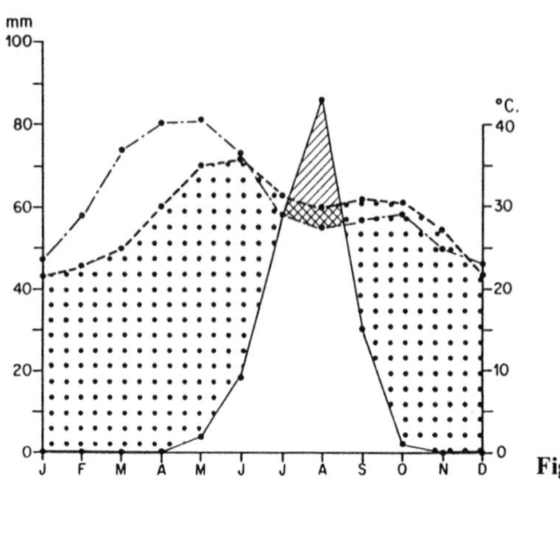

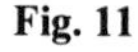
Fig. 11

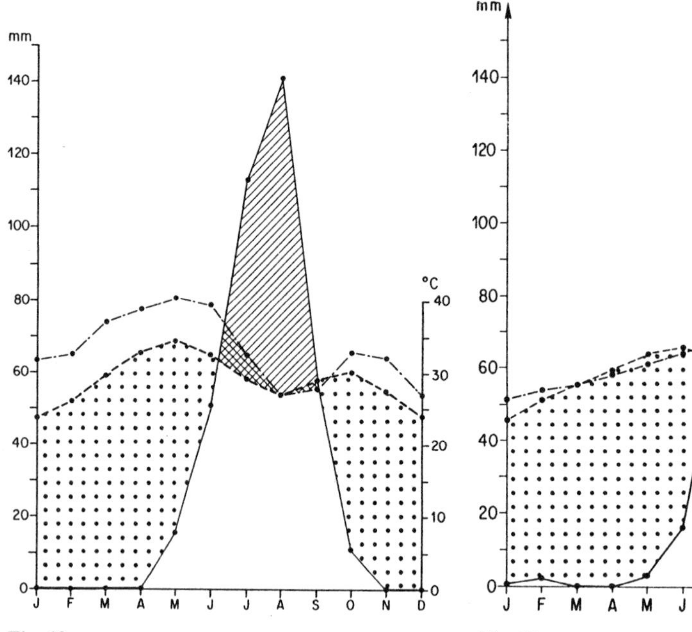

Fig. 12

Fig. 13

Fig. 11. *Timbuctoo (Mali)* Sahelian ecoclimatic zone, sensu stricto:
P = 250 mm; t = 29.0°C; Alt. = 264 m
2 t = 29 × 2 × 12 = 696 mm; PET (68.6 t) = 1989 mm
0.35 PET (Penman) = 740 mm; PET (Penman) = 2116 mm
Water deficit (2 t) = 696 – 250 = 446 mm
(0.35 PET) = 740 – 250 = 490 mm
Growing (rainy) season (P > 2 t) = 45 days
(P > 0.35 PET) = 40 days
Dry season (P < 2 t) = 320 days
(P < 0.35 PET) = 325 days

Fig. 12. *Tahoua (Niger)* Sahelian ecoclimatic zone, sensu stricto:
P = 385 mm; t = 28.8°C; Alt. = 387 m
2 t = 28.8 × 2 × 12 = 691 mm; PET (68.6 t) = 1976 mm
0.35 PET (Penman) = 802 mm; PET (Penman) = 2294 mm
Water deficit (2 t) = 691 – 385 = 306 mm
(0.35 PET) = 802 – 385 = 417 mm
Growing (rainy) season (P > 2 t) = 84 days
(P > 0.35 PET) = 81 days
Dry season (P < 2 t) = 281 days
(P < 0.35 PET) = 284 days

Fig. 13. *Podor (Senegal)* Sahelian ecoclimatic zone, sensu stricto:
P = 308 mm; t = 28.7°C; Alt. = 6 m
2 t = 28.4 × 2 × 12 = 682 mm; PET (68.6 t) = 1948 mm
0.35 PET (Penman) = 777 mm; PET (Penman) = 2218 mm
Water deficit (2 t) = 682 – 308 = 374 mm
(0.35 PET) = 777 – 308 = 469 mm
Growing (rainy) season (P > 2 t) = 67 days
(P > 0.35 PET) = 65 days
Dry season (P < 2 t) = 298 days
(P < 0.35 PET) = 300 days

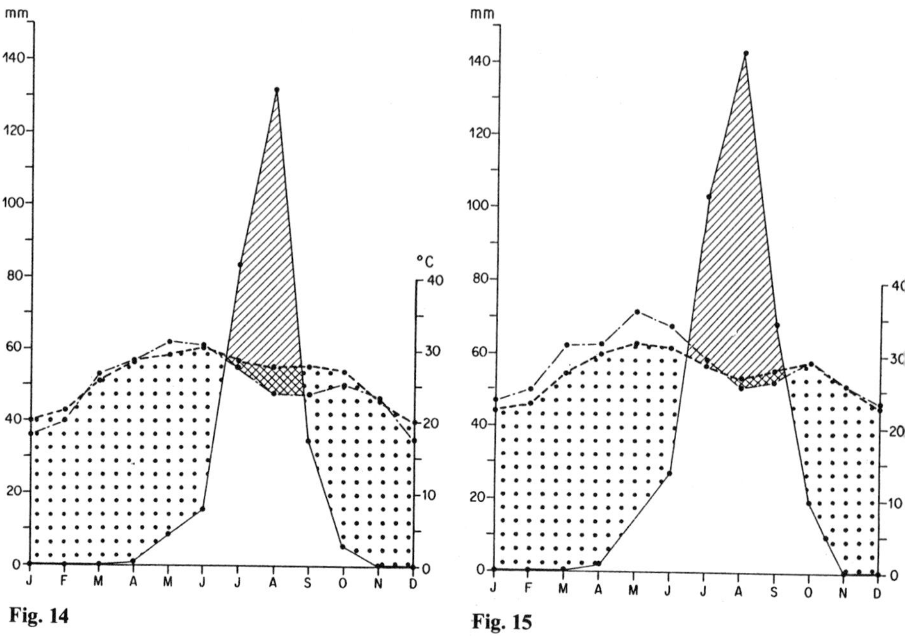

Fig. 14

Fig. 15

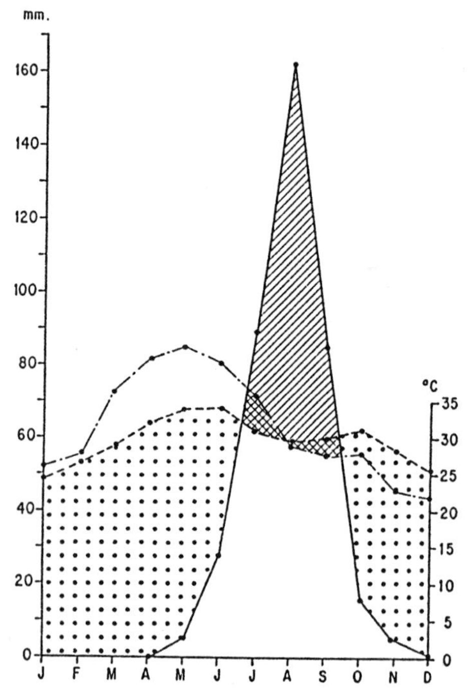

Fig. 16

Fig. 14. *El Fasher (Rep. of Sudan)* Sahelian ecoclimatic zone, sensu stricto:
P = 284 mm; t = 25.7°C; Alt. = 730 m
2 t = 25.7 × 2 × 12 = 617 mm; PET (68.6 t) = 1763 mm
0.35 PET (Penman) = 590 mm; PET (Penman) = 1687 mm
Water deficit (2 t) = 617 – 284 = 333 mm
(0.35 PET) = 590 – 284 = 306 mm
Growing (rainy) season (P > 2 t) = 63 days
(P > 0.35 PET) = 66 days
Dry season (P < 2 t) = 302 days
(P < 0.35 PET) = 299 days

Fig. 15. *El Obeid (Rep. of Sudan)* Sahelian ecoclimatic zone, sensu stricto:
P = 388 mm; t = 24.8°C; Alt. = 570 m
2 t = 24.8 × 2 × 12 = 595 mm; PET (68.6 t) = 1701 mm
0.35 PET (Penman) = 655 mm; PET (Penman) = 2168 mm
Water deficit (2 t) = 595 – 388 = 207 mm
(0.35 PET) = 655 – 388 = 267 mm
Growing (rainy) season (P > 2 t) = 84 days
(P > 0.35 PET) = 85 days
Dry season (P < 2 t) = 281 days
(P < 0.35 PET) = 280 days

Fig. 16. *Kaedi (Mauritania)* Sahelian ecoclimatic zone, sensu stricto:
P = 391 mm; t = 29.7°C; Alt. = 18 m
2 t = 29.7 × 2 × 12 = 713 mm; PET (68.6 t) = 2037 mm
0.35 PET (Penman) = 758 mm; PET (Penman) = 2168 mm
Water deficit (2 t) = 713 – 391 = 322 mm
(0.35 PET) = 758 – 391 = 367 mm
Growing (rainy) season (P > 2 t) = 81 days
(P > 0.35 PET) = 78 days
Dry season (P < 2 t) = 284 days
(P < 0.35 PET) = 287 days

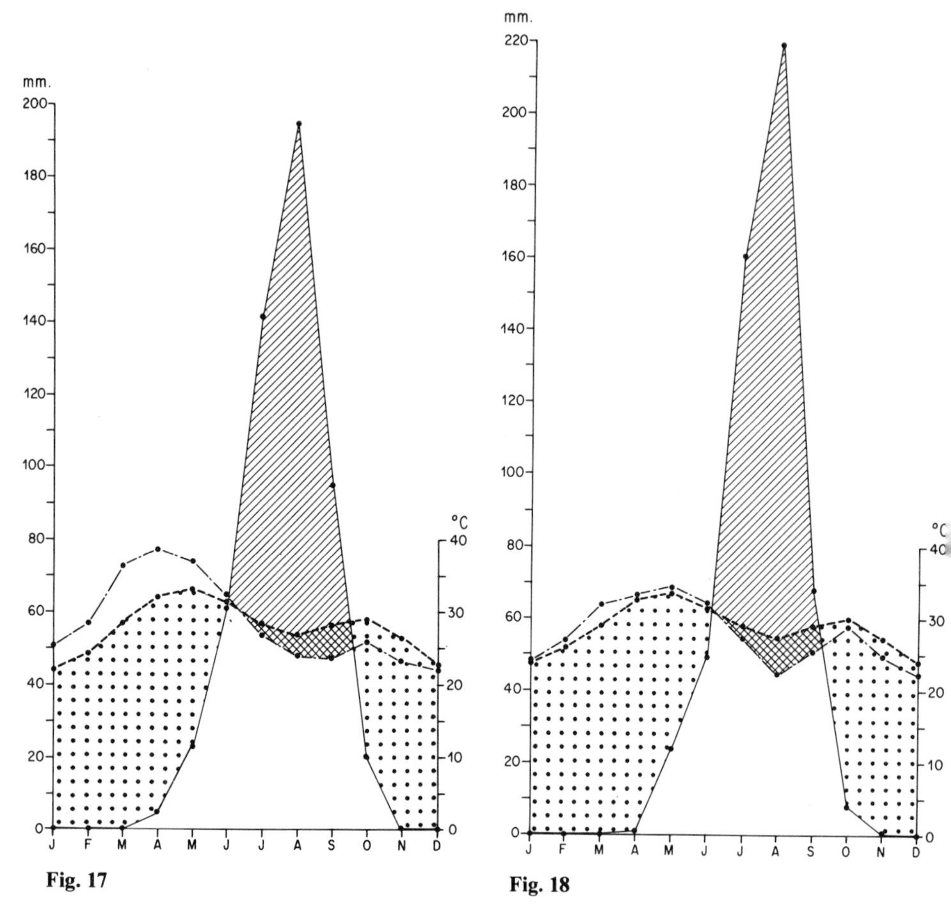

Fig. 17. *Mopti (Mali)* Sudano-Sahelian ecoclimatic zone:
P = 542 mm; t = 27.7°C; Alt. = 272 m
2 t = 27.7 × 2 × 12 = 665 mm; PET (68.6 t) = 1900 mm
0.35 PET (Penman) = 690 mm; PET (Penman) = 1973 mm
Water deficit (2 t) = 665 – 542 = 123 mm
(0.35 PET) = 690 – 542 = 148 mm
Growing (rainy) season (P > 2 t) = 102 days
(P > 0.35 PET) = 107days
Dry season (P < 2 t) = 263 days
(P < 0.35 PET) = 258 days

Fig. 18. *Zinder (Niger)* Sudano-Sahelian ecoclimatic zone:
P = 532 mm; t = 28.4°C; Alt. = 453 m
2 t = 28.4 × 2 × 12 = 682 mm; PET (68.6 t) = 1948 mm
0.35 PET (Penman) = 670 mm; PET (Penman) = 1916 mm
Water deficit (2 t) = 682 – 532 = 150 mm
(0.35 PET) = 670 – 532 = 138 mm
Growing (rainy) season (P > 2 t) = 93 days
(P > 0.35 PET) = 96 days
Dry season (P < 2 t) = 272 days
(P < 0.35 PET) = 269 days

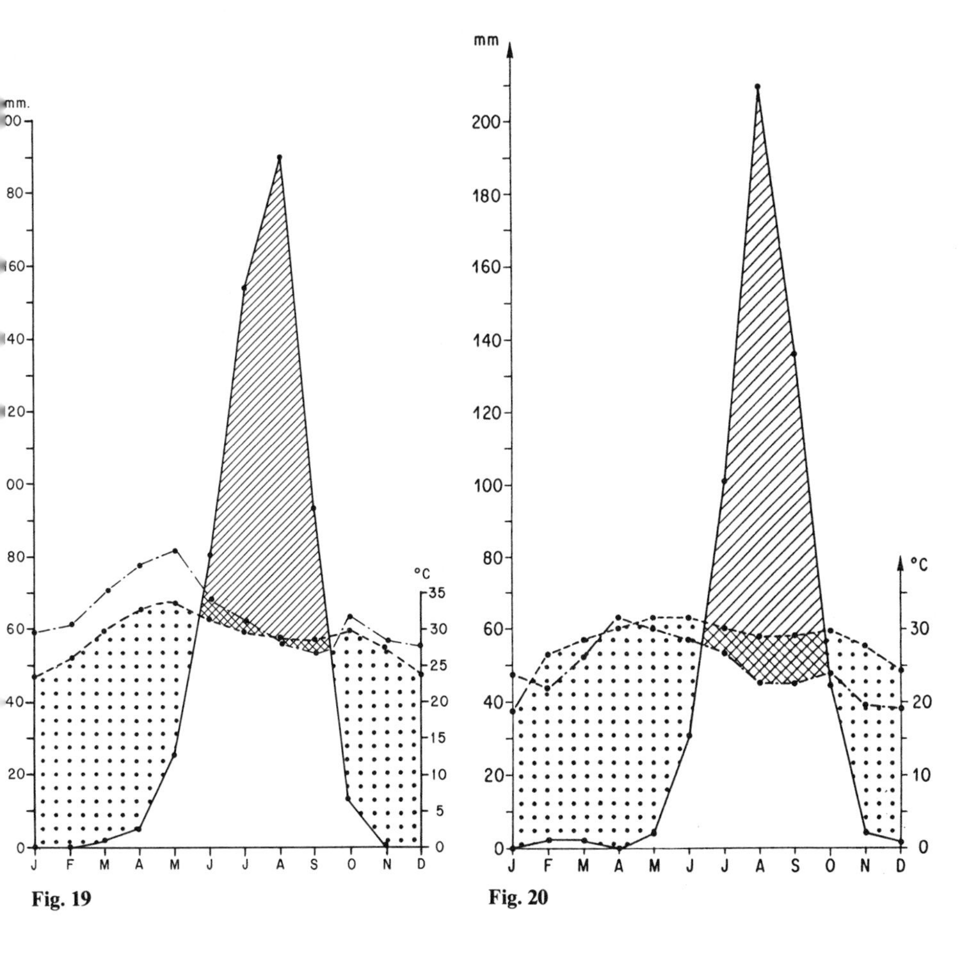

Fig. 19. *Dori (Burkina-Faso)* Sudano-Sahelian ecoclimatic zone:
P = 563 mm; t = 28.8°C; Alt. = 277 m
2 t = 28.8 × 2 × 12 = 682 mm; PET (68.6 t) = 1976 mm
0.35 PET (Penman) = 778 mm; PET (Penman) = 2225 mm
Water deficit (2 t) = 682 – 563 = 128 mm
(0.35 PET) = 778 – 563 = 215 mm
Growing (rainy) season (P > 2 t) = 114 days
(P > 0.35 PET) = 108 days
Dry season (P < 2 t) = 251 days
(P < 0.35 PET) = 257 days

Fig. 20. *Linguère (Senegal)* Sudano-Sahelian ecoclimatic zone:
P = 531 mm; t = 28.2°C; Alt. = 21 m
2 t = 28.2 × 2 × 12 = 677 mm; PET (68.6 t) = 1935 mm
0.35 PET (Penman) = 731 mm; PET (Penman) = 2095 mm
Water deficit (2 t) = 686 – 531 = 155 mm
(0.35 PET) = 731 – 531 = 200 mm
Growing (rainy) season (P > 2 t) = 102 days
(P > 0.35 PET) = 100 days
Dry season (P < 2 t) = 265 days
(P < 0.35 PET) = 257 days

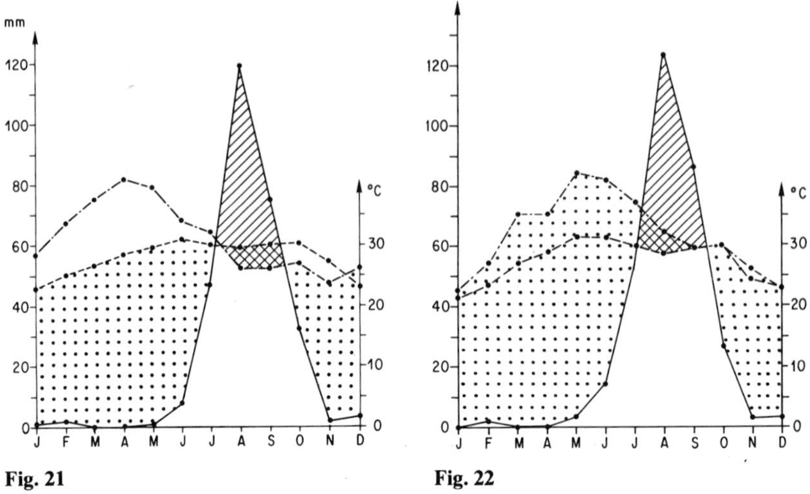

Fig. 21

Fig. 22

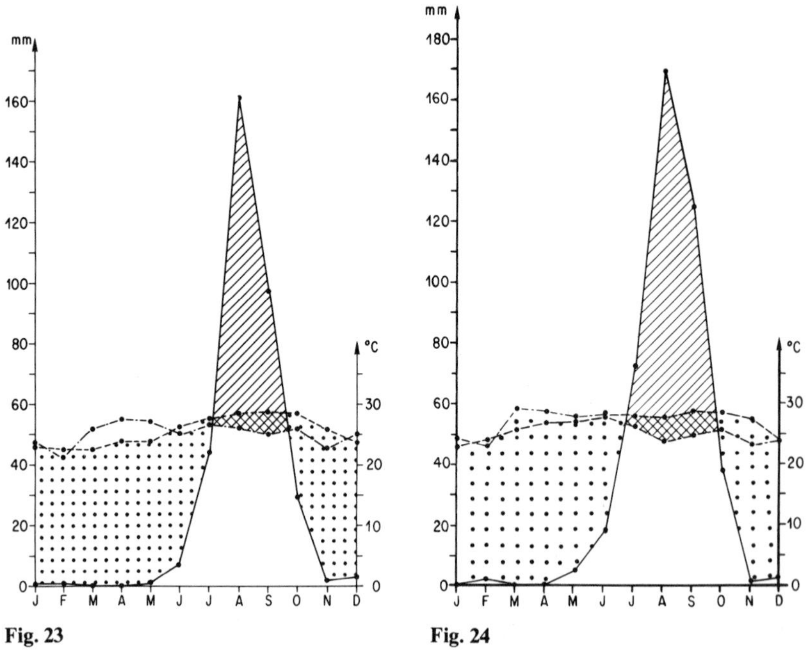

Fig. 23

Fig. 24

Fig. 21. *Rosso (Mauritania)* Sahelian ecoclimatic zone, sensu stricto:
P = 288 mm; t = 27.8 °C; Alt. = 7 m
2 t = 27.8 × 2 × 12 = 667 mm; PET (68.6 t) = 1907 mm
0.35 PET (Penman) = 752 mm; PET (Penman) = 2151 mm
Water deficit (2 t) = 677 – 288 = 379 mm
(0.35 PET) = 752 – 288 = 464 mm
Growing (rainy) season (P > 2 t) = 69 days
(P > 0.35 PET) = 72 days
Dry season (P < 2 t) = 296 days
(P < 0.35 PET) = 293 days

Fig. 22. *Dagana (Senegal)* Sahelian ecoclimatic zone, sensu stricto:
P = 309 mm; t = 28.7 °C; Alt. = 5 m
2 t = 28.7 × 2 × 12 = 689 mm; PET (68.6 t) = 1969 mm
0.35 PET (Penman) = 710 mm; PET (Penman) = 2031 mm
Water deficit (2 t) = 689 – 309 = 380 mm
(0.35 PET) = 710 – 309 = 401 mm
Growing (rainy) season (P > 2 t) = 75 days
(P > 0.35 PET) = 78 days
Dry season (P < 2 t) = 290 days
(P < 0.35 PET) = 287 days

Fig. 23. *St Louis (Senegal)* Coastal Sahelian ecoclimatic zone, sensu stricto:
P = 346 mm; t = 25.2 °C; Alt. = 4 m
2 t = 25.2 × 2 × 12 = 604 mm; PET (68.6 t) = 1729 mm
0.35 PET (Penman) = 603 mm; PET (Penman) = 1725 mm
Water deficit (2 t) = 604 – 346 = 258 mm
(0.35 PET) = 603 – 346 = 257 mm
Growing (rainy) season (P > 2 t) = 75 days
(P > 0.35 PET) = 78 days
Dry season (P < 2 t) = 290 days
(P < 0.35 PET) = 287 days

Fig. 24. *Kebemer (Senegal)* Subcoastal Sudano-Sahelian ecoclimatic zone:
P = 429 mm; t = 27.5 °C; Alt. = 40 m
2 t = 27.5 × 2 × 12 = 660 mm; PET (68.6 t) = 1887 mm
0.35 PET (Penman) = 710 mm; PET (Penman) = 2031 mm
Water deficit (2 t) = 660 – 429 = 231 mm
(0.35 PET) = 710–429 = 281 mm
Growing (rainy) season (P > 2 t) = 100 days
(P > 0.35 PET) = 99 days
Dry season (P < 2 t) = 265 days
(P < 0.35 PET) = 266 days

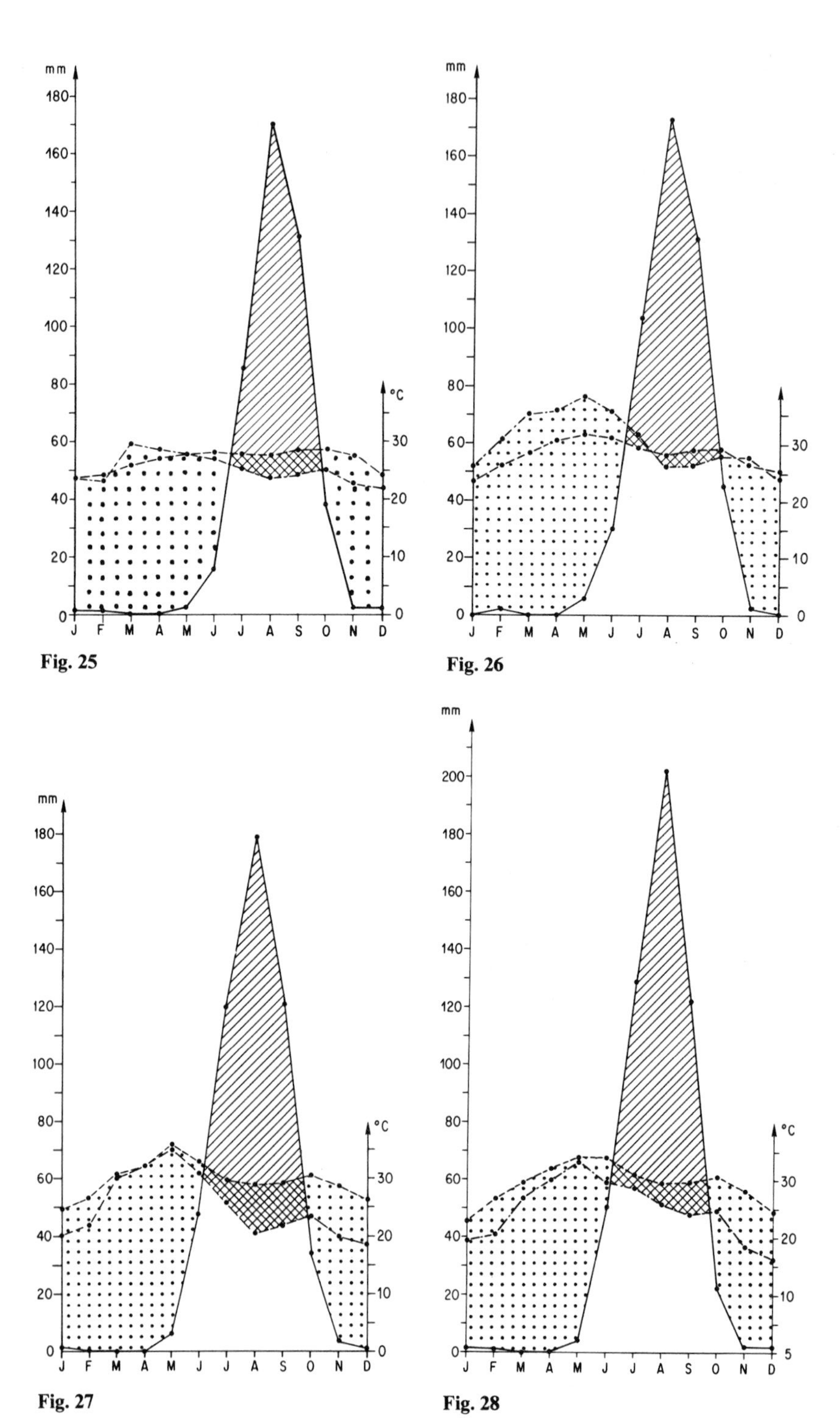

Fig. 25

Fig. 26

Fig. 27

Fig. 28

Fig. 25. *Louga (Senegal)* Subcoastal Sudano-Sahelian ecoclimatic zone:
P = 447 mm; t = 27.5°C; Alt. = 38 m
2 t = 27.5 × 2 × 12 = 660 mm; PET (68.6 t) = 1887 mm
0.35 PET (Penman) = 710 mm; PET (Penman) = 2031 mm
Water deficit (2 t) = 660 – 447 = 213 mm
(0.35 PET) = 710 – 447 = 263 mm
Growing (rainy) season (P > 2 t) = 100 days
(P > 0.35 PET) = 93 days
Dry season (P < 2 t) = 265 days
(P < 0.35 PET) = 272 days

Fig. 26. *Dahra (Senegal)* Sudano-Sahelian ecoclimatic zone:
P = 496 mm; t = 28.6°C; Alt. = 39 m
2 t = 28.6 × 2 × 12 = 686 mm; PET (68.6 t) = 1962 mm
0.35 PET (Penman) = 720 mm; PET (Penman) = 2059 mm
Water deficit (2 t) = 686 – 496 = 190 mm
(0.35 PET) = 720 – 496 = 224 mm
Growing (rainy) season (P > 2 t) = 102 days
(P > 0.35 PET) = 100 days
Dry season (P < 2 t) = 263 days
(P < 0.35 PET) = 265 days

Fig. 27. *Bakel (Senegal)* Sudano-Sahelian ecoclimatic zone:
P = 509 mm; t = 29.5°C; Alt. = 22 m
2 t = 29.5 × 2 × 12 = 708 mm; PET (68.6 t) = 2024 mm
0.35 PET (Penman) = 700 mm; PET (Penman) = 2002 mm
Water deficit (2 t) = 708 – 509 = 199 mm
(0.35 PET) = 700 – 509 = 191 mm
Growing (rainy) season (P > 2 t) = 117 days
(P > 0.35 PET) = 114 days
Dry season (P < 2 t) = 248 days
(P < 0.35 PET) = 251 days

Fig. 28. *Matam (Senegal)* Sudano-Sahelian ecoclimatic zone:
P = 535 mm; t = 29.5°C; Alt. = 17 m
2 t = 29.5 × 2 × 12 = 708 mm; PET (68.6 t) = 2024 mm
0.35 PET (Penman) = 686 mm; PET (Penman) = 1962 mm
Water deficit (2 t) = 708 – 535 = 173 mm
(0.35 PET) = 686 – 535 = 151 mm
Growing (rainy) season (P > 2 t) = 103 days
(P > 0.35 PET) = 109 days
Dry season (P < 2 t) = 262 days
(P < 0.35 PET) = 256 days

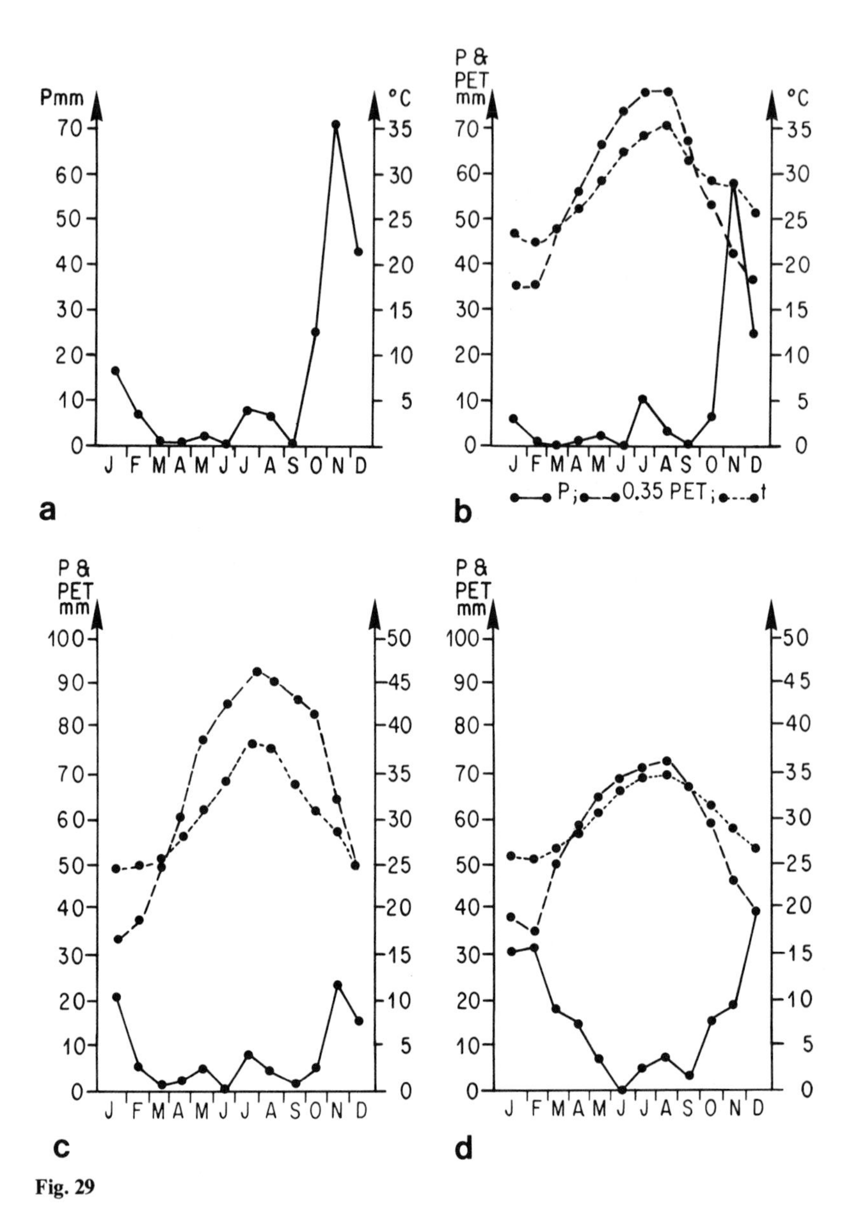

Fig. 29

Fig. 29a. *Suakin (Rep. of Sudan)* arid Mediterranean ecoclimatic zone, with hot winters:
P = 181 mm; t = 30°C; Alt. = 5 m
2 t = 30 × 2 × 12 = 720 mm; PET (68.6 t) = 2058 mm
Water deficit (2 t) = 720 − 181 = 539 mm

Growing (rainy) season (2 t) = 25 days
Dry season (2 t) = 340 days

b. *Port Sudan (Rep. of Sudan)* arid Mediterranean ecoclimatic zone with hot winters:
P = 111 mm; t = 28.3°C; Alt. = 2 m
2 t = 28.3 × 2 × 12 = 679 mm; PET (68.6 t) = 1941 mm
0.35 PET (Penman) = 662 mm; PET (Penman) = 1891 mm
Water deficit (2 t) = 679 − 111 = 568 mm
(0.35 PET) = 662 − 111 = 551 mm

Growing (rainy) season (P > 2 t) = 0 day
(P > 0.35 PET) = 15 days
Dry season (P < 2 t) = 365 days
(P < 0.35 PET) = 350 days

c. *Tokar (Rep. of Sudan)* hyperarid Mediterranean ecoclimatic zone with hot winters:
P = 90 mm; t = 29.9°C; Alt. = 19 m
2 t = 29.9 × 2 × 12 = 718 mm; PET (68.6 t) = 2051 mm
0.35 PET (Penman) = 812 mm; PET (Penman) = 2320 mm
Water deficit (2 t) = 718 − 90 = 628 mm
= 812 − 90 = 722 mm

Growing (rainy) season (P > 2 t) = 0 day
(P > 0.35 PET) = 0 day
Dry season (P < 2 t) = 365 days
(P < 0.35 PET) = 365 days

d. *Masawa (Ethiopia)* arid Mediterranean ecoclimatic zone with hot winters:
P = 181 mm; t = 29.8°C; Alt. = 10 m
2 t = 29.8 × 2 × 12 = 717 mm; PET (68.6 t) = 2044 mm
0.35 PET (Penman) = 677 mm; PET (Penman) = 1933 mm
Water deficit (2 t) = 717 − 181 = 534 mm
(0.35 PET) = 677 − 181 = 496 mm

Growing (rainy) season (P > 2 t) = 0 day
(P > 0.35 PET) = 10 days
Dry season (P < 2 t) = 365 days
(P < 0.35 PET) = 255 days

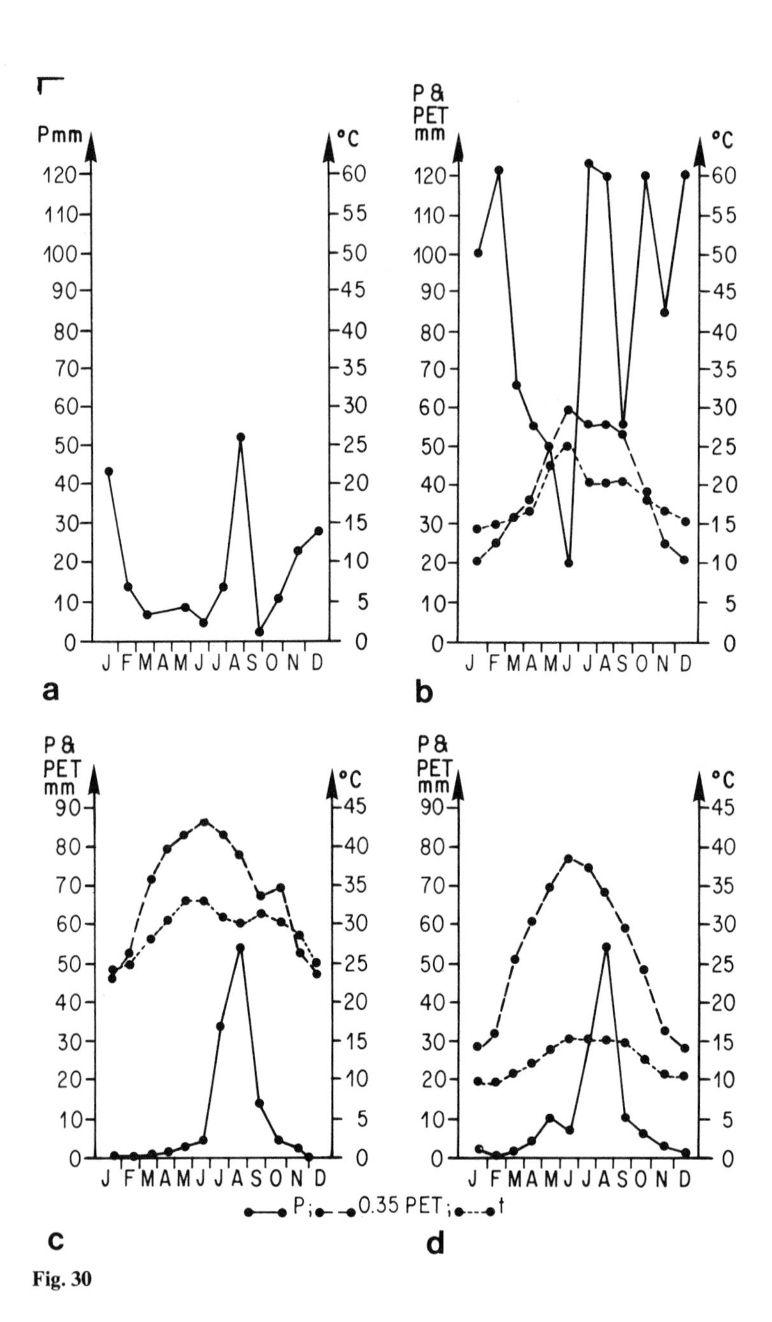

Fig. 30

Fig. 30a. *Erkwit (Rep. of Sudan)* montane arid Mediterranean-tropical transition ecoclimatic zone (bimodal rainfall regime: winter and summer maxima):

P = 218 mm; t = 20.0°C; Alt. = 1150 m
2 t = 20.0 × 2 × 12 = 480 mm; PET (68.6 t) = 1372 mm
Water deficit (2 t) = 480 – 218 = 262 mm

Growing (rainy) season (2 > 2t) = 10 + 15 = 25 days
Dry season (2 > 2t) = 172 + 168 = 340 days

b. *Faghena (Ethiopia)* montane subhumid equatorial, plurimodal ecoclimatic zone (4 rainfall maxima):

P = 1048 mm; t = 19.3°C; Alt. = 1760 m
2 t = 19.3 × 2 × 12 = 463 mm; PET (68.6 t) = 1324 mm
0.35 PET (Penman) = 487 mm; PET (Penman) = 1390 mm
Water balance (2 t) = 1048 – 463 = +585 mm
(0.35 PET) = 1048 – 487 = +561 mm

Growing (rainy) season (P > 2 t) = 330 days
(P > 0.35 PET) = 320 days
Dry season (P < 2 t) = 35 days
(P < 0.35 PET) = 45 days

c. *Derudeb (Rep. of Sudan)* midland arid Saharo-Sahelian ecoclimatic zone:

P = 117 mm; t = 29.2°C; Alt. = 518 m
2 t = 29.2 × 2 × 12 = 701 mm; PET (68.6 t) = 2003 mm
0.35 PET (Penman) = 812 mm; PET (Penman) = 2320 mm
Water deficit (2 t) = 701 – 117 = 584 mm
(0.35 PET) = 812 – 117 = 695 mm

Growing (rainy) season (P > 2 t) = 0 day
(P > 0.35 PET) = 0 day
Dry season (P < 2 t) = 365 days
(P < 0.35 PET) = 365 days

d. *Gebeit (Rep. of Sudan)* midland Saharo-Sahelian ecoclimatic zone:

P = 127 mm; t = 25.4°C; Alt. = 796 m
2 t = 25.4 × 2 × 12 = 610 mm; PET (68.6 t) = 1742 mm
0.35 PET (Penman) = 672 mm; PET (Penman) = 1920 mm
Water deficit (2 t) = 610 – 127 = 483 mm
(0.35 PET) = 672 – 127 = 545 mm

Growing (rainy) season (P > 2 t) = 40 days
(P > 0.35 PET) = 0 day
Dry season (P < 2 t) = 325 days
(P < 0.35 PET) = 365 days

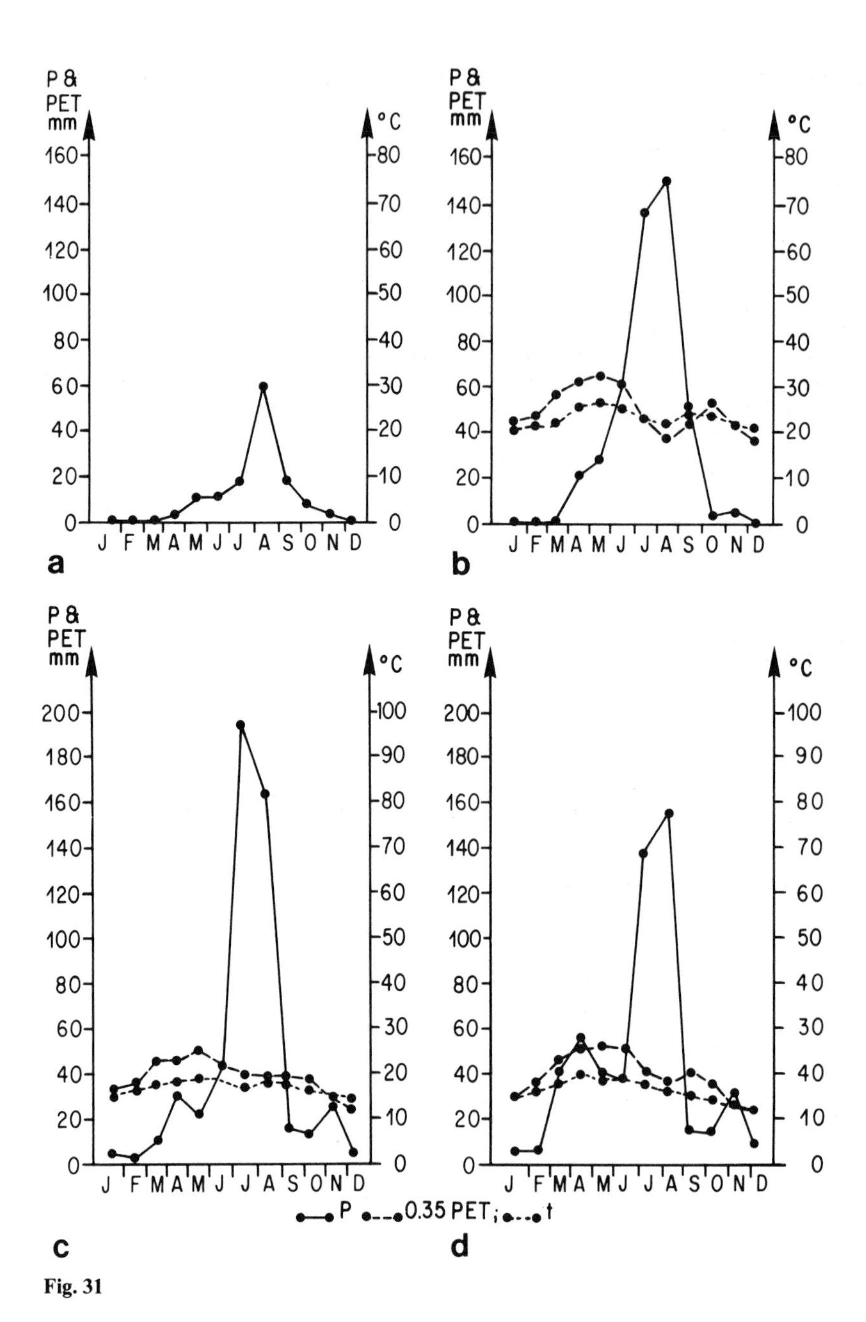

Fig. 31

Fig. 31a. *Sinkat (Rep. of Sudan)* midland Saharo-Sahelian ecoclimatic zone:

P = 127 mm; t = 27.0°C; Alt. = 840 m

2 t = 27.0 × 2 × 12 = 648 mm; PET (68.6 t) = 1852 mm

Water deficit (2 t) = 648 – 127 = 521 mm

Growing (rainy) season (> 2 t) = 15 days

Dry season (P > 2 t) = 350 days

b. *Keren (Ethiopia)* montane Sudano-Sahelian ecoclimatic zone:

P = 456 mm; t = 23.0°C; Alt. = 1460 m

2 t = 23.0 × 2 × 12 = 552 mm; PET (68.6 t) = 1578 mm

0.35 PET (Penman) = 587 mm; PET (Penman) = 1679 mm

Water deficit (2 t) = 552 – 456 = 96 mm

(0.35 PET) = 587 – 456 = 131 mm

Growing (rainy) season (P > 2 t) = 90 days

(P > 0.35 PET) = 90 days

Dry season (P < 2 t) = 275 days

(P < 0.35 PET) = 275 days

c. *Asmara (Ethiopia)* montane Sudano-Sahelian ecoclimatic zone:

P = 525 mm; t = 16.6°C; Alt. = 2325 m

2 t = 16.6 × 2 × 12 = 398 mm; PET (68.6 t) = 1139 mm

0.35 PET (Penman) = 465 mm; PET (Penman) = 1328 mm

Water balance (2 t) = 525 – 398 = +127 mm

(0.35 PET) = 525 – 465 = + 60 mm

Growing (rainy) season (P > 2 t) = 95 days

(P > 0.35 PET) = 90 days

Dry season (P < 2 t) = 270 days

(P < 0.35 PET) = 275 days

d. *Adigrat (Ethiopia)* montane semiarid equatorial ecoclimatic zone:

P = 552 mm; t = 16.0°C; Alt. = 2457 m

2 t = 16.0 × 2 × 12 = 384 mm; PET (68.6 t) = 1098 mm

0.35 PET (Penman) = 476 mm; PET (Penman) = 1360 mm

Water balance (2 t) = 552 – 384 = +168 mm

(0.35 PET) = 552 – 476 = + 76 mm

Growing (rainy) season (P > 2 t) = 180 days

(P > 0.35 PET) = 120 days

Dry season (P < 2 t) = 185 days

(P < 0.35 PET) = 245 days

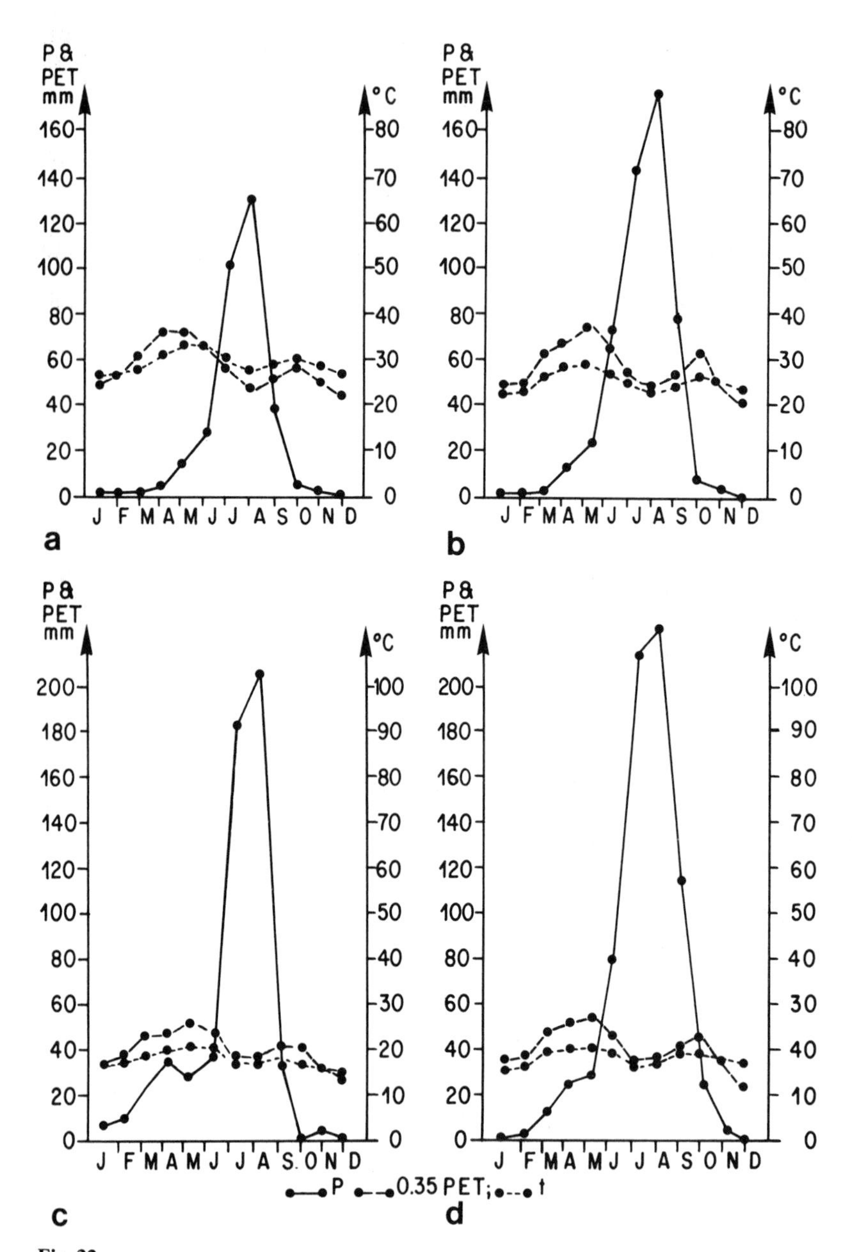

Fig. 32

Fig. 32a. *Agordat (Ethiopia)* midland Sahelian ecoclimatic zone:
P = 319 mm; t = 29.0° C; Alt. = 626 m
2 t = 29.0 × 2 × 12 = 696 mm; PET (68.6 t) = 1989 mm
0.35 PET (Penman) = 677 mm; PET (Penman) = 1934 mm
Water deficit (2 t) = 696 – 319 = 377 mm
(0.35 PET) = 677 – 319 = 358 mm
Growing (rainy) season (P > 2 t) = 75 days
(P > 0.35 PET) = 72 days
Dry season (P < 2 t) = 290 days
(P < 0.35 PET) = 293 days

b. *Barentu (Ethiopia)* submontane Sudano-Sahelian ecoclimatic zone:
P = 516 mm; t = 25.1° C; Alt. = 980 m
2 t = 25.1 × 2 × 12 = 602 mm; PET (68.6 t) = 1722 mm
0.35 PET (Penman) = 652 mm; PET (Penman) = 1862 mm
Water deficit (2 t) = 602 – 516 = 86 mm
(0.35 PET) = 652 – 516 = 136 mm
Growing (rainy) season (P > 2 t) = 120 days
(P > 0.35 PET) = 105 days
Dry season (P < 2 t) = 245 days
(P < 0.35 PET) = 260 days

c. *Mekele (Ethiopia)* montane Sudano-Sahelian ecoclimatic zone:
P = 563 mm; t = 18.0° C; Alt. = 2212 m
2 t = 18.0 × 2 × 12 = 432 mm; PET (68.6 t) = 1235 mm
0.35 PET (Penman) = 478 mm; PET (Penman) = 1365 mm
Water balance (2 t) = 562 – 432 = +131 mm
(0.35 PET) = 562 – 478 = + 85 mm
Growing (rainy) season (P > 2 t) = 100 days
(P > 0.35 PET) = 98 days
Dry season (P < 2 t) = 265 days
(P < 0.35 PET) = 267 days

d. *Adwa (Ethiopia)* montane subhumid Sudanian ecoclimatic zone:
P = 742 mm; t = 19.1° C; Alt. = 1980 m
2 t = 19.1 × 2 × 12 = 458 mm; PET (68.6 t) = 1310 mm
0.35 PET (Penman) = 506 mm; PET (Penman) = 1446 mm
Water balance (2 t) = 742 – 458 = +284 mm
(0.35 PET) = 742 – 506 = +236 mm
Growing (rainy) season (P > 2 t) = 135 days
(P > 0.35 PET) = 130 days
Dry season (P < 2 t) = 230 days
(P < 0.35 PET) = 235 days

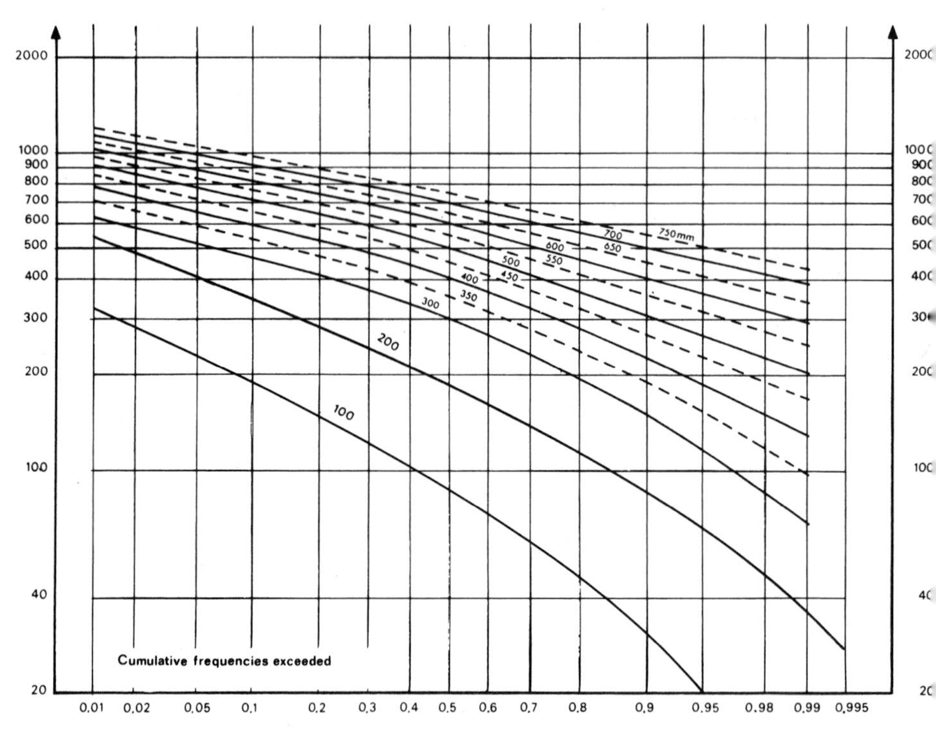

Fig. 33. Annual rainfall probability abacuses for the Sahel (Modified after Rodier 1982)

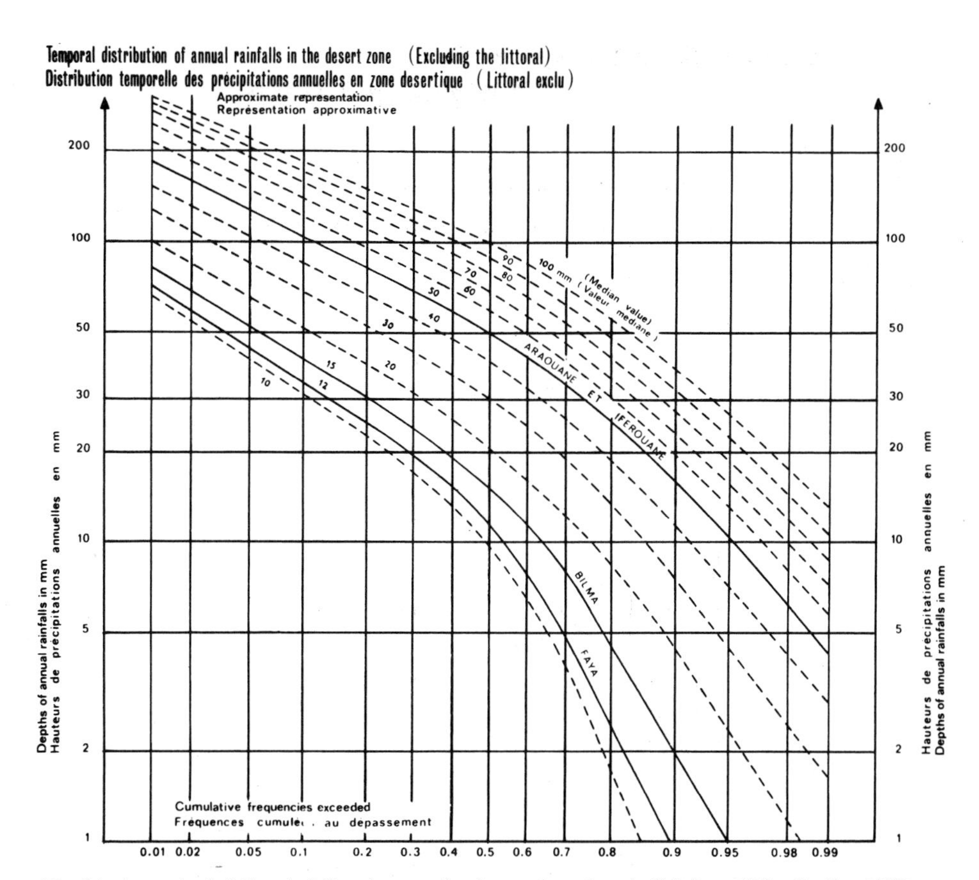

Fig. 34. Annual rainfall probability abacuses for the southern (tropical) Sahara (After Rodier, 1982)

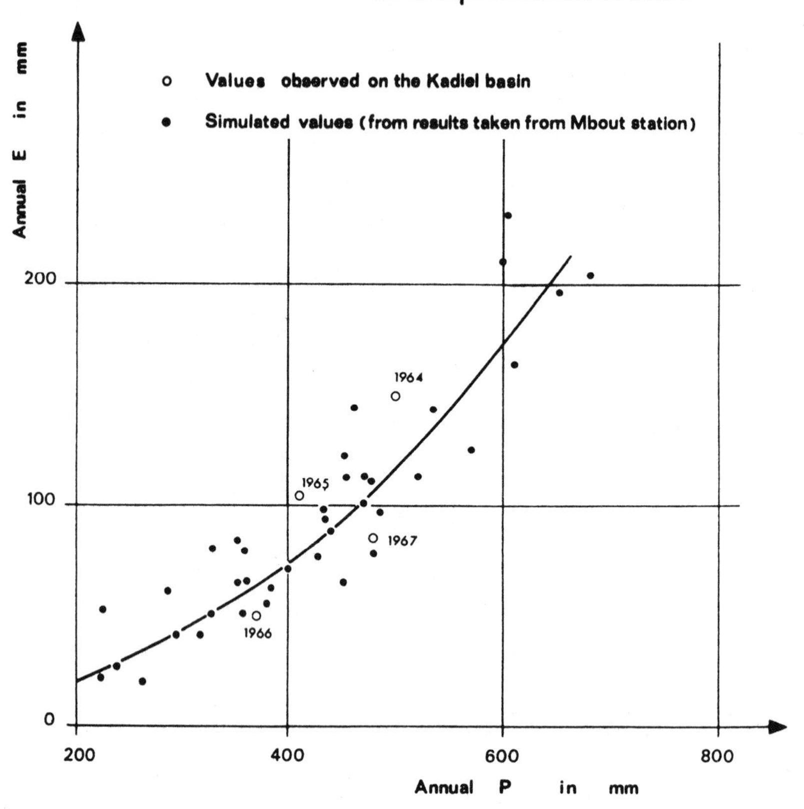

Fig. 35. Relationship between annual rainfall and annual runoff on a small watershed (25 km^2) on basement complex shists of the Ader Doutchi Plateau (p = 450 mm), Sahel Zone of Niger (After Rodier 1982)

Fig. 37. Rainy season precipitations (June-October); isohyets are based upon all records available for each station (Bellocq 1983)

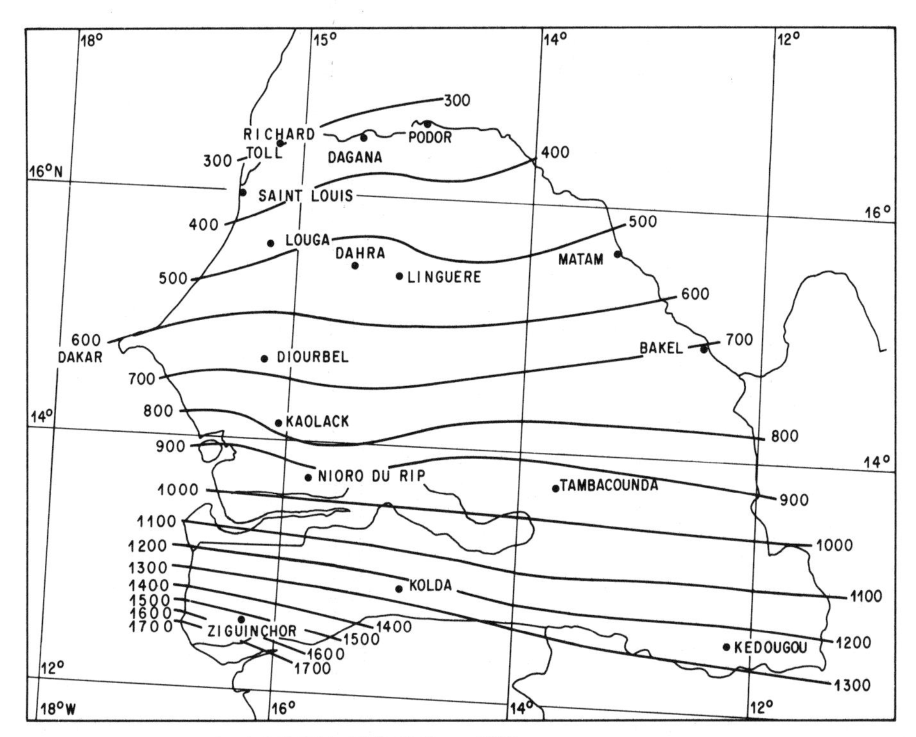

Fig. 36. Mean annual rainfall 1931–1960 (Bellocq 1983)

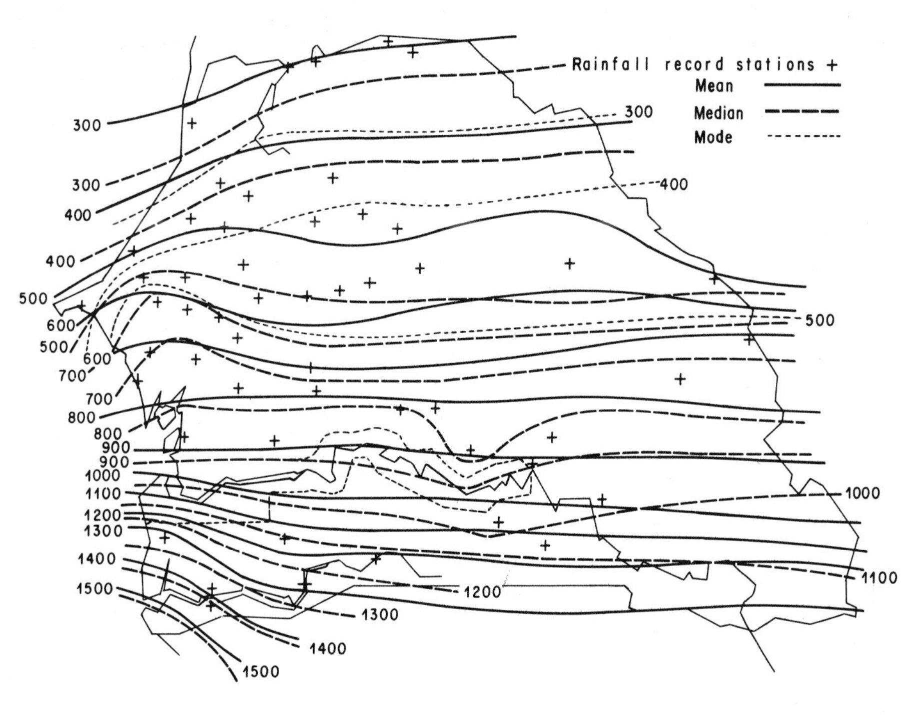

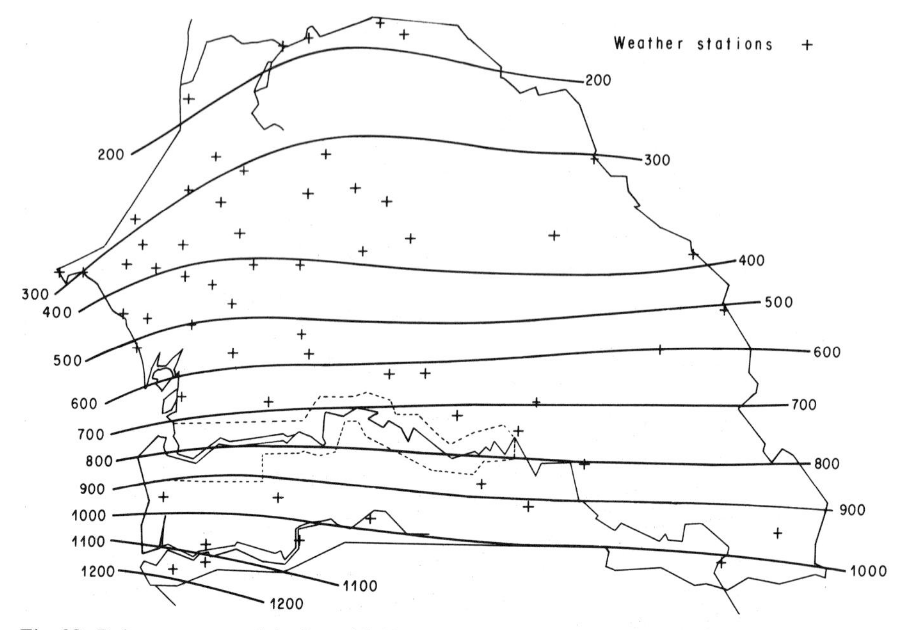

Fig. 38. Rainy season precipitation with 80% probability (Bellocq 1983)

Contrary to what has been often stated, the variability of rainfall in the Sahel is not higher than in other arid zones having a similar total average annual amount; rather the opposite, variability for a given isohyet is substantially greater in the Mediterranean arid zone (Le Houérou 1982/84), in the North American subtropics and in the Brasilian Nordeste (Le Houérou and Norwine 1988; Le Houérou 1989. Tables 4–8).

The peak of the rainy season occurs in August throughout the region; this is also the period in which monthly rain is the least unreliable (Le Houérou and Popov 1981).

Dependable annual rains, i.e. having a 75–80% probability of occurrence (Table 8; Figs 33–38) represent about 60–75% of the annual mean; the ratio increases with the mean from about 60–65% under 300 mm in northern Senegal to 80–85% under 1500 mm in sourthern Senegal (Dancette and Hall 1979). There does not seem to have been a long-term trend in rainfall evolution for the past 3000 years, but sequences of years with above and below long-term averages. The 1970–1985 mean is well below the long-term mean, perhaps by 30–40% or locally more (Le Houérou and Gillet 1985; Dancette 1985).

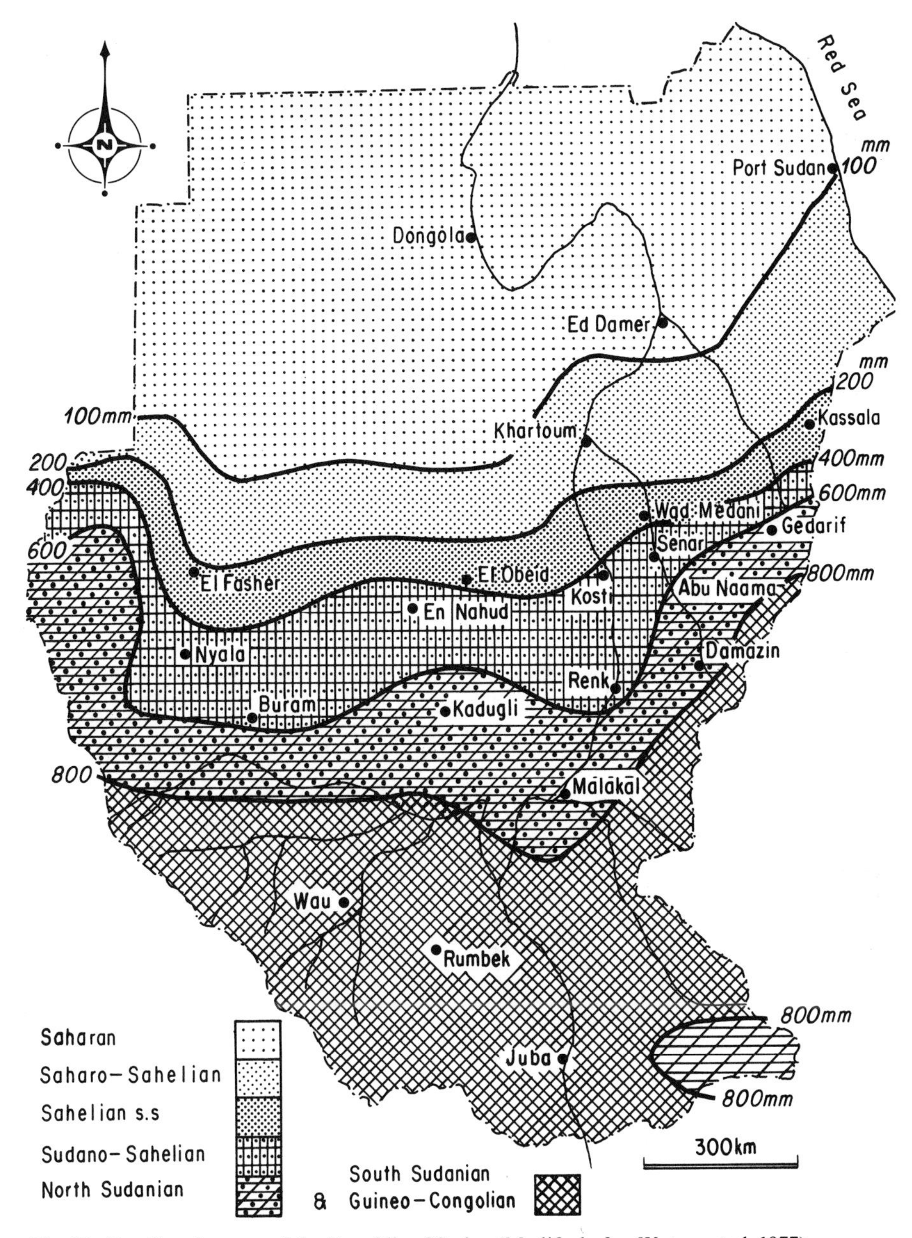

Fig. 39. Ecoclimatic zones of the Republic of Sudan (Modified after Watson et al. 1977)

Table 4. Variability of annual rainfall in the Sahara, Sahelian, Sudanian and Guinean ecological zones, north of the Equator (Le Houérou 1986)

Number of stations	Range of mean annual rain (mm)	Mean of the series (mm)	Mean coefficient of variation (C.V.) (%)	Standard error of the mean C.V. (%)
11	17–99	55	81	0.64
9	100–199	153	43	0.27
14	200–299	247	37	0.16
12	300–399	338	32	0.16
15	400–499	458	30	0.15
17	500–599	551	27	012
14	600–699	639	26	0.15
18	700–799	756	21	0.12
18	800–899	855	20	0.07
9	900–999	964	19	0.11
24	1000–1199	1105	17	0.04
26	1200–1399	1291	18	0.06
18	1400–3000	1842	17	0.02
205				

Table 5. Variability of annual rains in Africa north of the Sahara (Le Houérou 1982/84; data from 200 weather stations, in Algeria, Libya and Tunisia)

Mean annual precipitation (mm)	Mean coefficient of variation (%)	SE of the coefficient of variation (%)
0–50	106 (68–150)	0.8
50–100	68 (65–102)	0.7
100–200	57 (35–90)	0.6
200–300	36 (26–55)	0.4
300–400	37 (26–50)	0.4
400–600	31 (25–45)	0.6
600–800	28 (20–35)	0.3
800–1000	23 (20–25)	0.4
1000–1500	18 (15–20)	0.3

Table 6. Mean annual rainfall and 10% probability maximum rainfall in 24 h in the Sahel

Number of Stations	Range of mean annual rainfall	Mean of series $\bar{X}$	SE	Mean P max. in 24 h, 0.1 probability	SE	$\frac{P\ max}{\bar{X}}$
7	0–100	41.5	7.3	39.6	3.2	0.95
8	100–200	142.0	12.2	57.6	3.2	0.41
11	200–300	255.0	7.2	77.5	3.2	0.30
9	300–400	338.0	10.7	87.1	2.6	0.26
8	400–500	450.0	12.0	97.4	4.9	0.22
22	500–600	552.0	6.8	104.8	3.6	0.18
14	600–700	650.0	9.6	118.6	5.8	0.18

Table 7. Annual rainfall probabilities in the Sahel (Le Houérou 1986)

Mean annual rainfall (mm)	Probabilities								
	0.10	0.20	0.30	0.40	0.50	0.60	0.70	0.80	0.90
100	175	150	130	115	90	85	70	50	25
200	310	270	240	210	185	160	140	120	90
400	550	500	460	420	390	360	325	290	250
600	780	720	675	630	595	560	520	480	420

Table 8. Ratio of dependable rains (p 0.80) to the mean and to the median in the Sahel

Mean	DR/$\bar{X}$	DR/$\bar{m}$
100	50/100 = 0.50	50/90 = 0.55
200	120/200 = 0.60	120/185 = 0.65
400	290/400 = 0.72	290/390 = 0.74
600	480/600 = 0.80	480/595 = 0.81

2.1.3 Temperature (Figs. 8-32)

The mean annual temperature varies from 25° to 30°C, the mean annual minimum being 18–20°C and the mean maximum 35–38°C. The mean monthly minimum drops to 10–15°C in January; the mean monthly maximum increases to 38–44°C in May/June. The absolute minimum may drop to 0°C or slightly less in the northern Sahel, while the absolute maximum may rise above 50°C.

Some weather stations along the "thermic equator" of the 16°N parallel are among the rather rare places on earth where the mean annual temperature reaches 30°C: Nema, Kiffa in E. Mauritania; Kayes, Gao, Timbuktoo, Hombori and Menaka in Mali; Tillabery and Agades in Niger; Abeche in Chad; Shendi, Shambat, Khartoum and Kessala in the Republic of Sudan. The Sahel is thus probably the hottest, broad ecological region on earth and certainly hotter than the Sahara when the mean annual temperature is taken as a criterion.

Of course, temperatures are somewhat milder in a strip some 50 km wide along the Atlantic Ocean; partly due to the upwellings associated with the Stream of the Canaries; the eastern part, east of Lake Chad, is somewhat hotter than the western half.

2.1.4 Solar Radiation (Landsberg et al. 1965)

Global radiation varies from 160 Klangleys yr^{-1} to the south-west to 200 Klangleys to the north-east. It is about 400–500 Klangleys day^{-1} in December-January and increases to 600–700 in May-June. The daily mean insolation is 8–11 h day^{-1}, i.e. 70–90% of the astronomic potential during the dry season and dropping to 50–60% for the rainy season.

2.1.5 Wind (Cochemé and Franquin 1967)

Mean annual wind speed averages 3 m s^{-1} (2.5–5.0). It is higher in the Cape Verde peninsula (Dakar: 7.1 m s^{-1}) and in the Niger River Bend (Timbuctoo and Gao: 4–5 m s^{-1}). The mean monthly maximum of 4–5 m s^{-1} occurs usually during the second half of the dry season when the NE continental trade winds, locally called "Harmattan", occur from January to May, with a very low air moisture, while the minimum occurs during the second part of the rainy season and the early dry season, from August to November.

Rainy season winds come from the SW, with the monsoon, i.e. from the Gulf of Guinea.

2.1.6 Air Humidity and Saturation Deficit (Leroux 1983)

Air moisture is fairly uniform across the region for a given season; the monthly mean minimum and mean maximum air humidity is shown below (Leroux 1983):

	J	F	M	A	M	J	J	A	S	O	N	D	Yr
Minimum	20	15	10	10	20	30	40	60	40	30	20	20	25
Maximum	30	25	20	25	40	60	70	85	70	60	40	30	45
Average	25	20	15	17	30	45	55	72	55	45	30	25	35

Again, these values are among the lowest on earth for a broad ecological region; particularly lower, in average, than in most parts of the Sahara (Le Houérou 1982/84).

The vapor pressure deficit is 5–10 millibars at night during dry season and 30 to 45 mb in the daytime; in August it drops to 0–3 mb at night and 15–30 in daytime (Davy et al. 1976).

2.1.7 Evaporation and Evapotranspiration (Davy et al. 1976; Riou 1975; Dancette 1979; Frère and Popov 1984) (Figs. 8-32)

Annual evaporation over Lake Chad is 2350 mm, (Tetzlaff and Adams 1983) while computed potential evapotranspiration at N'Djamena is 1800 mm (Cochemé and Franquin 1967; Riou 1975; Davy et al. 1976; Dancette and Hall 1979). PET in the Sahel varies from less than 1800 mm yr^{-1} in the south to over 2200 to the north (Cochemé and Franquin 1967; Davy et al. 1976; Le Houérou 1988c) as computed via Penman's formula and/or experimentally measured in lysimeters (the difference between the figures from experiments and computed from Penman's formula differ by less than 4% (Riou 1975). Calculated or measured PET is about 60% of class A pan on a year-round basis (30% in the rainy season and 70% in the dry season) (Riou 1975). Daily PET is around a low 4 mm day^{-1} in January and a high 6.5 mm day^{-1} in May-June. PET is naturally substantially lower along the ocean shore on a strip about 50 km wide where it is 100–300 mm (i.e. 5–15%) lower than the zonal inland values (Davy et al. 1976). Conversely, it may be higher than the zonal value in areas which are subject to stronger winds, such as the Bend of the Niger (Timbuctoo, Gao).

2.1.8 Seasons (Le Houérou 1980b)

There are four main seasons in the annual cycle:

1. The rainy season from mid-June to mid-September;
2. The "deferred season" from September to November when temperatures are still high, air moisture relatively high, some moisture may still remain in some soils in depressions. Most ponds and superficial water resources dry out. This is the time when transhumants and nomads return to their dry season quarters;
3. The cool dry season from November to February;
4. The hot dry season from March to May.

A fifth season is sometimes singled out, quite appropriately: the prerainy season in May-June when air moisture steadily increases as the ITCZ marches northward with the monsoon of the Gulf of Guinea; the first thunderstorms break off with numerous aborted rains (small rains evaporate before they reach the ground).

Men and animals, exhausted by the long, dry and hot season and the high level of static electricity, have become nervous and excitable (this state is sometimes referred to by local Europeans as the time of "Sudanitis", a disease characterized by a quarrelsome mood). This season is also characterized by the "precession of foliation" in trees and shrubs which is an early break of dormancy before the onset of the rains: many species of trees and shrubs develop new leaves 15–60 days before the first rains. The precession of foliation is a very important phenomenon for domestic herbivores and wildlife since green herbage has long since disappeared and the only feed available is dried grass, i.e. with no vitamins, no carotene, no protein and virtually no phosphorus (Clanet and Gillet 1980; Lawton 1980; Walker 1980; Le Houérou 1978, 1979c, 1980b). Herbivores are then able to balance their diet from this new leave flush and recover their good health by the onset of the rains. The precession of foliation seems to be triggered by the steady increase of air humidity resulting from the northward march of the ITCZ and the monsoon of the Gulf of Guinea, contrasting sharply with the latter conditions of very dry and hot air blown from the NE by the "Harmattan" (continental trade wind). The development of new leaves does not seem to involve any absorption of moisture and nutrients from the environment: water, protein and carbohydrates transmigrate from roots and stem to the twigs; transpiration seems to be prevented by a thick wax cuticle which disappears later, in younger leaves, when the rains are well established (Walker 1980). The precession of foliation does not therefore constitute growth as such; although these new leaves constitute a highly nutritive feed very rich in protein, phosphorus and vitamins, at a time when these are most needed since most herbivores are then at a late pregnancy stage. This is also the time when herds will start to move towards their rainy season pastures, as the ponds begin to fill up and good browse is available along the trail, before the grass begins to grow.

The hot dry season (March-May) is an extremely harsh time: temperatures rise daily above 40 °C and do not drop below 35 °C at night, while air humidity drops below 10% every day to rise to 20–30% at night. Ponds have long since dried out, most trees and shrubs have shed their leaves; the landscape is parched by the intense radiation and scorched by numerous bushfires that sweep across 15 to 20% of the country.

The rainy season, conversely, is the time of bounty and festivity, water and milk are plentiful (unless there is a generalized drought).

2.2 Edaphic Factors

2.2.1 Geology (Furon 1950)

The overall geological structure of the region is fairly simple; the basement complex of the African Plate is covered with a thin and discontinuous mantle of sediments with two marine and two continental episodes coming in an alternate succession: the Paleozoic transgressions, from the Cambrian to the Gotlandian, left large deposits, especially to the west; there are virtually no deposits from the Upper Paleozoic and Lower Mesozoic; then continental deposits are found from the Middle Mesozoic (Continental Intercalary); again marine transgressions at the Upper Cretaceous (Cenomanian to Danian) are followed by Paleogene transgressions, then new continental deposits (Continental Terminal) from Oligocene to the Holocene.

The Precambrian basement complex of the African Plate is made of three more or less metamorphosed, folded and faulted systems, with occasional eruptive seams. The older cycle, the most intensely metamorphosed (Suggarian, Dahomeyan) is essentially composed of granites and gneisses. The Middle Precambrian (Pharuzian, Birrimian), less intensely metamorphosed, includes granites, gneisses, crystalline shists, quartzites, greywackes. The Upper Precambrian (Nigritian, Tarkwaian, Falemian) is still less crystallized (or not crystallized for the latter); it includes shists, quartzites, cipolins, shales, arkoses, sandstones and conglomerates. The Precambrian peneplaned chains and ranges outcrop over huge areas, particularly east of Lake Chad. These outcrops represent overall about one-third of the Sahel, i.e. 1 million km^2; naturally they constitute the core of some of the Sahelian mountains: Adrar of Iforas, Air, Nuba Mounts.

The sedimentary mantle is rather thin, except, of course, in the sedimentary basins of the Senegal River, the Niger River, the Nile and its tributaries.

The Cambrian transgression covered most of Africa and left rather thin deposits, particularly in the NW part of our region (mainly limestones and shales) in Senegal and Mauritania.

The Ordovician sandstones and ferrugineous sandstones outcrop over huge areas particularly to the W and NW: Senegal, Mauritania, Mali and Burkina-Faso (formerly called Upper-Volta); they constitute the Mandingue Plateau, the Bandiagara Plateau, the Adrar of Mauritania, the Walata and Nema Plateaux, the Gourma Pediplain.

The Continental Intercalary (from Jurassic to Middle Cretaceous). The Upper Paleozoic being virtually absent, the Mesozoic continental deposits rest directly on the Cambro-Silurian or on the Basement Complex. The C.I. includes the Nubian sandstones occupying huge areas in the middle and lower reaches of the Nile Basin. The Continental Intercalary outcrops over more than 0.5 million km^2, i.e. 20% of the Sahel, particularly east of the Niger River. The C.I. includes variegated shales, gypsum, clays, sandstones, lacustrine limestones, comparable to the West European "Weald" facies of the continental Lower Cretaceous.

The Cenomanian transgression flooded Central West Africa and the Sahara linking the Ocean to the Thetys via a northern expansion of the Gulf of Guinea on both sides of the Ahaggar mountains, thus achieving the junction with the southern expansion of the Mesogaea. The Upper Cretaceous outcrops E and NE of the Niger Bend in the Adrar of Iforas and Tamesna (Senonian-Maestrichtian, Danian); it is also known from boreholes in Senegal and NW of the Republic of Niger, as the Maestrichtian is, in both cases, the main deep aquifer. Limited outcrops occur also in the SE part of the Republic of Niger on the small plateaux of Ader Doutchy, Damergou and Tegama.

The Paleocene transgression (Montian-Ypresian), following the Danian regression, reached north of the Niger Bend and the inland of Senegal and left deposits of limestones and a thin layer of rock phosphate (Taiba, in Senegal and Tilemsi in Mali: ca. 12% P).

The Continental Terminal. At the end of the Paleocene, the sea withdrew from the continent, never to come again into our region. A series of fluvio-lacustrine deposits from Oligocene-Miocene age topped by Plio-Quaternary laterites and lateritic hardpans (ferralites) usually overlain by a thick ferrugineous duricrust, which fossilizes the whole series, occur.

The Continental Terminal is referred to as the Um el Rawaba (also written Umm al Ruaba) formation in the Republic of Sudan, where it outcrops over some 300 000 km^2, south of the 14° parallel in the middle and upper Nile Basin. The Umm al Ruaba series is composed of unconsolidated sands, sandstones, gravels, clays and shales.

Pleistocene and Holocene. Continental deposits of fluvio-lacustrine and alluvial nature sedimented in thick layers of several 100 m in the Chad Basin, the Inland Delta of the Niger River and the Jezira of the Sudan. Sand dune formations cover huge areas, particularly in the northern half of the Sahel. Marine deposits occupy small surfaces in Senegal and Mauritania.

Eruptive rocks and lava flows of Tertiary and Quaternary ages occupy limited area in the Adrar of Iforas, Air, Tibesti, J. Gurgeil, J. Marra and J. Oda (mainly trachytes and basalts).

The overall importance of sandstones in the sedimentary series [Ordovician, Devonian, Continental Intercalary (Nubian sandstone), Continental Terminal], in addition to the proximity of the Sahara and its extensive outcrops of sandstones (Tassilis), explain the huge areas covered in the Sahel by sandy geomorphic units and sandy soils (Mainguet 1972–1985).

2.2.2 Geomorphology (Michel 1973; Hunting Technical Services 1964; Pias 1970; Mainguet 1972–85) (Fig. 40, 41)

Sand dune formations cover huge areas in the northern half of the Sahel and parts of the southern half. The northern limit of drifting sands and "live dunes" is approximately the 150-mm isohyet of long-term mean annual rainfall (Mainguet et al. 1980). This limit has shifted southwards recently due to desertization processes resulting from the conjunction of drought and land abuse. This will be discussed later. The southern limit of the "clad", "fixed dunes" is usually around the 600-mm isohyet; but it expands to 1000 mm [Bongor, Sarh (formerly: Ft Archambault)], at 15–17° E longitude and 8–11° N latitude. These dune formations date back to upper Pleistocene and Holocene; there are four systems including the present-day live dunes. The three "fossil ergs" date back to periods when the Sahara extended further south than it does presently; this southern expansion was 200 to 500 km beyond the present limit of "loose sands" (Pias 1970; Michel 1973; Maley 1973, 1977, 1980, 1981, 1982, 1983a, b; Wickens 1975; Boulet 1978; Daveau 1970; Elouard 1976; Grove 1958; Mainguet 1972–85; Hunting Technical Services 1964; Leprun 1978b).

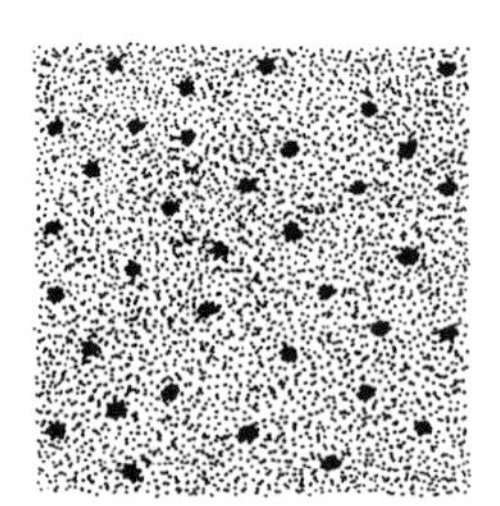

Aerial view of a typical sahelian vegetation: continuous herbaceous cover dotted with a sparse ligneous stratum.

Fig. 40. Sketch of Sahelian landscape and vegetation physiognomy of a Mimosaceae thorn scrub (After Naegelé 1969)

The older dune system is anterior to the "Inchirian" wet period, i.e. older than 40 000 BP; the second system is dated 15 000 to 20 000 BP: the so-called Ogolian dune system. The "pre-Inchirian" system extended further south than the Ogolian, down to the 8° parallel in SE Chad. But the two other dry periods, much shorter and of lesser aridity, occurred 6000–7000 BP and 2000–3000 BP during which some superficial reshaping and reshuffling occurred within the Ogolian system. The latter became vegetated and fixed during the wet periods of "Chadian" (7000–11 000 BP) and Nouakchottian (3000–6000 BP). The four dune systems may be recognized by the color of the sand (the older being more reddish and the younger paler) and the shape of the dunes.

The older dunes or *"longitudinal dunes"* have an almost symmetrical shape; the distance from crest to crest is 2 to 3 km; the height from crest to hollow is 8–20 m and the length is 30–60 km; the color of the sand is light reddish/yellowish. The orientation of the parallel ridges is in the N-S direction. The intermediate age dunes or "*transverse dunes*" are asymmetric in cross-section with a steeper slope facing the west, on the lee side of the dune-forming winds. The height of the dunes from top to bottom varies from 5–10 m for small dunes to 20–30 m for larger ones; the ridges are sinuous, contrary to the rather straight longitudinal dunes, and unlike the latter, the ridges and the valleys are not flattened. The length of individual ridges may reach 20 to 40 km. The presently mobile dunes are of various types, often barchans or barchanoids. There are also huge surfaces of sand sheets and sand veils with a flat or gently undulating surface. The thickness of the dune systems varies from a few centimeters in some sand veils up to 50–100 m in the thickest parts of the pre-Inchirian and Ogolian systems. The dune systems are locally called Erg, Edeyen or Qoz (Republic of Sudan) which have become part of the scientific terminology.

Pediplains are rather flat grounds developed in a structural surface and often coated with a sand or silt veil. They occur in the foothills of the mountains and plateaux, inselbergs from the basement complex or from igneous extrusions. Outliers and table mountains develop along the cliffs of the plateau escarpments, particularly in the Ordovician sandstones.

Plateaux and structural surfaces of the subhorizontal sedimentary deposits are usually topped by thick Plio-Quaternary iron hardpans, and associated soils.

Clay plains and alluvial flood plains are flattened surfaces of accumulated weathered material, locally interrupted by inselbergs; they develop vertisols and occupy large areas in the Inland Delta of the Niger and the Jezira, between the two Niles.

Fluvio-lacustrine deposits of wetter periods are found in many parts including in the interdunal depressions.

Fossil valleys and wadis, remains from wetter periods, are numerous around the mountain areas of Adrar, Air, Ennedi, etc.; the Tilemsi valley is thought to have been the upper course of the lower Niger River. The "Dallols" in the Rep. of Niger are dried out Pleistocene valleys originating in the Air mountains.

2.2.3 Topography (Fig. 41)

The whole Sahel is a rather flat, gently rolling country of low elevation (200–600 m), with the exception of the Chadian-Sudanese border where elevation reaches 1000 m and above. Most of the mountains are located at the border of the Sahara and the Sahel: Adrar of Mauritania, Adrar of Iforas, Air, Tibesti, Ennedi; the only exceptions are J. Gourgeil and J. Marra in the western part of the Rep. of Sudan rising to 2300 and 3100 m respectively. The Tibesti (3400 m), the Gourgeil and Marra are of volcanic nature. Air (1900 m) is mainly metamorphic with some volcanism; Adrar of Iforas (900 m) is part of the Basement Complex topped by sedimentary deposits, while Ennedi (1500 m), Adrar of Mauritania (700 m) and the Bandiagara plateau (800 m) are entirely made of paleozoic sediments (mainly sandstones). The Inland Delta of the Niger and Lake Chad are 260 and 250 m above sea level respectively, while the bottom of the Taoudeni syncline is only at 200 m elevation.

2.2.4 Hydrology (Rodier 1964, 1975, 1982; Roose 1977) (Fig. 42)

Most of the region is virtually arheic; the main rivers are all allochtonous: the Nile, the Chari and Logone, the Niger and the Senegal and Gambia, i.e. originating in the mountains beyond the limits of our region. Local valleys, such as the Tilemsi, Azaouak and Dallols, are all dried out remnants of wetter Pleistocene climates. Some local runoff naturally occurs during the rainy season and fills numerous, local, temporary ponds, the catchment of which is usually a few hectares in surface. Some larger, subpermanent ponds exist in the Niger Bend, on the Basement Complex: Gossi, Oursi, etc. The Inland Delta of the Niger from Mopti to Timbuctoo feeds many lakes.

Until the early Holocene, the upper Niger River flooded into the Araouane depression; then it was captured at Tosseye, east of Timbuctoo, by a tributary of the Tilemsi coming from the Adrar of Ifora mountains (Chudeau 1909, 1919;

→

Fig. 41. Geology and soils in the Sahel zone of the northern Ferlo, in NE Senegal. **A** Geological structure and geomorphic relief (Michel 1973). *1* Continental terminal; *2* Lutetian; *3* Lower Eocene; *4* Paleocene; *5* Maestrichtian; *6* marine sediments and alluvia; *7* Late Pliocene embankment surface (?); *8* boreholes. **B** Soil catena at Fété Olé, northern Ferlo, from dune top to central interdunal hollow (Poupon 1980). *a* Top of dune and upper slope; *b* midslope; *c* step and downslope; *d* interdunal hollow. *1* Humiferous topsoil; *2* leached (eluvial) horizon; *3* ferruginized accumulation (illuvial) horizon; *4* clay horizon with a prismatic structure; *5* iron spots, incrustations and concretions; *6* clay horizon with pseudogley; *7* light sand (parent material)

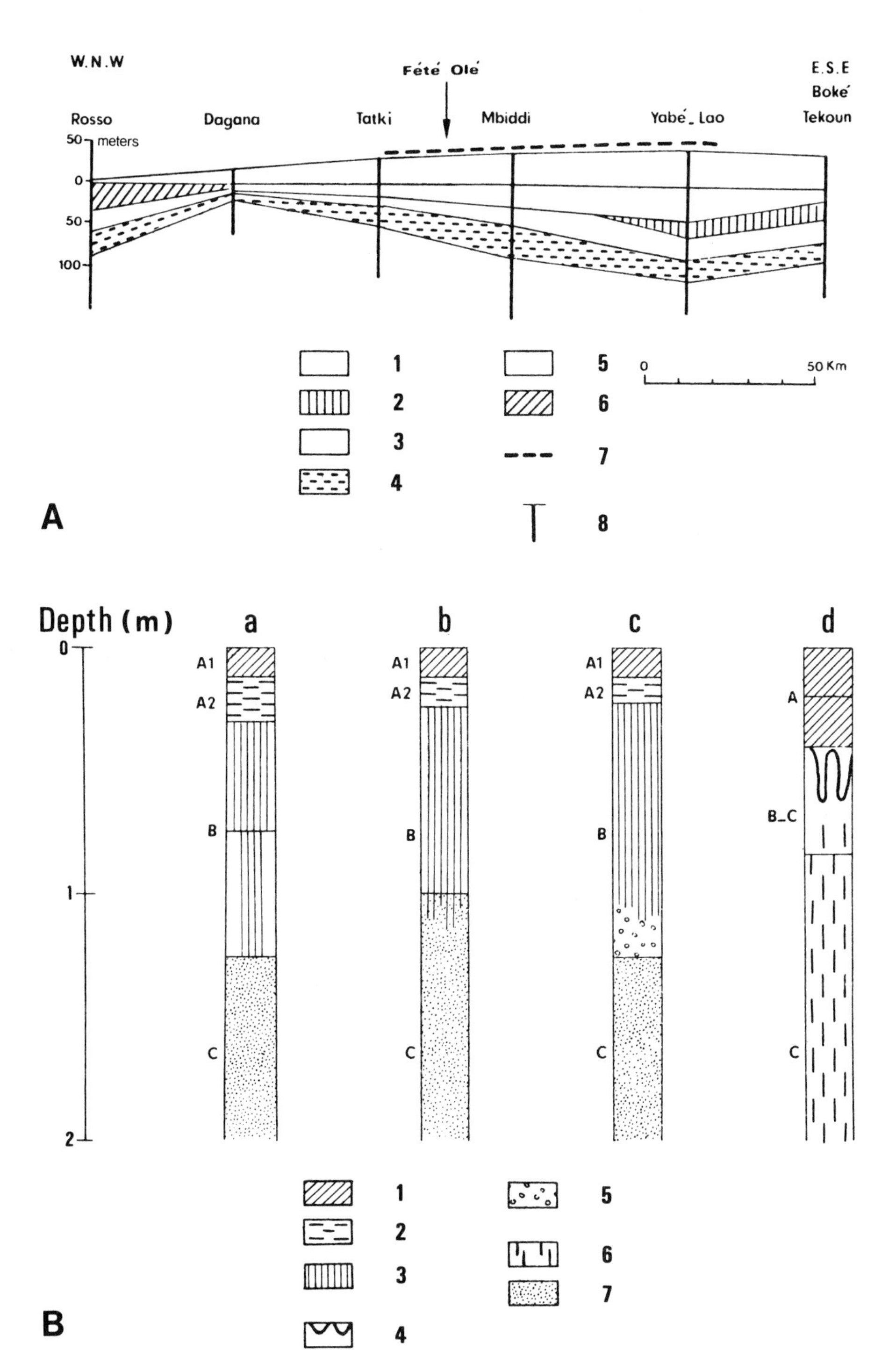

W.N.W
Féte Olé
E.S.E
Boke
Rosso
Dagana
Tatki
Mbiddi
Yabé_Lao
Tekoun
50 meters
0
50
100
1
2
3
4
5
6
7
8
0
50 Km
A
Depth (m)
a
b
c
d
A1
A2
B
C
A
B_C
0
1
2
1
2
3
4
5
6
7
B

Fig. 41

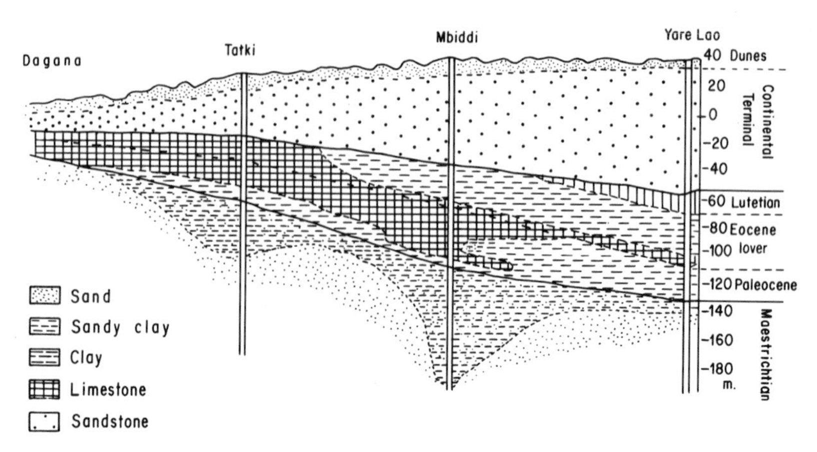

Fig. 42. Geological cross-section of the Ferlo from NW to SE (Naegelé 1971a)

Furon 1929; Urvoy 1942; Baulig 1950; Tricart 1965, 1969; Blanck 1968; Beaudet et al. 1977; Delibrias et al. 1984; Petit-Maire and Gayet 1984).

Groundwater is scarce except for the Maestrichtian aquifers of northern Senegal and north-western Niger where boreholes can discharge many liters per second. Otherwise, deep groundwater seems to exist only in very small and local aquifers in the weathered rock on the upper part of the Basement Complex along the faults and cracks, they only discharge a few m^3 per hour. This is enough for man and cattle, but insufficient for any irrigation.

2.2.5 Soils

(Fauck 1973; Hunting Technical Services 1964; FAO/UNESCO 1973; Boulet 1978; Leprun 1978b; Maignien 1965; Pias 1970)

According to the US classification (USDA 1975), the main taxonomic units (orders, suborders and groups) are as follows in the Sahel:

1. Entisols (no pedogenic horizon): Torripsamments, Torrifluents, Torriorthents, Quartzipsamments, Ustipsamments, Usorthents.
2. Inceptisols (soils with weakly differentiated horizons showing alteration of the parent material): Ustropepts (fluventic, aquic, vertic, lithic and oxic).
3. Vertisols (black cracking clays, black cotton soils): Torrerts, Chromusterts (typic, entic, aquic), Pellusterts (typic, entic).
4. Aridisols (soils with pedogenic horizons, but poor in organic matter, usually dry): Camborthids (typic, lithic), Hapargids (typic, lithic).
5. Alfisols (soils with a brown surface horizon, a medium to high base supply and a subsurface clay accumulation horizon): Haplusalfs (typic, lithic, udic, psammentic, arenic, rhodic), Paleustalfs, Rhodulstalfs.

Sandy soils occupy huge areas, probably more than half the area, i.e. 1.5 million km^2. They are slightly acidic with a pH from 6.0 to 6.5 and are poor in organic matter, nitrogen and phosphorus. In the FAO classification they are labelled Cambic and Luvic Arenosols which correspond to the Psamments and Aridisols in the US taxonomy, and to the brown subarid soils in the French classification. These are the zonal soils in the northern half of the Sahel. In the southern half soils are somewhat more evolved towards ferric Luvisols, i.e. Ustipsamments, Plinthic Paleustalfs and Plinthustalfs or tropical ferrugineous soils in the French taxonomy. These may be deep, shallow (iron duricrust) or sandy. They constitute the zonal soils in the Sudano-Sahelian ecoclimatic subzone and in the Sudanian zone. They are poor in organic matter, nitrogen and phosphorus and fairly acidic (pH: 5–6). Other important soil groups are the alluvial, i.e. Fluvisols, Torrifluvents and alluvial deposit soils. Other important azonal groups are Gleysols = Aquents, Tropaquepts and Aquults = hydromorphic gley soils. Vertisols are also very important in the sedimentary basins of the Niles, Lake Chad and Niger perhaps as much as 0.25 million km^2, much of which is in the Rep. of Sudan (S. Kordofan, Jezira, Kessala).

Saline and alkaline soils are negligible in extent, contrary to many other arid zones of the world; however, they play an important role in transhumant and nomadic animal husbandry as husbandmen take the herds and flocks to the limited areas with saline soils during the rainy season. The animals lick the ground for a few days or weeks for their annual supply of minerals; this is locally known as the "salt cure", which is one of the motivations of transhumance and nomadism.

2.2.6 Runoff, Erosion (Roose 1977, 1981; Roose and Piot 1984; Rodier 1982; Chevallier et al. 1985)

Runoff. In the sandy parts of the Sahel there is virtually no runoff on large or intermediate watersheds; there are no autochtonous rivers, only local, temporary ponds with small watersheds of a few hectares to a few km^2. The situation is different on other substrata such as the Basement Complex, the Cretaceous/Eocene sediments, the Continental Intercalary and the Continental Terminal.

Runoff and water erosion depend to a large extent on climatic aggressivity, particularly on rainfall intensity (Wischmeier 1959). The heaviest annual shower in the Sahel represents 5 to 15% of the average long-term annual precipitation, i.e. in an area with 400 mm mean annual precipitation, the heaviest annual rain with a 50% probability will reach around 40 mm in 24 h (Rodier 1964, 1975; Roose 1977; Brunet-Moret 1967). Rain intensities of 0–10 mm day^{-1} represent 50 to 60% of the total amount of annual precipitation. As in other arid zones of the world this proportion of light rains is inversely related to the annual mean, and therefore increases with aridity. The heaviest shower in 10 years reaches the values shown in Table 6, computed from Rodier (1964) for 71 Sahelian and south Saharan stations.

The instantaneous intensity with a 10-year return period at Tahoua (Rep. of Niger; 400 mm mean annual rainfall) is as follows, according to Heusch (1975):

10 min	150 mm h^{-1}
30 min	90 mm h^{-1}
1 h	60 mm h^{-1}
5 h	10 mm h^{-1}
24 h	5 mm h^{-1}

On small basins (1.0–10.0 ha) in the Ader Doutchy Hills (east of Tahoua, Rep. of Niger), under 400 mm annual rainfall, on Cretaceous and Paleocene sedimentary substratum, experimental runoff data averaged 10 to 50% of annual rains from 1965 to 1967, depending on the size of the watershed, slope, soil type, land use and vegetative cover (Vuillaume 1968; Delwaulle 1973b; Heusch 1975; Roose 1981).

An 8-year survey on the Oursi Pond watershed in northern Burkina-Faso: catchment 263 km^2 on the Basement of Complex, with a mean slope of 1.5%, a mean rainfall of 375 mm yr^{-1} for the period, a pond surface of 20 km^2 and a storage capacity of 25 million m^3, gave a mean runoff rate of 20% (Chevallier et al. 1985), while 25% was found in the northern rim of the "dead" Inland Delta of the Niger River in Central Mali (De Vries and Djiteye 1982) on a smaller scale of a few hectares, on rather flat, silty alluvial soils.

On gently sloping sandy soils of the Oursi Pond watershed in northern Burkina-Faso runoff averaged 5 to 6% on small experiment basins of 3000 m^2 in 1977–79 (Piot and Millogo 1980), under a mean annual rainfall of 400 mm. In other experiments in the northern Sudanian zone of Burkina-Faso on gently sloping, sandy loam cultivated soils, on the Basement Complex, Forest and Poulain (1978) found runoff rates of 3–4% on totally protected and vegetated plots; up to 38% on bare tilled Wischmeir test plots and 18–20% under traditional cropping practices.

It would seem that on the sandy grasslands of the Sahel the runoff rate at the individual field or individual paddock scale remains below 5%; it may reach 10% on cultivated sandy soils and up to 30–50% on medium- to fine-textured soils, and perhaps more on those having a bare and sealed surface.

At the peak of the rainy season, in August, at the Fete-Ole IBP research station in northern Senegal, with 20–30 depressions km^{-2} covering some 8–12% of the landscape, temporary ponds may cover some 5% of the land (Poulet 1982).

Runoff is considerably increased by wild fires, particularly late dry-season fires as shown below (Roose 1978; Roose and Piot 1984). Locations: Ganse, N Burkina-Faso, 12°22′ N–1° 19′ W alt. 300 m; Sahelo-Sudanian tree savanna, mean annual precipitation 691 mm, plot size 250 m^2, 0.5% slope, 7 years of data: 1968–74 (Roose and Piot 1984):

Treatment	Average runoff % of rains	Maximum runoff % of rains	Erosion (t ha^{-1})
Total protection	0.25	1.0	0.03
Discontinuous vegetation cover	3.00	10.0	0.15
Early fires	2.60	9.3	0.10
Late fires	13.30	60.5	0.35

At Saria in Central Burkina-Faso, 12°16′ N 2°09′ W, alt. 300 m, on ferrugineous pisolithic iron hardpan, on 100–250 m^2 plots, 1.4% slope, P = 643 mm, 1971-1974:

Treatment	Average runoff % of rains	Maximum runoff % of rains	Erosion ($t\ ha^{-1}$)
Bare tilled soil	39	70	20
Earthed up sorghum	27	60	6
Recent fallow	10	30	0.5
Old fallow	3	5	0.15
Soil erosion index K = 0.23			

At Linoghin, 12°22′ N–0°30′ W, on vertic tropical brown soil (central Burkina-Faso):

Treatment	Average runoff % of rains	Erosion ($t\ ha^{-1}$)
Cultivated	18.5	3.6
Cultivated and ridged	7.0	0.6
Burned savanna	19.0	0.8
Unburned savanna	3.7	0.1
Wischmeier test plot (bare, tilled)	47.0	14.1
Soil susceptibility index K = 0.14		

At Gampela, Burkina-Faso, 12°25′ N–1°21′ W, alt. 280 m, slope 0.8%, rainfall 714 mm, RUSA[2] 319., gravelly ferrugineous soils on hardpan, duration: 1967–72:

Treatment	Average runoff % of rains	Maximum runoff % of rains	Erosion ($t\ ha^{-1}$)
Cultivated, ridged	10.4	31	3.2
Cultivated, unridged	23.0	41	5.0
Wischmeier test plot (bare, tilled)	–	–	16.0
K erosion index = 0.20			

Erosion. We could not find any quantified data on wind erosion in the literature, neither for the Sahel nor for the southern Sahara. Water erosion has been experimentally measured on a few small plots under various land-use simulations in the Sahelian and Sudanian ecoclimatic zones of the Rep. of Niger and

[2]RUSA: mean annual rainfall erosivity index (Wischmeier and Smith 1978), $t\ ha^{-1}\ yr^{-1}$.

Burkina-Faso (formerly, Upper-Volta) (Delwaulle 1973a, b; Roose 1977, 1981; Roose and Piot 1984; Piot and Millogo 1980). Surprisingly enough, the climatic aggressivity factor "R" of Wischmeier is fairly constant throughout the Sahelian, Sudanian and Guinean zones. It approximates Ram = 0.5, Pam ± 0.05, as shown by Roose (1977), where Ram is the mean annual "R" index and Pam is long-term mean annual rainfall, Ram is expressed in potential erosivity in t ha^{-1} yr^{-1} in the case of uncoherent material with a 9% slope. In other words, the climatic erosion potential in metric t ha^{-1} is equal to half the mean annual precipitation in mm throughout the region.[3]

The soil erosivity index of Wischmeier increases from about 0.10 in the Guinean zone to 0.20–0.30 in the northern Sudanian zone and up to 0.30–0.60 in the Sahel (Roose 1977, 1981).

On the moderately sloping sandy soils of the Sahel, water erosion may vary from 200 kg ha^{-1} yr^{-1} on land fully and permanently covered with natural vegetation, to 1000–3000 kg ha^{-1} yr^{-1} on cultivated soils (pearl millet, sorghum, peanut) and up to 18 000 kg ha^{-1} yr^{-1} on Wischmeier test plots (Piot and Millogo 1980). On other soil types, erosion on cultivated land may reach 8000–10 000 kg ha^{-1} yr^{-1} (Delwaulle 1973a; Heusch 1975).

2.3 Anthropozoic Factors: Man and Land Use (Tables 9–12)

Unlike the populations of the Guinean and Congolian ecoclimatic zone, further south, the people of the Sahel have a long and complex history starting with the Christian Kingdoms of the Mid-Nile Valley of the 6–14th centuries, the Ghana Kingdom (4–12th cent.), the Mali Empire (14th cent.), the Songhai Empire (15th cent.), the Hausa States (10–18th cent.), the Kanem-Bornu Empire (12–17th cent.), the Mossi Kingdoms (14–19th cent.), the Kebbi Empires (15–16th cent.), the Fulani Establishments (12–16 and 19th cent.). These populations have been in indirect contact with the outside world through trans-Saharan trade throughout history, through the Berber and Almoravide invasions of the 4th and 11th centuries, the marine trade since the late 15th century, before the area opened to European penetrations of the 18th and 19th centuries. Archaelogical evidence bear witness to the strong human impact of the environment for over 2000 years (Mauny 1961, 1978; McIntosh and McIntosh 1981a, b, 1983a, b; Monod and Toupet 1961).

The present population densities in the Sahel are rather low in absolute terms, but they have sharply increased since about World War II. Densities are also very variable in space depending on climate, soils and water availability: 15–75 inhabitants km^{-2} in W Senegal and along the Niger-Nigeria border, around the

[3]Wischmeier equation: $R = Ec. I_{30}$; $Ec = 1214 + 890 \log I$; in metric units: $R = (1/1735.6)\ Ec.\ I_{30}/100$; Ec = kinetic energy in t km^{-2} mm^{-1} rain. I = rain intensivity in mm h^{-1}; I_{30} = maximum intensivity in 30 min (Wischmeier 1959; Wischmeier and Smith 1978).

Table 9. Mean national human population density (Inh km^{-2})

Country	Country surface (10^3 km^2)	1950	1970	1983
Burkina-Faso	274	12.7	19.8	27.3
Chad	1284	2.1	2.8	3.7
Mali	1240	2.8	4.3	4.6
Mauritania	1031	0.6	1.2	1.7
Niger	1267	1.7	3.2	4.6
Senegal	196	2.7	21.8	31.3
Sudan	2506	3.4	5.6	8.1
Overall	6767	3.5	5.6	7.7

Table 10. Evolution of the human population in the Sahel countries (FAO production yearbooks; numbers in 1000)

Country	1950		1970		1983		$\frac{1983-1950}{1950}$
	Total	Rural (%)	Total	Rural (%)	Total	Rural (%)	
Burkina-Faso	3490	87	5413	87	7483	79	114.4
Chad	2642	92	3643	90	4743	81	79.4
Mali	3445	91	5362	91	5747	85	66.8
Mauritania	660	89	1245	88	1781	81	169.9
Niger	2120	96	4008	93	5819	86	174.5
Senegal	2480	75	4267	80	6131	73	147.2
Sudan	8615	78	14 090	82	20 200	75	134.5
Total/mean	23 452	86.9	38 028	87.3	51 909	80.0	121.3
Overall increase	-	-	62%	-	121%	-	-
Mean annual increase		-	2.7%	-	2.7%	-	-
Livestock to people ratio (TLU/Inh.)	0.89		0.90		0.90		

Table 11. Areas of cropland in the Sahel countries (FAO production yearbooks) (surfaces in 1000 ha)

Country	1975	Total country (%)	1983	Total country (%)	Increase 1975–1983 (%)
Burkina-Faso	2509	9.2	2633	9.7	4.9
Chad	3007	2.3	3155	2.5	4.9
Mali	1900	1.5	2058	1.7	8.3
Mauritania	188	0.2	208	0.2	10.6
Niger	2496	2.0	3605	2.8	46.2
Senegal	5000	25.5	5227	26.7	4.5
Sudan	12 160	4.9	12 448	5.0	2.5
Overall	27 261	4.0	29 379	4.3	7.78

Mean annual increase 1.0%

Table 12. Human population densities and supporting capacities in Africa (Kassam and Higgins 1980; Le Houérou and Popov 1981)

Ecoclimatic zones	Length of growing season (days)	Present population densities (inhabitants km^{-2})	Potential population supporting capacities (inh. km^{-2})	
			Under low inputs	Under intermediate inputs
Saharan	0–15	3	1	2
Saharo-Sahelian	15–80	6	3	4
Sahelian sensu stricto	80–120	16	7	20
Sudano-Sahelian	120–180	20	32	154
N. Sudanian	180–210	16	49	215
S. Sudanian	210–300	19	82	390
Guinean and Congolian	> 300	21	146	561

Jebel Marra, along the Senegal, Niger, Chari and Nile rivers; 10–15 inhabitants km^{-2} in the Sudano-Sahelian zone; 5–10 inhabitants km^{-2} in the Sahel proper; about 5 inhabitants km^{-2} in the Saharo-Sahelian subzone and, naturally, still less in the Sahara (oases and mountains).

The rate of demographic growth is 1.5 to 2.5% per annum among the pastoralists and 3.0 to 3.5% among the settled farmers (Le Houérou 1985). The rural element represents over 80 and often over 90% of the overall population, as there is virtually no industry and very little mining. The remainder is essentially composed of trading and civil service with some local fishing in Senegal, in the Inland Delta of the Niger River and other rivers and lakes. The smoking of some 40 000 metric tons of fish every year in the Inland Delta of the Niger has had a rather strong impact on the woody vegetation of the neighbourhood for several decades.

The population lives in villages and towns; the larger cities are mainly located on the southern fringe at the confines of the Sudanian ecoclimatic zone or within it: Dakar, Bamako, Segou, Mopti, Ouagadougou, Niamey, Maradi, Zinder, Sokoto, Kano, Katsina, Maiduguri, N'Djamena. There are only a few smaller towns in the Sahel proper: St. Louis, Nouakchott, Podor, Kaedi, Matam, Kayes, Nioro, Timbuctoo, Gao, Tahoua, Agades, Abeche, El Fasher, El Obied. The only large city in the Sahel is Khartoum (2 million) located at the junction of the White and Blue Niles.

Rural populations in the Sahel are far beyond the estimated human supporting potential capacities under conditions of low or intermediate inputs as shown in Table 12.

The Saharan, Saharo-Sahelian and Sahelian zones are mainly occupied by nomadic and transhumant pastoralists, from west to east: Moors, Fulani, Tuaregs, Zaghawa, Kabbabish, Baggara. These keep different kinds and breeds of animals, cameleers: Moors, Tuaregs, Kabbabish, Zaghawa; cattle husbandmen: Fulani, some Tuaregs, Baggara; all keep small stock to various degrees. Some are long-range nomads such as the Moors, Kabbabish, some Tuaregs and some Fulani; others are short-range transhumants such as some Tuaregs, some Fulanis, etc. In terms of numbers the transhumants (short- and longe-range) are dominant, moving from the dry season ranges in the Sudano-Sahelian and northern Sudanian zones to the rainy season ranges in the Sahel and Sahelo-Saharan zones (Bernus 1981; Bonte et al. 1979; Tubiana and Tubiana 1977; Gallais 1967, 1975; Dupire 1962). However, short-range transhumance and nomadism may be locally dominant as by the Tuaregs of the Niger Bend practising the so-called pastoral endodromy (Barral 1974, 1977).

The Sudano-Sahelian and northern Sudanian zones are predominantly inhabited by settled farmers with dry season incursions of transhumant pastoralists, particularly Fulani. The settled farmers belong to the following main ethnic groups, from west to east: Toucouleurs, Wolof, Serere, Malinke, Sonninke, Bambara, Dogon, Mossi, Songhai, Hausa, Kanuri, Baguirmi. In addition to these traditional farmers, some are recently settled pastoralists, often freed, former dependent or slave nomads, e.g. the Moor's Harratines, the Tuareg's Buzus or Bellas, the Fulani's Rimaibe and Farfaru, etc. Both sedentary farming and semi-nomad farming systems are based on subsistence staple food crops: millet, cowpea, some groundnuts, some Bambara groundnuts, some cassava and sorghum in the clay depressions. These systems are based on the alternation of crop and fallow; after the harvest in September, crop residues are grazed (and manured) by resident or nomadic herds. But, as population grows, the pressure on the land increases, less and less land is left to fallow or the period of fallow is shortened or completely cancelled. As a result, fertility and yields decrease, therefore, more and more land is needed per unit of food harvested.

Some ethnic groups, however, practice agroforestry since time immemorial, using *Faidherbia albida* at a density of 10–50 trees ha^{-1} (CTFT 1988; Charreau and Vidal 1965). This allows for the maintenance of soil productivity on a long-term sustained basis without any fallow. As the tree sheds its leaves for the rainy season, there is no shading problem. Another traditional agroforestry system is used in the Rep. of Sudan consisting in a crop rotation of *Acacia senegal* (spontaneous or planted) with sorghum or millet in "Qoz" regions of Kordofan and Darfur between the isohyets of 250 and 800 mm of annual rainfall. The cultivation part of the cycle lasts 4–10 years, *Acacia* regeneration 5–8 years, and the gum arabic tapping period 6–10 years. The whole cycle thus lasts 15–20 years (Kassas 1970; Seif El Din 1965; Seif El Din and Mubarak 1971). However, due to the fast-growing human population, this well-balanced system is on the decline (Kassas 1970); it was almost vestigial in Kordofan in 1988.

The southern half of the Sahel is subject to increasing cultivation and competition between pastoralism and farming. The latter has increased at a pace similar to population growth: 2–3% per annum, for the past 3 decades (Le Houérou 1979b; Haywood 1981; Le Houérou and Gillet 1985). Confronted with this acute land-use problem, the Rep. of Niger set up a legal northern limit to cultivation along the 15° N lat., i.e. the isohyet of 350–400 mm of annual rainfall.

This ordinance, proclaimed in 1961, could never be enforced (Bernus 1981). At present the actual northern limit of cropping is some 100 km further north along the 250-mm isohyet. This subject will be discussed further in Section 3.2.5.2.

2.4 Ecoclimatic Classification (Le Houérou and Popov 1981) (Figs. 2, 3, and 39)

As mentioned above, the Sahel is usually subdivided into three subzones:

	Saharo-Sahelian	100–200 mm, perennial grass steppe;
Sahel	Sahelian	200–400 mm, Mimosacéae thornscrub, annual grasses;
	Sudano-Sahelian	400–600 mm, Combretaceae Savanna, annual grasses;
Sudan	North Sudanian	600–900 mm, mixed savanna, perennial grasses.

In addition to these zones there are limited areas of montane and Afro-Alpine ecoclimatic zones geographically included in the Sahel: the Ennedi montane area (Gillet 1968), the Jebel Gourgeil montane area (Quezel 1968), the Jebel Marra montane and Afro-Alpine area (Wickens 1976) and the Air montane area (Bruneau de Miré and Gillet 1956). Montane areas are understood as those above 1000 m of elevation: the lower montane zone, 1000–1800 m and upper montane zone, 1800–2800 m. The Afro-Alpine zone is only known in Jebel Marra above 2800 m. The lower limit of 1000 m corresponds to high risks of frost in December-January (Gillet 1968), while the Afro-Alpine zone undergoes regular night frosts year-round (Le Houérou and Popov 1981).

3 Flora and Vegetation

3.1 Flora (Wickens 1976; Brenan, 1978; Lebrun 1976, 1981; White 1983; Le Houérou 1988d) (Table 1, Fig. 43)

The flora of the Sahel is rather poor, i.e. approximately 1500 species of flowering plants for an area of 3 million km^2. The flora is typically paleotropical. The richest families are: Poaceae, Cyperaceae, Leguminoseae, Capparidaceae, Malvaceae, Convolvulaceae, Zygophyllaceae, Euphorbiaceae, Asclepiadaceae, Acanthaceae and Solanaceae. The rate of endemism is very low: some 40 species, i.e. 3% (White 1983). There are some 200 species confined to the intertropical African arid zone of West, East and South Africa. Many species are shared with the Sudanian zone, which, in the Sahel, are restricted to watercourses, around ponds and areas subject to flooding.

There is a N-S gradient of floristic composition from the southern Sahara to the Sudanian zone, similar to the rainfall gradient. There are also some dif-

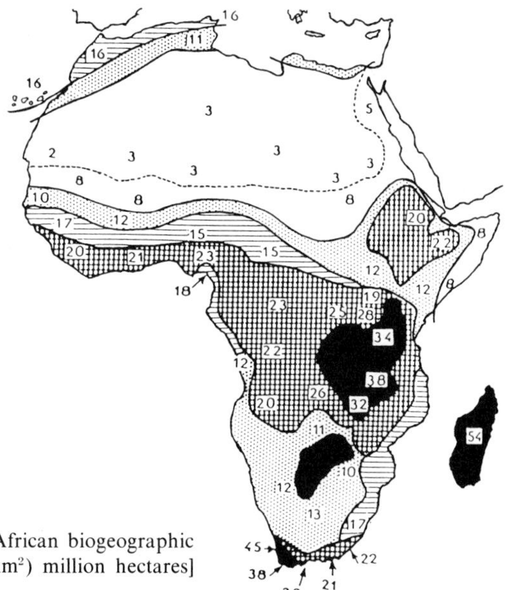

Fig. 43. Floristic richness of the main African biogeographic zones [number of species per (10 000 km^2) million hectares] (After Lebrun 1960)

ferences between the eastern Sahel and the western Sahel, the transition zone lying between the Niger Bend and Lake Chad. Typical eastern species are:

Acacia laeta
Acacia mellifera
Acacia nubica
Albizia amara subsp. *sericocephala*
Combretum cordofanianum
Combretum hartmannianum
Commiphora quadricincta
Crotalaria thebaica
Dobera glabra
Geigera acaulis
Farsetia longisiliqua
Ochradenus baccatus
Terminalia brownii
etc.

Among the common western species we have:

Andropogon penguipes
Butyrospermum paradoxa
Caralluma retroscipiens subsp. *tombuctuensis*
Diheteropogon hagerupii
Elyonurus elegans
Euphorbia balsamifera
Indigofera senegalensis
Jatropha chevalieri
Panicum laetum
Parkia clappertoniana
Pennisetum violaceum
etc.

Montane species include dominant shrubs and perennial grasses:

Albizia amara subsp. *sericocephala*
Andropogon distachyus
Aristida adoensis
Botriochloa pertusa
Cenchrus ciliaris
Chrysopogon plumulosus
Euphorbia candelabrum
Euphorbia nubica
Ficus ingens
Ficus populifolia
Ficus salicifolia
Heteropogon contortus
Hyparrhenia hirta
Olea laperrini
Rhus vulgaris
Terminalia brownii
etc.

Typical and common Saharo-Sahelian species are:

Acacia ehrenbergiana (= *A. flava* auct.)
Aristida sieberiana
Capparis decidua
Cornulaca monacantha
Crotalaria saharae
Crotalaria thebaica
Cymbopogon shoenanthus
Danthonia forsskalei
Fagonia spp.
Farsetia spp.
Grewia tenax
Lasiurus hirsutus
Leptadenia pyrotechnica
Monsonia nivea
Morettia canescens
Neurada procumbens
Ochradenus baccatus
Panicum turgidum
Salsola baryosma
Pulicaria crispa
Schouwia thebaica
Stipagrostis pungens
etc.

Common Sudano-Sahelian species are:

Adansonia digitata
Andropogon penguipes
Andropogon pseudapricus
Anogeissus leiocarpus
Bombax costatum
Butyrospermum paradoxum
Celtis integrifolia
Combretum glutisonum
Combretum ghazalense
Combretum micranthum
Combretum molle
Combretum nigricans
Combretum racemosum
Ctenium elegans
Diheteropogon hagerupii
Diospyros mespiliformis
Faidherbia albida
Guiera senegalensis
Khaya senegalensis
Lannea acida
Loudetia togoensis
Maytenus senegalensis
Mitragyna inermis
Parkia biglobosa
Parkia clappertoniana
Pennisetum pedicellatum
Piliostigma reticulata
Pterocarpus lucens
Slerocarya birrea
Stereospermum kunthianum
Terminalia avicennoides
etc.

which may also be found in the Sahel proper, but in depressions and around ponds only, in the latter case.

Pan-Sahelian species are, among trees and shrubs:

Acacia ataxacantha
Acacia laeta
Acacia nilotica s.l.
Acacia senegal
Acacia seyal
Acacia tortilis s.l.
Balanites aegyptiaca
Bauhinia rufescens
Boscia senegalensis
Cadaba farinosa
Calotropis procera
Combretum aculeatum
Commiphora africana
Cordia sinensis
Dichrostachys cinerea
Grewia bicolor
Hyphaene thebaica
Maerua crassifolia
Ziziphus mauritiana
etc.

Pan-Sahelian annual grasses are:

Aristida funiculata
Aristida mutabilis
Cenchrus biflorus
Cenchrus prieurii
Chloris prieurii
Chloris virgata
Dactyloctenium aegyptium
Eragrostis tremula
Panicum laetum
Tragus recemosus
Schoenefeldia gracilis
etc.

For the distribution of common species across the ecoclimatic zones between the Sahara and the equator, see Table 1.

3.2 Vegetation

3.2.1 Physiognomy and Structure (Figs. 40, 44)

The Sahelian vegetation has been named in many different ways, often in an unsatisfactory manner: steppe, pseudosteppe, scrub, savanna, and even "prairie" or "meadow" or "tropical meadow". To the present author, none of these terms is completely satisfactory. The northern open perennial grass formation, virtually treeless outside the topographical depressions, could be labelled "steppe"; the other terms being obviously inadequate. The Mimosaceae formation with an almost continuous layer of annual grasses could be named "thorn scrub" as it has sometimes been in the Rep. of Sudan; it could not be appropriately called "savanna" which would imply tall perennial grasses. The Sudano-Sahelian Combretaceae formation is sometimes called the "Combretaceae savanna",

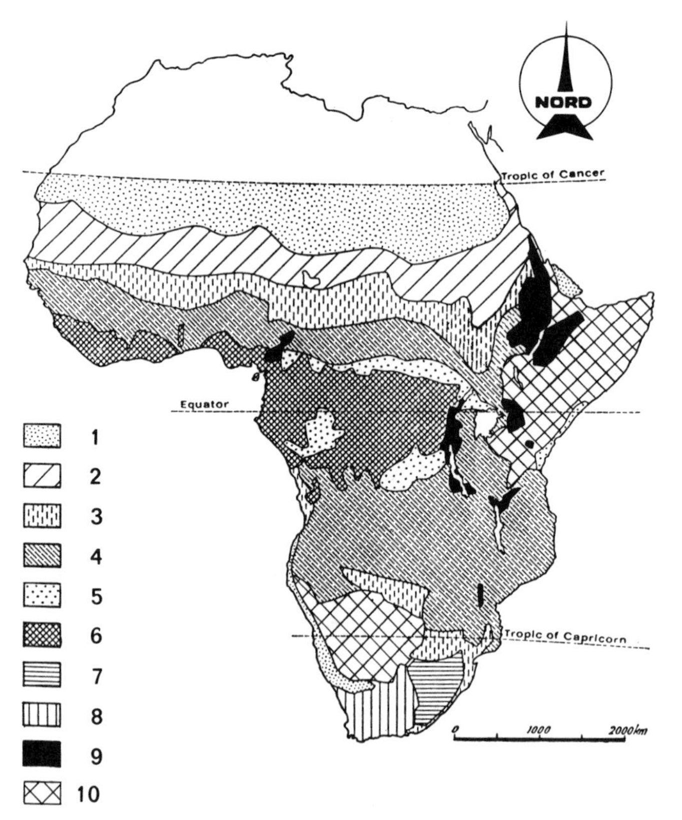

Fig. 44. Sketch of the vegetation of Africa. *1* Desert; *2* Sahel *Mimosoideae* Scrub & *Combretaceae* Savanna; *3* Northern Sudanian Savanna and austro-african grasslands; *4* Southern Sudanian Savanna and Miombo woodland; *5* Mixed Forest-Woodland-Savanna; *6* Guinean and Congolian Rain-Forest; *7* Subtropical and Temperate Grassland; *8* Mixed Shruband and Grassland, Mediterranean Shrubland; *9* Montane and Afro-Alpine mixed formations; *10* *Acacia-Commiphora* scrubland (perennial grasses) of East Africa and the Kalahari

which is also inappropriate because of the annual grass layer. According to the observations made in Central Mali each percent of canopy cover of shrubs and trees corresponds to 70 ± 10 "trubs" ha^{-1} in old fallows (computed from Wilson 1982b).

Actually, each percent of canopy cover corresponds to:

Shrubs (ha^{-1})	Mean diameter (m)
509	of 0.50
128	1.00
32	2.00
14	3.00
8	4.00
5	5.00
1.3	10.00

Vegetation structure depends to a large extent on:

1. The ecoclimatic subzone concerned;
2. The physiography and soil;
3. Land use and anthropozoic pressure.

In the Sahel zone of Kordofan (Rep. of Sudan) under mean long-term rainfall of 300–600 mm, using remotely-sensed data (aerial photography and Landsat MSS imagery) combined with ground truth checking, Olsson (1985) found the following values for *tree cover* (shrubs excluded) in the *Acacia senegal* Gum Belt around El Obeid:

Land-use type	No. of test plots	No. of stems km^{-2} ($\bar{x} \pm SE$)	Percent canopy cover ($\bar{x} \pm SE$)	No. of stems km^{-2} for each percent of canopy cover ($\bar{x} \pm SE$)	Wood biomass (kg Dm ha^{-1})
Arable land and bush fallow	22	2023 254	5.9 0.8	357 15	2800
Forest reserves	14	8042 564	18.4 1.1	441 25	11 000
Overall area	36	4363	10.7	380	6030

For mapping purposes, woody biomass was divided into nine classes from 0–1850 to over 15 000 kg dry wood ha^{-1}, the median class being 7550–9300 kg ha^{-1}. In other words, each percent of tree canopy cover equalled an average of 3.8 stems ha^{-1} (3.6 in arable land and 4.4 in forest reserves), i.e. an average canopy cover of 26.3 m^2 (23 and 28) per individual tree, with a mean canopy diameter of 5.8 m and a mean biomass of 138.3 kg dry wood per tree.

Annual grassland of *Aristida mutabilis*, with *Acacia senegal* in NE Mali. Deep sandy soil; long-term mean annual rainfall 400 mm. Standing crop biomass and production at the end of the rainy season (mid-September): 120 g m^{-2} (Photo Le Houérou)

1. *In the Saharo-Sahelian subzone (100–200 mm),* on sandy soils ("clothed" and stabilized sand dunes, sand sheets and sand veils), the pristine vegetation is a perennial grass steppe dominated by highly xeromorphic species such as *Panicum turgidum*, with a canopy cover of 5–20%. This is a "diffuse" type of steppe, i.e. covered with sparse but regularly distributed tussocks. On stony and gravel plains and pediments there may be no perennial vegetation, but only small ephemeral plants covering 0 to 50% of the ground during the short (1–2 months) rainy season. Silt and clay valleys have a layer of short (e.g. *Aristida hordacea*) or tall (e.g. *Sorghum aethiopicum*) annual grasses; but they may be locally dominated by a sparse shrub layer 0.5–2.0 m high:

   Boscia senegalensis *Cordia sinensis*

   and a still sparser layer of small trees 2.0–5.0 m high:

   Acacia tortilis *Balanites aegyptiaca*
   Acacia ehrenbergiana

   The woody canopy layer may cover 0 to 10% of the ground with 0–200 stems ha^{-1}. Denser stands of trees and shrubs may occur in depressions and along wadis having a deep water table (underflow) with the following phreatophytes:

   Acacia nilotica *Hyphaene thebaica*
   Faidherbia albida *Salvadora persica*

2. *In the Sahel proper subzone (200–400 mm),* the "Mimosaceae thorn scrub" has three main vegetation layers with an overall canopy cover of 30–80%:

a) *A herbaceous layer* composed of annual grasses and forbs covering 20–80% of the ground, with an average height of 40–80 cm. Common dominant grasses are (Tables 13–14):

Aristida funiculata — *Cenchrus biflorus*
Aristida mutabilis — *Schoenefeldia gracilis*

Common dominant forbs are:

Blepharis lineariifolia — *Limeum viscosum*
Borreria radiata — *Mollugo nudicaulis*
Gisekia pharnaceoides — *Zornia glochidiata*

Legumes are usually neither common nor dominant, *Zornia* excepted:

Alysicarpus spp. — *Indigofera* spp.
Crotalaria spp. — *Tephrosia* spp.

except in the "piospheric" zones around villages, wells, etc.:

Cassia spp. — *Zornia glochidiata*

Table 13. Botanical composition (frequency) and specific contribution to canopy cover and aerial phytomass at the time of Maximum Standing Crop (mid-September) in some exclosures of Sahelian grazing ecosystems of northern Burkina-Faso (Oursi Watershed) (Grouzis 1979)[a]

Range types	Ams		Cep		Sgr		Sgl		Spt		Averages	
Edaphic conditions	Vegetated sand dune system		Sandy pedilain		Silty pediplain		Shallow silty pediplain		Loamy depression		------	
Botanical composition (%)	B C		B C		B C		B C		B C		B C	
Specific contributions (%) of canopy cover		Si C		Si C		Si C		Si C		Si C		Si C
Annual grasses	27.0	85.0	26.8	83.6	50.0	99.9	42.1	97.2	35.3	60.1	36.2	85.2
Sedges	2.1	0	7.3	0	0	0	0	0	5.8	9.6	3.0	1.9
Legumes	23.0	13.5	17.2	10.6	13.3	0	13.1	0	35.3	30.2	20.4	10.9
Forbs (Misc. Families)	48.3	1.2	48.4	5.8	36.7	0	44.4	2.6	23.4	0	40.2	1.9
Bare ground (%)	21.0		27.6		33.0		50.0		0.0		26.3	

[a] *B C* Botanical composition in percent of all species present;
Si C species contribution to frequency and canopy cover in %.
Plant communities:
Ams = *Heliotropium strigosum/Zornia glochidiata;*
Cep = *Cenchrus biflorus/Alysicarpus ovalifolius;*
Sgr = *Schoenefeldia gracilis/Panicum laetum;*
Sgl = *Schoenefeldia gracilis/Urochloa trichopus;*
Spt = *Panicum laetum/Aeschynomene indica.*

Table 14. Herbaceous primary production in the Ferlo (1979–81). Mean values computed from the data published by Boudet (1983a)

Range types[a]	1	2	3	4	5	6	7	8	Xa[b]	Xp[c]
No. of sites	5	9	24	21	20	20	7	8	114	114
MSC (kg DM ha^{-1} yr^{-1})	369	1026	682	755	1259	769	1112	1166	955	980
G3 % MSC[d]	50	54	53	39	47	35	53	45	47	46
G2	16	18	25	25	18	29	25	19	22	21
G1	2	3	4	5	2	8	3	1	3.5	4
Total G	68	75	82	69	67	72	81	65	72	71
L2	27	21	11	22	24	16	11	27	20	20
L1	0.3	1.7	0.5	2	1	1	1	1	1.2	1
Total L	27	23	12	24	25	17	12	28	21	21
M2	0.4	0.3	2	3	3	2	2	3	2	2
M1	0.3	1	4	2	4	4	4	3	3	3
Total M	1	1	6	5	7	6	6	6	5	5
U	4	1	1	2	3	5	1	1	2	2
Bare ground (%)	55	19	28	25	5	30	24	14	25	27
Forage value index[e]	77	83	73	74	76	69	81	80	77	75
Rain-use efficiency (RUE) (kg DM ha^{-1} yr^{-1} mm^{-1})	1.1	3.1	2.1	2.3	3.8	2.3	3.4	5.0	2.9	2.8
Overall RUE[f]: 2.9										

[a] Range types: *1* low potential on iron hardpan; *2* low potential downslope; *3* low potential upslope; *4* medium potential more or less flat ground; *5* medium potential downslope; *6* medium potential upslope; *7* high potential interdunal depressions; *8* high potential downslope.
[b] Xa = Arithmetic mean.
[c] Xp = Weighted mean.
[d] MSC = Maximum Standing Crop, end of rainy season. G3 = Good grazing value grasses; G2 = medium value grasses; G1 = poor value grasses; L2 = medium value legumes; L1 = poor value legumes; M2 = medium gazing value forbs; M1 = poor gazing value forbs; U = unpalatable and/or toxic.
[e] Forage value, see Table 15, pp 111–112.
[f] RUE = Rain-use efficiency = $\frac{\text{MSC kg DM ha}^{-1}\text{ yr}^{-1}}{\text{Rainfall mm}}$.

The herbaceous layer may be subdivided into *an upper stratum* dominated by medium-size grasses, 40–80 cm high: typical representatives of which are:

Annual grasses:

Andropogon pseudapricus
Aristida funiculata
Aristida mutabilis
Cenchrus biflorus
Cenchrus prieurii
Chloris pilosa
Chloris prieurii
Chloris virgata
Ctenium elegans
Diheteripogon hagerupii
Elionorus elegans
Eragrostis tremula
Loudetia togoensis
Pennisetum pedicellatum
Pennisetum violaceum
Schizachryrium exile
Schoenefeldia gracilis

and *a lower stratum* 0–40 cm high with grasses and forbs:

Annual grasses:

Aristida adscensionis
Aristida hordacea
Dactyloctenium aegyptium
Eragrostis cilianensis
Eragrostis tenella
Leptothrium senegalense
Microchloa indica
Tetrapogon cenchriformis
Tragus racemosus
Tragus berteronianus

Annual forbs:

Blepharis lineariifolia
Boerhavia spp.
Borreria spp.
Cleome spp.
Commelina spp.
Gisekia pharnaceoides
Glossonema spp.
Ipomeaea spp.
Limeum viscosum
Mollugo cerviana
Tribulus terrestris
Zornia glochidiata

b) *A shrub layer* 0.5–3.0 m high whose canopy may cover 0–20% of the ground with the following spiny species being usually dominant:

Acacia ehrenbergiana
Acacia laeta
Acacia senegal
Acacia torilis
Balanites aegyptiaca
Ziziphus mauritiana

together with some non-spiny species:

Boscia senegalensis
Cordia sinensis
Euphorbia balsamifera
Guiera senegalensis

c) *A very sparse layer of small trees* 3–6 m high with a canopy cover of 1–5%, the usually dominant species are:

Acacia tortilis subsp. *raddiana*
Balanites aegyptiaca
Commiphora africana
Grewia bicolor

The number of trees and shrubs may reach 500 ha^{-1} or more; under and immediately around the canopies a sciaphilous and nitrophilous vegetation of annual grasses and forbs develop with:

Achyranthes aspera
Andropogon pinguipes
Brachiaria xantholeuca
Chloris prieurii
Gynandropsis gynandra
Panicum laetum
Pennisetum pedicellatum
Setaria pallide-fusca

3. *In the Sudano-Sahelian subzone (400–600 mm),* the structure is similar to the one found in the Sahel proper, with, however, some noticeable differences:

a) Perennial grasses may occasionally occur, particularly in depressions (*Andropogon gayanus*).

b) The herbaceous layer tends to be continuous and the grasses somewhat taller (60–120 cm):

Andropogon pseudapricus | *Loudetia togoensis*
Diheteropogon hagerupii

c) Increased importance of the tree-shrub layers with up to 20–35% canopy cover and 500–1500 individuals per hectare.

d) Dominance of broad-leaved species, as opposed to the Sahel proper and to the Sahelo-Saharan subzones where spiny, microphyll species are dominant. Common and dominant broad-leaved species are:

Adansonia digitata
Bombax costatum
Combretum aculeatum
Combretum ghazalense
Combretum glutinosum
Combretum micranthum
Combretum nigricans
Pterocarpus lucens
Sclerocarya birrea
Sterculia setigera

e) Development of vegetation arcs (the so-called tiger bush), particularly on shallow soils on iron hardpans, showing alternating stripes of bush and bare ground in contour patterns. The Sudano-Sahelian subzone vegetation thus physiognomically differs from the one found in the Sahel proper. The Sudano-Sahelian and Sudanian *Acacia seyal* shrublands, developing on medium-to fine-textured soils are, however, fairly similar to the Sahelian "Mimosaceae scrub", but usually denser, with over 1000 stems ha^{-1}.

3.2.2 Plant Communities (Figs. 45–46)

1. *In the Saharo-Sahelian subzone*

a) *On fixed sand dunes, sand sheets, sandy pediplains*

The main woody species are:
Acacia ehrenbergiana
Acacia tortilis subsp. *raddiana*
Balanites aegyptiaca
Capparis decidua
Commiphora africana
Euphorbia balsamifera
Leptadenia pyrotechnica
Maerua crassifolia

Perennial grasses and graminoids:
Aristida pallida
Aristida papposa
Cymbopogon schoenanthus
Cyperus jeminicus
Lasiurus hirsutus
Panicum turgidum

Annual grasses:
Aristida mutabilis
Cenchrus biflorus

Forbs:
Indigofera sessiliflora
Tribulus terrestris

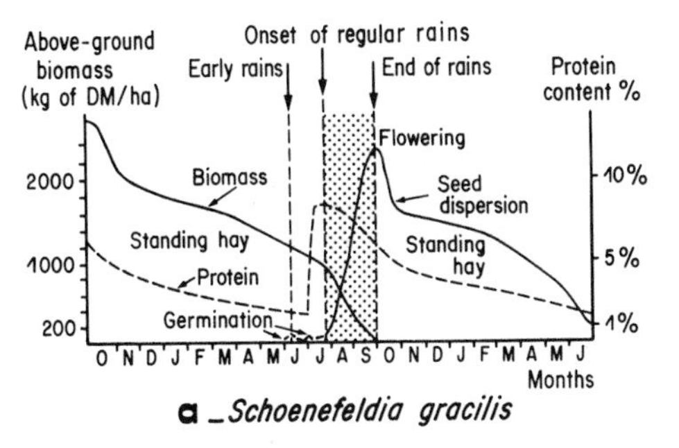

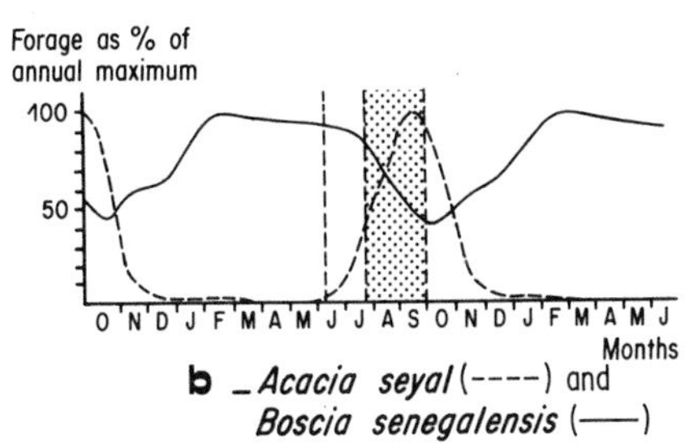

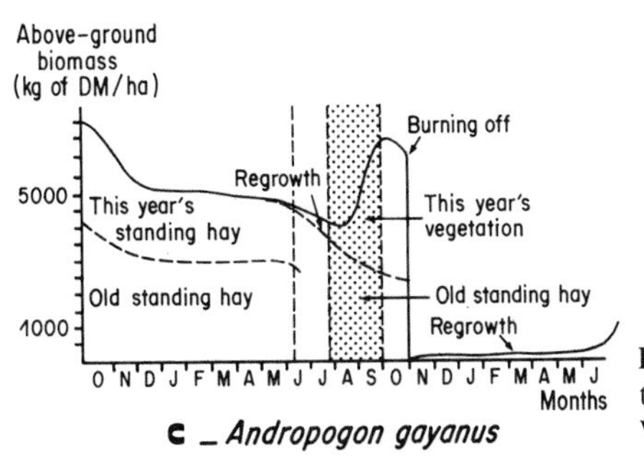

Fig. 45. Seasonal variation in phytomass and protein content of some forage plants of the Sahel (After Wilson 1982)

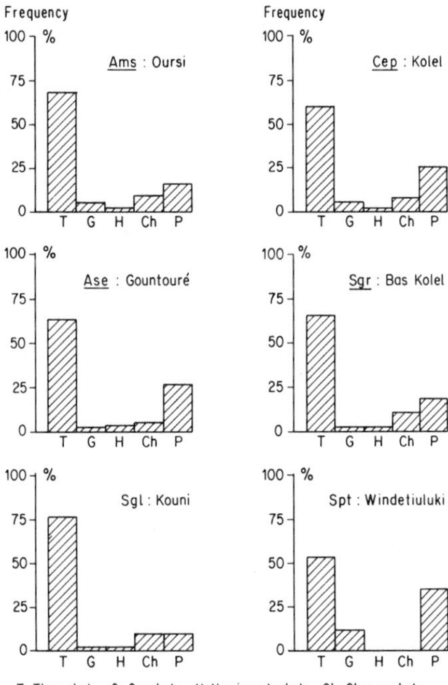

Fig. 46. Biological spectra in six grazing lands ecosystems of the Sahel of northern Burkina-Faso (After Grouzis 1979). Plant communities (dominant species) (Toutain 1976; Grouzis, 1979): **Ams** *Heliotropium strigosum/Zornia glochidiata;* **Cep** *Cenchrus ciliaris/Alysicarpus ovalifolius;* **Ase** *Schoenefeldia gracilis/Tripogon minimus;* **Sgr** *Schoenefeldia gracilis/Panicum laetum;* **Sgl** *Schoenefeldia gracilis/Urochloa trichopus;* **Spt** *Panicum laetum/Aeschynomene indica*

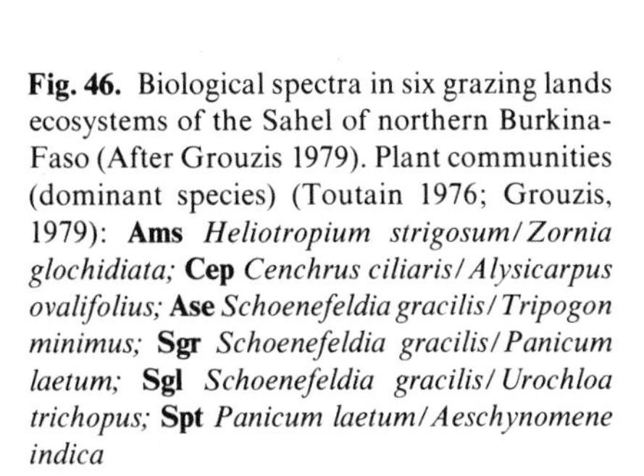

b) *On silty pediplains*

Woody species:
Acacia erhenbergiana
Acacia tortilis subsp. *raddiana*
Balanites aegyptiaca
Boscia senegalensis
Commiphora africana
Cordia sinensis

Annual grasses:
Aristida adscensionis
Aristida funiculata
Aristida hordacea
Panicum laetum
Schoenefeldia gracilis
Sorghum aethiopicum

Forb:
Tribulus terrestris

2. *In the Sahel proper subzone*

a) *On sandy soils*

Woody species:

Acacia laeta
Acacia senegal
Acacia tortilis subsp. *raddiana*
Balanites aegyptiaca
Commiphora africana

Perennial grasses:
Aristida longiflora
Cymbopogon proximus

Annual grasses:
Aristida funiculata
Aristida mutabilis
Cenchrus biflorus
Eragrostis tremula

Forbs:
Alysicarpus ovalifolius
Blepharis leariifolia
Tephrosia purpurea
Tribulus terrestris

b) *On silty pediplains*

Woody species:
Acacia ehrenbergiana
Acacia laeta
Acacia mellifera (eastern Sahel)
Acacia nubica (eastern Sahel)
Acacia senegal
Balanites aegyptiaca
Bauhinia rufescens
Boscia senegalensis
Gewia bicolor
Maerua crassifolia

Annual grasses:
Aristida funiculata
Aristida mutabilis
Cenchrus biflorus
Panicum laetum
Schoenefeldia gracilis
Tetrapogon cenchriformis

Annual legumes:
Alysicarpus ovalifolius
Zornia glochidiata

Annual grassland of *Schoenefeldia gracilis* in a hollow on silty soil. Standing crop biomass and production: 250 g m^{-2}. Tree layer: *Acacia tortilis, A. nilotica, Anogeissus leiocarpus, Balanites aegyptiaca.* Mean long-term rainfall 350 mm; Ferlo region of the Sahel of N. Senegal (Photo A. Cornet)

c) *On loam and clay pediplains*

Woody species:
Acacia ehrenbergiana
Balanites aegyptiaca
Boscia senegalensis
Combretum aculeatum
Cordia sinensis
Grewia bicolor

Annual grasses:
Aristida funiculata
Panicum laetum
Pennisetum mollissimum
Schoenefeldia gracilis

Annual legumes:
Zornia glochidiata

d) *On loamy shallow soils (iron pans, rocks)*

Woody species:
Acacia ataxacantha
Acacia ehrenbergiana
Acacia laeta
Acacia macrostachya
Acacia mellifera (eastern Sahel)
Acacia nubica (eastern Sahel)
Balanites aegyptiaca
Boscia senegalensis
Commiphora africana
Euphorbia balsamifera
Maerua crassifolia

Perennial grass:
Cymbopogon proximus

Annual grasses:
Aristida adscensionis
Aristida funiculata
Aristida mutabilis
Panicum laetum
Schoenefeldia gracilis
Tetrapogon cenchriformis

Annual legume:
Zornia glochidiata

3. *Sudano-Sahelian subzone*

a) *On sandy soils*

Woody species:
Acacia senegal
Balanites aegyptiaca
Boscia senegalensis
Combretum glutinosum
Commiphora africana
Guiera senegalensis
Piliostigma reticulatum
Slerocarya birrea

Perennial grasses:
Andropogon gayanus
Aristida longiflora
Cymbopogon giganteus
Hyparrhenia dissoluta
(= *Hyperthelia d.*)

Annual grasses:
Andropogon pseudapricus
Aristida mutabilis
Cenchrus biflorus
Ctenium elegans
Diheteropogon hagerupii
Elionurus elegans
Eragrostis tremula
Loudetia togoensis
Pennisetum pedicellatum
Schoenefeldia gracilis
Schizachryrium exile

Annual legumes:
Alysicarpus ovalifolius
Zornia glochidiata

b) *On silty and loamy pediplains*

Woody species:
Acacia mellifera
Acacia senegal
Acacia seyal
Acacia tortilis subsp. *raddiana*
Balanites aegyptiaca
Bauhinia rufescens
Boscia senegalensis
Combretum aculeatum
Combretum glutinosum
Grewia bicolor
Guiera senegalensis
Sclerocarya birrea
Ziziphus mauritiana

Perennial grass:
Andropogon gayanus (rare)

Annual grassland of *Schoenefeldia gracilis, Chloris prieurii,* with a tree and shrub layer of *Acacia senegal.* Mean long-term rainfall 350 mm; Ferlo region of the Sahel of northern Senegal. Flat sandy soil, poor growth due to shortage of rain in that particular year, with patches of bare soil. Standing crop biomass and production: 80–100 g m^{-2} (Photo A. Cornet)

Annual grasses:

Andropogon pseudapricus
Aristida adscensionis
Aristida mutabilis
Diheteropogon hagerupii
Elionurus elegans
Loudetia togoensis
Pennisetum pedicellatum
Schoenefeldia gracilis

Annual legumes:

Aeschynomene indica
Alysicarpus ovalifolius
Crotalaria spp.
Psoralea plicata
Zornia glochidiata

c) *On loamy shallow soils (iron pans)*

Woody species:

Acacia ataxacantha
Acacia seyal
Boscia senegalensis
Combretum glutinosum
Combretum micranthum
Combretum nigricans
Grewia bicolor
Guiera senegalensis
Pterocarpus lucens
Ziziphus mauritiana

Annual grasses:

Andropogon pseudapricus
Aristida adscensionis
Loudetia togoensis
Microchloa indica

Cenchrus biflorus
Ctenium elegans
Diheteropogon hagerupii
Elionurus elegans
Pennisetum pedicellatum
Schoenefeldia gracilis
Tropogon minimus

Annual legume:
Zornia glochidiata

4. *Flood plain vegetation*

Many meadow types of grassland could be differentiated according to the average length of the flood period.

a) *1-Month flooding or less:*

Andropogon gayanus
Cynodon dactylon
Panicum anabaptistum
Oryza barthii
Vetiveria fluvibarbis
Vetiveria nigritana

b) *2 to 3-Month flooding:*

Acroceras amplectens
Brachiaria mutica
Echinochloa pyramidalis
Eragrostis barteri
Oryza longistaminata
Vossia cuspidata

c) *2-Month flooding or more:*

Echinochloa pyramidalis
Echinochloa stagnina
Oryza longistaminata
Vossia cuspidata

5. *Around ponds and main depressions*

Woody species:
Acacia ataxacantha
Acacia macrostachya
Acacia nilotica
Acacia sieberiana
Anogeissus leiocarpus
Celtis integrifolia
Crataeva adandoni
Dalbergia melanoxylon
Diospyros mespiliformis
Faidherbia albida
Feretia apodanthera
Hyphaene thebaica
Lannea acida
Mitragyna inermis
Terminalia avicennoides

Perennial grasses:
Andropogon gayanus
Cymbopogon maximus
Cynodon dactylon
Hyparrhenia dissoluta
Panicum anabaptistum
Vetiveria nigritana

Annual grasses:
Andropogon pseudapricus
Panicum laetum

Diheteropogon hagerupii
Elionurus elegans
Pennisetum pedicellatum (Shade)
Schoenefeldia gracilis

6. *Montane communities (> 1000 m)*

Woody species:
Acacia mellifera
Albizia amara
Anogeissus leiocarpus
Balanites aegyptiaca
Carissa edulis
Commiphora africana
Commiphora quadricincta
Dichrostachys cinerea
Euphorbia candelabrum
Ficus ingens
Ficus populifolia
Ficus salicifolia
Ficus teloukat
Grewia flavescens
Grewia tenax
Grewia villosa
Olea laperrini
Rhus vulgaris
Sterculia pilifera
Stereospermum kunthianum
Terminalia brownii
Ziziphus mauritiana

Perennial grasses:
Andropogon schirensis
Botriochloa pertusa
Cenchrus ciliaris
Chrysopogon plumulosus
Cymbopogon commutatus
Cymbopogon schoenanthus
Dichanthium annulatum
Enneapogon spp.
Anthephora hochtetteri
Eremopogon foveolatus
Heteropogon contortus
Hyparrhenia hirta
Sporobolus helvolus
Sprobolus ioclados
Themeda triandra

Annual grasses:
Aristida spp.
Brachiaria spp.
Dactyloctenium aegyptium
Eragrostis spp.
Tetrapogon cenchriformis
Tragus berteronianus
Tragus racemosus

Forbs:
Barleria spp.
Blepharis linariifolia
Blepharis persica
Boerhevia spp.
Borreria spp.
Cleome spp.
Commelina spp.
Crotalaria spp.
Forskalea tenacissima
Geigera acaulis
Gisekia pharnaceoides
Heliotropium spp.
Indogofera spp.
Ipomea spp.
Lavandula coronopifolia
Limeum spp.
Mollugo spp.
Oldenlandia spp.
Rynchosia spp.
Salvia aegyptiacaia
Sesbania spp.
Solanum spp.

Glinus lotoides	*Tephrosia* spp.
Glossonema spp.	*Tribulus* spp.
Gynandropsis gynandra	*Vernonia* spp.

3.2.3 Seasonal Aspects, Short-Term Dynamics (Figs. 45, 47)

Seasonal aspects of vegetation are sharply contrasted for climatic reasons. The rainy season is short and growth may be intense since the total annual growth is achieved in 1 to 3 months. Daily increments are thus very high, particularly at the peak of the growing season, in August. This point will be discussed with some detail later.

During a normal rainy season, water and green grass are plentiful everywhere; then for a period of 9–11 months the landscape is parched, only trees and shrubs provide some green spots.

The botanical composition of the herbaceous layer may change considerably from one year to the next depending on early rains and on the rainfall events that had occurred the previous years. This history of rain events and grass-layer composition strongly influences seed production and seed stocks in the soils. A given species may be dominant in a specific place for one or several years and then almost disappear for a number of years for no obvious reasons. This dynamism is very complex and far from fully understood; there seems to exist complex

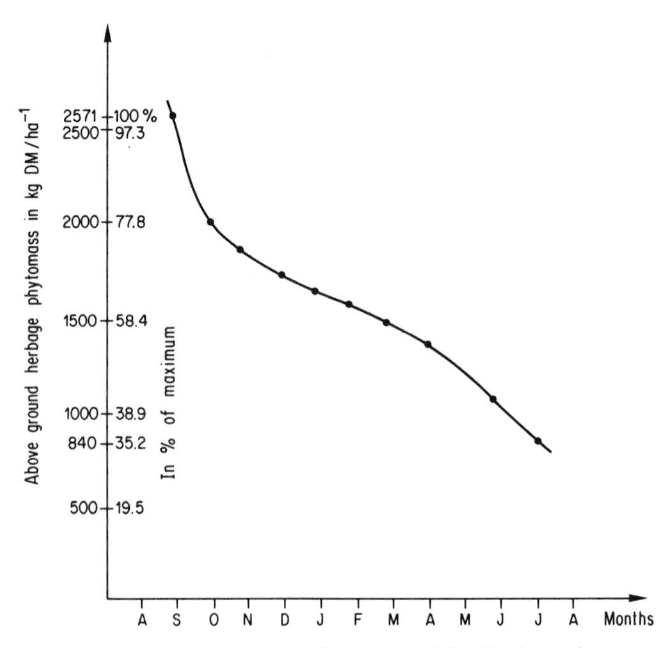

Fig. 47. Decay of maximum standing crop (MSC) of herbage phytomass of a *Schoenefeldia gracilis* grassland, under conditions of total protection from livestock, on sandy soils, at Niono, Mali (After Diarra 1976)

interactions between rainfall events and history, seed production, consumption and storage by granivores (mainly arthropods, birds, rodents), and seed hardness. With respect to hardness, seeds may be classified as "soft" (*Cenchrus biflorus*), "semi-soft" (*Dactyloctenium aegyptium*), "semi-hard" (*Zornia, glochidiata*) or "hard" (*Borreria radiata*) (De Vries and Djiteye 1982; AM Cissé 1986). When several rain events interspersed with dry spells occur at the beginning of the rainy season, most soft and semi-soft seed species may germinate and die; vegetation under these circumstances will be dominated by semi-hard or hard seed species. One may thus, to some extent, predict the composition of the annual herbaceous layer in a given year, knowing the seed stock and the early rainy season rainfall events.

Cissé (1986) studied the germination and establishment of a number of herbage species in the Sudano-Sahelian ecoclimatic zone of central Mali (Ranch of Niono). The establishment of seedlings appears to result from a combination of various ecophysiological characteristics: seed hardness, germination speed, length of growing cycle, seed production, size of species, existence of one or several germination pulses and drought tolerance in early establishment phases. The following table emerges from this research:

Germination and establishment characteristics of some annual herbage species of the Sahel (A.M. Cissé 1986).

Species	Germination		Drought tolerance of seedlings[c]	Vegetative growth			Reproductive growth		
	Speed[a]	Pulses[b]		Carbo-xylation pathway	Cycle length[d]	Size[e]	Seeds Percent of MSC	Weight 1000 seeds	Hardness[f]
Diheteropogon hagerupii	F	+	+ + +	C4	L	T	3	0.8	S
Pennisetum pedicellatum	F	+	+	C4	L	T	–	0.4	S
Cenchrus biflorus	F	+ +	+ + +	C4	S	I	10	2.1	S
Schoenefeldia gracilis	F	+	+ + +	C4	S	I	3	0.2	S
Loudetia togoensis	F	+	+ + +	C4	S	I	3	2.5	S
Blepharis linariifolia	F	+	+ + +	C4	L	I	18	13.9	S
Elionurus elegans	F	+ + +	+ +	C4	S	I	–	0.4	S
Borreria stachydea	S	+ + +	+	C3	L	I	16	7.0	H
B.chaetocephala	S	+ + +	+	C3	L	S	–	1.0	H
B.radiata	S	+ +	+	C3	L	S	13	0.7	H
Zornia glochidiata	H	+ + +	+ +	C3	S	S	14	1.9	H
Indigofera prieureana	H?	+ + ?	+ +	C3?	L	T	–	–	–
Cassia mimosoides	H	+ +	+ +	C4	L	I	–	3.6	H
Eragrostis tremula	S	+ + +	+ +	C4	S	I	–	0.05	H
Tribulus terrestris	F	+	+ + +	C4	S	S	23	1.1	S

[a] Germination speed: F, fast; S, slow; H, variable.
[b] Pulses: +, one pulse; + +, a few pulses; + + +, several pulses.
[c] Drought tolerance of seedlings: +, weak; + +, intermediate; + + +, strong.
[d] Cycle length: L, long; S, short.
[e] Size: T, tall; I, intermediate; S, small.
[f] Seed hardness: H, hard; S, soft.

From the above table it appears that drought-tolerant species that would tend to dominate during a sequence of drought years (characterized by irregular and infrequent early rains and interrupted by lengthy drought spells) would have soft seeds, a quick germination, a short biological cycle and drought-tolerant seedlings able to withstand competition. Conversely, during good rainy years, with regular, early rainfalls the species having the opposite characteristics (hard seeds, slow germination, long biological cycle, tall size and mediocre drought tolerance of their seedlings) would be granted a selective advantage over their competitors. It would also seem that typical Sahelian species tend to belong to the former category, whilst Sudano-Sahelian and Sudanian species belong to the latter, as one could have expected.

This research also showed that the establishment of seedlings depends strongly on the texture of the top soil, particularly when rainfall deficiency occurs in the early growing season, since it takes twice as much rain to wet a silty soil from hygroscopic water potential (-200,000 to -100,000 kPa, depending on air moisture) to permanent wilting point (-1,600 kPa) as to moisten a coarse, sandy soil to the same water potential.

A survey on 36 permanent quadrats in the Sudano-Sahelian zone of Mali (Hiernaux et al. 1978) over a period of 3 years (1976–78) showed that the coefficient of similarity (Sørensen 1948) was about 60% for 2 consecutive years (57–63%) and only 45% for the 3 years. These years were not particularly wet or dry and there was no fire or grazing interference. One could therefore expect that the variation in botanical composition would still be greater under the usual conditions, i.e. with fire and grazing interference. One can only say that less than 50% of the species found at a given site in a given year are to be found there permanently. On the same plots during the same period of 3 years, variation in biomass production was 37%, substantially less than in the botanical composition. Similar facts have been reported by various authors, but no in-depth study of the phenomenon has been attempted to date, to our knowledge, except for the dynamics of productivity in Central Mali (De Vries and Djiteye 1982).

3.2.4 Biological Recovery After Drought

Herbaceous Layer. Biological recovery of herbaceous grasses and forbs is quite good in the second or third year consecutive to multi-annual drought when weather returns to normal and the seed stock has been restored (Bille 1977; Cornet 1981; Boudet 1977a; Wilson 1982b; Barry et al. 1983; Barral et al. 1983; De Vries and Djiteye 1982; Grouzis 1984, 1988).

This, however, is only true if the anthropozoic pressure is moderate. In heavily stocked areas (3 ha TLU^{-1} yr^{-1}, or more) such as around permanent ponds of the Oudalan of northern Burkina-Faso (Barral 1977; Sicot and Grouzis 1981; Grouzis 1984) or in the "transition zone" on the northern edge of the Inland Delta of the Niger River in Central Mali (Wilson 1982b; Wilson et al. 1983), large extents of permanently bare ground with a sealed surface may develop, even on sandy soils, with or without biological encrustation of *Cyanophyceae* (Dulieu et al. 1977; Barbey and Conte 1976). In such cases there is little or no recovery unless human intervention takes place and the cause of degradation is discontinued (Le Houérou 1979b; Peyre de Fabrègues and De Wispelaere 1984; Grouzis 1982, 1983; Toutain 1977a).

Woody Layer. Recovery in the woody layer, even in "brousse tigrée" (vegetation arcs) has been reported after the 1970–73 drought (Boudet 1976). In that particular case new seedlings developed and survived, while apparently dead stumps and root stocks came back to life. Among the regenerating species were excellent browse plants showing usually poor regeneration, such as *Pterocarpus lucens.* But in heavily stocked areas there may be no regeneration. In moderately stocked regions a species turnover may take place in the ligneous populations towards more xerophilous species such as:

Acacia spp.
Balanites aegyptiaca
Boscia senegalensis
Capparis decidua
Cordia sinensis
Dichrostachys cinerea
Leptadenia pyrotechnica
Orchradenus baccatus
etc.

which tend to replace more mesic species such as the Combretaceae:

Anogeissus
Combretum
Guiera
Terminalia

and others:

Adansonia
Bauhinia
Bombax
Grewia
Lannea
Piliostigma
Pterocarpus
Sclerocarya
Sterculia
etc.

(Bille 1977, 1978; Poupon 1980; Piot et al. 1980; Grouzis 1979, 1984). Rodent pullulation after the great drought has also sometimes jeopardized the biological recovery of the woody layer (Poulet and Poupon 1978). Naturally, recovery depends also on soil and topographical conditions; in depressions and areas subject to periodical rainy season flooding recovery is faster and more complete than in dry sites, hill tops and shallow soils (Haywood 1981; Courel 1985). In northern Senegal a detailed study of the woody vegetation population dynamics and demography concluded that it would take some 30 years to recover from the 1970–73 drought under total protection conditions (Bille 1978; Poupon 1980). This means that under the current conditions of increasing anthropozoic pressure, the Sahel will probably never recover from the recent drought of 1970–84 (Peyre de Fabrègues 1985; Peyre de Fabrègues and De Wispelaere 1984; Le Houérou 1980b; Le Houérou and Gillet 1985).

Heavy mortalities in the woody layer during the 1983–85 drought have been reported, particularly north of the 16° parallel (Barry et al. 1983; Boudet 1984a, 1985; Peyre de Fabrègues 1985; Peyre de Fabrègues and De Wispelaere 1985; Le

Houérou and Gillet 1985; Courel 1985) in Mauritania, Senegal, Mali, Burkina-Faso, Niger and Chad. At the time of writing it is still too early to know whether recovery will take place or to what extent.

Most woody species shed their leaves during the second half of the dry season (March-June). Many of them produce new leaves 15 to 60 days before the onset of the rains, as discussed above (Sect. 2.1.7). Some rare species, however, such as *Faidherbia albida*, shed their leaves at the onset of the rains and remain leafless throughout the rainy season. This bears important consequences on the centuries-old agroforestry techniques of cropping millet and groundnuts in *Faidherbia* orchards, with fairly high yields and no fertilization; this point will be discussed later.

3.2.5 Long-Term Dynamics: The Role of Wildfires

Two different questions should be addressed here:

1. What is the primeval vegetation of the Sahel?
2. What is the present trend?

3.2.5.1 Primeval Vegetation

Primeval vegetation is understood as one which is little affected by man and his livestock, in an attempt to avoid the word "climax", an abstraction that has been the subject of too many speculations and cannot have any real existence since climate has changed on several occasions in the past; and since man has been present with his livestock for more than 7500 years, while some of these changes were taking place.

1. *In the Saharo-Sahelian Transition Subzone*

   There is little doubt that primeval vegetation is a perennial grass steppe, probably similar to what can still be seen in remote places, far away from any permanent water. However, the botanical composition may have been substantially different from what we know today. More palatable perennial grasses such as:

   Andropogon schirensis
   Antephora hochstetteri
   Botriochloa pertusa
   Cenchrus ciliaris
   Chrysopogon plumulosus
   Dichanthium annulatum
   Enneapogon brachystachyus
   Enneapogon scaber
   Eremopogon foveolatus
   Heteropogon contortus
   Hyparrhenia hirta
   Lasiurus hirsutus
   Sporobolus ioclados
   Themeda triandra
   etc.

   which have become very rare and restricted to rocky montane sites, might have played a much more important role before the impact of man became severe.

This zone has probably never borne much woody vegetation outside depressions and sites benefitting from run-in or having a water table.

2. *In the Sudanian Zone*

Primeval vegetation is obviously a dry deciduous forest, often degraded into open woodland or various types of savannas due to wildfires, woodcutting, clearing for cultivation followed by abandonment (shifting cultivation). This point is subject to little arguing, as dynamic sequences are still easy to observe though there may be some minor points of disagreement between authors on the interpretation of some terms of the dynamic sequences.

3. *In the Sudano-Sahelian Transition Subzone*

The present day *Combretaceae* savanna seems to result from an open woodland with perennial grasses such as:

Andropogon gayanus
Andropogon tectorum
Aristida longiflora
Diheteropogon amplectens
Hyparrhenia dissoluta

Andropogon gayanus may still be found in protected areas under 400–600 mm average annual rainfall (Boudet and Leclerc 1970; Grouzis, pers. communication 1985). In such protected areas the woody vegetation canopy may still cover up to 30% of the ground or more, with forest galleries in the depressions and along watercourses (Hiernaux 1980). In most of those places *Andropogon gayanus* died during the 1969–73 drought and did not come back even under light grazing or total protection conditions.

4. *In the Sahelian Zone Proper*

The present herbaceous layer is almost entirely annual. One may wonder why, since perennial grasses prevail to the north in the Saharo-Sahelian subzone and to the south in the Sudanian zone and used to play a substantial role in the Sudano-Sahelian transition subzone. The facts are that perennial grasses such as:

Aristida pallida
Aristida papposa
Cenchrus ciliaris
Chrysopogon plumulosus
Cymbopogon proximus
Dichanthium annulatum
Schmidtia pappophoroides
Sehima ischaemoides
etc.

are still present in the foothills of the mountains of Air, Ennedi, Gourgeil, on shallow rocky soils under mean annual rainfalls of 100–200 mm. One cannot therefore see any climatic reason why these perennial grasses could not exist in the Sahel (some actually do in pinpoint locations: *Aristida pallida, Cymbopogon proximus*, etc.). The reason is probably fire.

Sahelian annual grassland shortly after the occurrence of an early-season sweeping bushfire. The shrub layer consists of *Acacia tortilis, Balanites aegyptiaca* and *Boscia senegalensis*. The burnt area will remain unproductive and subject to erosion forces until the next rainy season, some 9 months later. Region of Oursi, Burkina-Faso (Photo M. Grouzis)

In the Saharo-Sahelian zone there is not enough fuel load (200–500 kg DM ha^{-1}) and the tussocks of perennial grasses are too sparsely distributed to carry fire over large acreages.

In the Sahel proper, where the biomass of dry grass usually varies between 800 and 3000 kg DM ha^{-1}, fire may spread over huge areas (e.g. 850 000 ha in the N. Ferlo area of Senegal in 1964–65). The combined effect of fire with a very long and harsh dry season may be sufficient to explain the elimination of perennial grasses from the Sahel (Le Houérou and Naegelé 1972).

In fact, in the areas of Louga and Dahra in northern Senegal, under long-term mean annual rainfalls of 300–400 mm, *Andropogon gayanus,* var. *bisquamulatus* was still fairly common in old fallows and along field edges in 1986, right after the intense and prolonged droughts of 1983 and 1984. Similar cases have been reported by Grouzis (1987) and Benoit (1984) from northern Burkina-Faso where rather dense stands of *A. gayanus* still survive in little disturbed areas under long-term rainfalls of 300–400 mm, without any additional runoff water and away from any water table.

One possible explanation of the rareness of perennial grasses in the Sahel Sensu stricto and in the Sudano-Sahelian ecozones is that the Sudanian *Andopogoneae*, which are extremely fire-tolerant, are there at the dry limit of their geographical area of extension and therefore sensitive to the combination of drought and fire. The Saharan and sub-Saharan species, on the other hand, which are extremely drought-tolerant, but sensitive to fire, also cannot withstand the combination of fire and drought. Among these dought-tolerant, fire-sensitive species, one may quote the following:

Asthenatherum forskhalei
Aristida sieberiana
Botriochloa spp.
Chrysopogon plumulosus
Cenchrus ciliaris
Cenchrus setigerus
Cymbopogon schoenanthus
Desmostachya bipinnata
Dichanthium annulatum
Enneapogon spp.
Eremopogon foveolatus
Lasiurus hirsutus
Panicum turgidum
Pennisetum divisum
Pennisetum elatum
Pennisetum setaceum
Stipagrostis spp.
Stipagrostis pungens

The Sahel ecozone *sensu stricto* would thus be a kind of no-man's-land between northern Saharan perennial grasses and southern Sudanian perennial *Andropogoneae* (A. Cornet, pers. communication, 1988).

In the Sahelian zone of the Republic of Sudan, in N. Kordofan and N. Darfur provinces wildfires used to sweep across the country burning 15–20% of the land area each year (Wickens, pers. communication, 1985). Similar or higher estimates have been made for the Sahel zone of Senegal (Naegelé 1971a, p. 107) and Mali (Le Houérou 1980f, p. 43). In the Sahelian region of Ferlo, in northern Senegal, a 5-year study using satellite imagery, low altitude reconnaissance flights (LARF) in conjunction with ground checking found that 10 to 20% of the range bore traces of burning at the end of the dry season in those years when rainfall was close to or higher than the long-term average, i.e. whenever there was enough fuel to carry fire (Vanpraet 1985; Le Houérou 1988b). Several 100 000 ha were burned in the Ferlo of northern Senegal in the early 1988/1989 dry season, following a good 1988 rainy season, and therefore resulting in considerable fuel buildup. An official document of the Republic of Sudan (1977) estimated that wildfires burned every year some 35% of the range resource; for the mainly Sahelian provinces of Darfur, Kordofan, Blue Nile, Khartoum and Kessala, the mean percentage of range burned every year was estimated to average 21%, that is 30, 30, 15, 10 and 20% respectively (Dept. of Range Management, Khartoum, 1977; Baasher 1961; Shepherd and Baasher 1966; Shepherd 1968).

Because of range depletion and reduction in biomass, fires are not as destructive as they used to be; but they may still be very harmful in "good years" when the fuel load exceeds 1000 kg DM ha^{-1} and the dead grass constitutes an almost continuous mat for 9–11 month dry season. It is estimated that an average biomass of about 1000 kg DM ha^{-1} is necessary to carry fire over any substantial surface (Gillet 1967).

3.2.5.2 Present Trend, Desertization

Desertization occurred on several occasions during the Pleistocene (cf. Sects. 2.2.2 and 2.2.4). In the Holocene, up to about 3000–3500 BP the Niger river flooded some 250 km further north from Timbuctoo to Araouane (the channels are still conspicuous on air photographs and satellite images over some 80–100 km) in NE Mali, at the border of Mauritania (Petit-Maire 1982; Petit-Maire et al. 1983 a, b), over an area of some 50 000 km^2 which is presently a sandy desert. Whether the desiccation of this area is due to a decrease in rainfall or to the capture of the upper Niger (locally called Joliba) by the Tafassasset river (a tributary of the Tilemsi) at Tessaoua (also called Tosseye) at about 3000 BP is unclear and subject to heated controversy.

During historical times, until ca. 1100 AD, the Inland Delta of the Niger extended 150–200 km further north than its present reaches, throughout the Méma, far into Mauritania to the west of Lake Faguibine in the so-called dead Delta. Channels are known to have been still functioning west of Lake Faguibine in the late 19th century (Chudeau 1909, 1918, 1919; Furon 1929, 1950). The southward drift of the "live Delta" seems to have resulted from a slight subsidence or monoclinal dip of some 5 m or so, which resulted in the drying out of the Fala of Molodo and of a formerly flooded area of some 25 000 km^2, with numerous archaeological remains in topographical situations similar to the present-day "Togguéré" of the live Delta (Le Houérou, in Haaland, 1979; Haywood 1981).

Present-day desertization processes are of a different nature from those suggested above: chiefly the combination of a 15-year drought (1970–1984) with a rapidly increasing anthropozoic pressure on the land and on the ecosystems.

The Sahel has just undergone a 15-year drought; a period of an unprecedented length (1970–1984) since the beginning of instrumental records in the mid-19th century (St. Louis City 1855; Dakar City 1887). The mean annual rainfall for the period 1970–1984 was about 60% of that of the 1900–1969 period, with extreme droughts in 1972–1973 and in 1983–1984, when rains averaged only 40 to 50% of the long-term mean (Le Houérou and Gillet 1985). The Niger River went dry at Niamey in May 1985, for the first time in living memory; the Chari River at N'Djamena also went dry in May 1985, a fact which only happened once in the past, in 1914, since the beginning of the records.

At the same time human and livestock populations have both increased by a factor of about 2.3 between 1950 and 1983 in the seven Sahelian countries of Mauritania, Senegal, Mali, Burkina-Faso, Niger, Chad and Republic of Sudan (FAO production yearbooks). The annual rate of demographic growth in stock and human populations has thus been a geometric progression of 2.7% per annum, i.e. a doubling period of 26 years or a 16-fold increase per century. The average population density in these seven countries was 17 inhabitants km^{-2} in 1983. The ecological consequences of these human and stock population increments have been as follows.

Depleted annual grassland with *Acacia seyal.* Clay soil on pediplain overlaying Micashist and Gneiss Basement Complex. Herbage crop virtually nil in that particular year. Mean long-term annual rainfall 400 mm region of Oursi, Burkina-Faso (Photo M. Grouzis)

1. *The Expansion of Cropping*

Crops expand over rangelands at an average rate of about 1% per annum according to official statistics from the seven above mentioned countries; but at a much faster pace (2.0–2.5%) according to a number of surveys based on remote sensing. In a test zone of 60 000 km^2 in the Sahel of Central Mali a survey showed an increase of rain-fed cropping of 2.3% per annum between 1952 and 1975 (dates of the aerial photo coverages) (Le Houérou 1976b, 1979a; Haywood 1981). Similar findings have been reported from other countries, particularly from Niger, Chad, Senegal, Republic of Sudan, Burkina-Faso (Barral et al. 1983; Barry et al. 1983; De Wispelaere 1980a, b; De Wispelaere and Toutain 1976, 1981; Lamprey 1975; Gaston 1975a, 1981; Peyre de Fabrègues and De Wispelaere 1984).

In the Republic of Sudan an official government report states that rain-fed cropping in Darfur and Kordofan provinces increased five-fold between 1960 and 1975, with a sharp concurrent decrease in yields of groundnuts and sesame (70–75%) (Le Houérou 1976a). In N. Darfur and N. Kordofan the southern limit of the Sahara moved some 80–100 km southward between 1958 and 1975, due essentially to overgrazing (Lamprey 1975).

2. *Reduction or Suppression of the Crops/Fallow Rotation*

Under traditional conditions cropping occurred for 2–3 years and the land was then left fallow for a decade or more in order to restore fertility. Due to pressure on the land, resulting from demographic growth, this period of fallow is shortened to 1–3 years or even completely cancelled. Such a practice results in a sharp decline of yields and therefore new clearings have to occur on soils and in areas that are less and less suited for cropping. The northern limit of cropping in Niger during the 1950's used to be the 15° N parallel, i.e. the 400-mm isohyet. Since then cultivation has moved some 150 km northward to 16° 20′ N under the isohyet of 250 mm with a high rate of crop failure and, naturally, the concurrent destruction of large tracks of good grasslands (Bernus 1981).

3. *Overstocking and Overgrazing*

Due to the sharp increase in stock numbers as shown in Tables 10–15, in conjunction with prolonged drought, the Sahel grasslands have undergone an extremely severe degradation for the past 15 years. Desertization has freely expanded along the Saharan margins over huge areas. In the Republic of Sudan, the southern limit of the Sahara shifted some 80–100 km southward between 1958 and 1975, i.e. over an area of about 150 000 km^2 (Lamprey 1975). The impact of overgrazing and overstocking in the Sahel may be described as follows (Le Houérou and Gillet 1985).

Impact on the herbaceous Layer. Elimination of perennial grasses such as:

Andropogon gayanus
Aristida longiflora
Aristida pallida
Aristida papposa
Cymbopogon proximus
etc.

Replacement of mesic annual grasses of good fodder value such as:

Brachiaria spp.
Digitaria spp.
Panicum spp.
Pennisetum spp.
Setaria spp.
Sorghum spp.
Urochloa spp.

by more xeromorphic and less palatable and nutritious species such as:

Aristida spp.
Cenchrus spp.
Chloris spp.
Eragrostis spp.
Stipagrostis spp.
etc.

Invasion of less productive, palatable and nutritious forbs:

Blepharis spp.
Borreria spp.
Cassia spp.
Gisekia pharnaceoides
Limeum spp.
Mollugo spp.
Tribulus spp.
Zornia glochidiata
etc.

Foreground: detail of a sealed soil surface on a sandy soil. This sealing process is a potent factor of desertization as the ground surface is made almost impervious by a biological crust of Cyanophyceae (*Scytonema* sp.). The woody layer of *Acacia seyal* and *Pterocarpus lucens* is in very poor shape, trees and shrubs die off; the vegetation is laid out in contour stripes (vegetation arcs = "tiger bush"). Long-term mean annual rainfall 550 mm; region of Niono, central Mali, Sudano-Sahelian eco-climatic zone (Photo Le Houérou)

and one ubiquitous pantropical, ruderal, shrubbish Asclepiadaceae: *Calotropis procera* and a New World pest the tobacco shrub: *Nicotiana glauca*.

Denudation occurred over large tracts of land around villages, ponds, wells, boreholes and rivers which, on silty or gravelly soils become totally bare and barren. In Central Mali, for instance, on the test zone mentioned above the areas of bare ground increased from 4 to 26% between 1952 and 1975 (Le Houérou 1979b; Haywood 1981). In the Sahel of Chad vegetal cover decreased by 32% between 1954 and 1974, while erosion increased by 28% over the same period (Gaston 1975a, 1981), and the southern limit of drifting sands moved 50 km southward.

Impact on the Woody Layer. The impact of human and livestock pressure is still more catastrophic on the woody layer. Remote-sensing surveys coupled with ground truth checking in several Sahelian countries have shown that the tree and shrub layer cover has receded by about 1% per year since the 1950's (Gaston 1975a, 1981; Le Houérou 1980b; Haywood 1981; De Wispelaere and Toutain 1980; Barral et al. 1983; De Wispelaere 1980; Peyre de Fabrègues 1985). Under pristine Sahelian conditions, the canopy cover of woody vegetation may vary from 5 to 30% depending on local factors such as mean annual rainfall, soil, topography, etc. One could assume an average of 15% with 100 to 1000 trees and shrubs per hectare (Poupon 1980; Le Houérou 1980b; Hiernaux 1980; Bille 1977, 1978). In depleted areas this "trub" cover may be reduced to 1–2% or less (Piot et al. 1980; Menaut, in Toutain et al. 1983).

Generally, the regression of woody species is particularly acute on the upper part of the topography, on hard ferrugineous pans, on surface-sealed silty soils, gravel plains and pediments. But, because of increased runoff (40–60% on small watersheds of these kinds), a temporary increase in ligneous vegetation may occur in the depressions, around ponds and along water-courses (Haywood 1981). Conversely, for the same reasons, ligneous populations in depressions may be killed by radicular asphyxia resulting from prolonged floodings.

At the same time there is a turnover of species in tree/shrub vegetations towards more xerophytic genotypes; that is the expansion of:

a) Deciduous spiny microphyll species such as *Acacia* spp., *Dichrostachys cinerea*;
b) Ephedroid species: *Capparis decidua, Leptadenia pyrotechnica, Ochradenus baccata*;
c) Evergreen mesophyllic, highly sclerophyllous species such as *Boscia senegalensis, Cordia sinensis, Balanites aegyptiaca* having a sclerophyllous index of 0.8–1.5 cg (0.008–0.015 g) DM cm^{-2} (Poupon 1980; Le Houérou 1988b).

These tend to replace broad-leaved species such as Combretaceae:

Anogeissus leiocarpus | *Terminalia* spp.
Combretum spp. | *Guiera senegalensis*

and representatives of other families:

Slerocarya birrea | *Grewia bicolor*
Sterculia setigera | *Bombax costatum*

Denuded ground in sandy and silty soils may become covered on the surface by a biological encrustation of Cyanophyceae: *Scytenoma* spp. which renders the surface almost totally impervious and therefore sterile for flowering plants. The process of sealing of the soil surface in conjunction with the development of vegetation arcs ("tiger bush") is also a potent factor of desertization particularly on silty-loamy soils on iron hardpans (Boudet 1972b). The process may end in totally denuded and sterile areas over huge surfaces, particularly in the southern half of the Sahel. Then the "tiger bush" in turn is destroyed by overbrowsing, excessive woodcutting and other mismanagement practices. The depletion of the woody vegetation is extremely serious since this is the only source of protein carotene, phosphorus and minerals for livestock through the 9–11 month dry season. This means that when the woody vegetation is destroyed, no animal husbandry can be maintained in the dry season unless with costly and uneconomic concentrates (Le Houérou 1979c).

The phenomenon of sealing of soil surface on a silty soil in the same region as shown in the photo on p. 93. The dark patches correspond to places previously occupied by shrubs and therefore richer in organic matter and not yet sealed. The clear parts, in contrast, correspond to sealed silty soil surface with a biological incrustation as mentioned in the photo on p. 93. Region of Niono, Central Mali (Photo Le Houérou)

3.2.6 Inventory and Mapping

Some 1.5 million km^2 of Sahelian and Sudanian grasslands have been surveyed and mapped at various scales (1:50 000 to 1:500 000) for the past 25 years. The French IEMVT (Institut d'Elevage et de Médecine Vétérinaire des Pays Tropicaux), for one, surveyed more than 850 000 km^2 in West Africa from 1960 to 1975 (Boudet 1975b). Regrettably, this type of study, most active in the 1960's, has slowed down in the 1970's and nearly stopped in the 1980's, with the exception of pinpoint or local diachronic studies. A new type of study using advanced very high resolution radiometer (AVHRR) on board the National Oceanic and Atmospheric Administration (NOAA) satellite Nos. 6, 7 and 8 was initiated in the 1980's (Tucker et al. 1985a, b), which will be discussed further in Section 5.6.

The Sahelian grasslands are thus fairly well documented from the purely descriptive viewpoints of botanical composition, range types, productivity, carrying capacities, production systems, management strategies and tactics and monitoring.

From the viewpoint of ecosystem functioning, the only serious attempts so far in the Sahelian and Sudanian zones were carried out by French teams of ORSTOM at Fete-Ole in northern Senegal from 1970 to 1978 (Bourlière 1978), in northern Burkina-Faso from 1975 to 1985 (Grouzis 1984, 1988) and in Central Mali by a Dutch team from 1977 to 1980 (De Vries and Djiteye 1982) and an international ILCA team (Wilson et al. 1983). This subject, again, will be discussed further in Sections 5.6 and 7.3.

3.2.7. Primary Production

In addition to the surveys mentioned above, which are generally pinpointed in time, a few in-depth productivity studies extending over a number of years have been carried out in various countries on both herbaceous layers and woody vegetation (Bille 1977; Poupon 1980; De Vries and Djiteye 1982; Sicot and Grouzis 1981; Hiernaux et al. 1978; Piot et al. 1980; Gaston 1981; Tucker et al. 1985a, b). Regional syntheses were published by the present author (Le Houérou 1980a; Le Houérou and Hoste 1977; Le Houérou and Gillet 1985) and by Boudet (1975a, 1984b).

Productivity is directly tied to rainfall for a given plant community or for a given geographical area, over a number of years; but the amount of annual rainfall is not a good predictor of herbage yield for a given, specific year. The latter seems to depend as much on the seasonal distribution as on the total amount, but this is not peculiar to the Sahel.

Over a number of years the variation of seasonal distribution and production is "averaged". The correlation between mean annual rainfall and mean annual production is thus very highly significant with correlation coefficients above 0.8 and a null hypothesis probability of no relationship below 0.001 (Le Houérou and Hoste 1977; Le Houérou 1982; Sicot and Grouzis 1981).

3.2.7.1 The Herbaceous Layer (Figs. 48, 49)

1. *Herbage Production*

   Aerial and underground phytomass represent 60 and 40% of the net primary production (NPP) respectively (Bourlière 1978). The NPP is 25–36% above the maximum standing crop (MSC) due to fast decay (fungi, bacteria, insects) during the rainy season (Bille 1978; Le Houérou 1980f; De Vries and Djiteye 1982), as shown in Figs. 45–47.

   The MSC in September-October, in the Saharo-Sahelian transition zone, averages 150–600 kg DM ha^{-1} yr^{-1}, i.e. a rain-use efficiency (RUE) of 1.0 to 4.0 kg DM ha^{-1} yr^{-1} mm^{-1} (Table 14).

   In the Sahel proper, MSC is 500 to 2500 kg DM ha^{-1} yr^{-1} depending on soil and range type, i.e. a RUE factor of 1.7 to 8.0. In clay depressions, flooded meadows (50–150 cm deep for 2–5 months) of *Echinochloa stagnina*, locally known as "Bourgou", may reach maximum standing crops of 5000 to 15 000 kg DM ha^{-1} yr^{-1}; RUE is naturally meaningless in that particular case.

   In the Sudano-Sahelian transition subzone, MSC may vary from 800 kg DM ha^{-1} yr^{-1} on shallow silty soils on iron hardpans to over 3000 kg in clay depressions, i.e. a RUE factor of 1.5 to 6.0. As mentioned above, the sum of increments is 25 to 36% above the MSC.

Biomass reduction during the dry season is quite sharp: 40 to 65% (Figs. 45–47) due to the action of ants, termites, grasshoppers, locusts, rodents, birds and weathering. The latter is particularly acute in the second half of the dry season when the DM content of the aerial biomass in the grass layer is 95% or above, due to very low air moisture in the atmosphere (10% or less every afternoon from March to May). Dry grass is then very fragile, broken into pieces by the least shock (wind, treading animals, etc.) and blown away. This sharp decrease in standing crop throughout the dry season has often been underestimated in the assessment of carrying capacities, deduced from MSC figures.

The overall geographical productivity figure of an RUE factor of 2.66 for the herbaceous layer in the Sahel, found by Le Houérou and Hoste in 1977, remains a valid order of magnitude ($Y = 2.643\,X^{1.001}$; Y denoting mean annual MSC and X denoting mean annual rainfall). This RUE factor may naturally be somewhat higher or lower locally according to the dynamic status of the ecosystems (Le Houérou 1984).

The correlation between production and "infiltrated rains" *for a given year* is excellent (Cornet 1981; Cornet and Rambal 1981). But the measurement of water intake is difficult to apply over large areas under field conditions since it needs rather sophisticated or labour-intensive (hence, costly) methods and the figures obtained can only be extrapolated to rather small areas due to extreme heterogeneity in the patterns of rain distribution in space. The figures are therefore very site-dependent.

The correlation with "useful rains" is still quite good (Cornet 1981; Boudet 1984b), with coefficients of determination ranging from $r^2 = 0.75$ to $r^2 = 0.95$. Useful rains are those that are equal or superior to 0.35 PET (Le Houérou and Popov 1981) or 0.50 PET (Boudet 1984b), on a weekly basis, i.e. rains which are equal or superior to the following values, in mm day^{-1}:

P > 0.35 PET (Penman)	P > 0.50 PET (Penman)	
2.8	4.0	in May
2.5	3.5	June
2.1	3.0	July
1.7	2.5	August
1.5	2.2	September
2.1	3.0	October

Boudet (1984b) thus found four correlation curves between production and useful rains for the Ferlo Province of northern Senegal for the 1979 and 1980 seasons (Figs. 48 and 49):

i) Very high yielding grasslands:
$\log Y = 0.0041\,X + 0.085$
$URE = 6.0;\ r^2 = 0.973;\ n = 5$

ii) High yielding grasslands:
$\log Y = 0.00387\,X - 0.1422$
$URE = 4.0;\ r^2 = 0.86;\ n = 18$

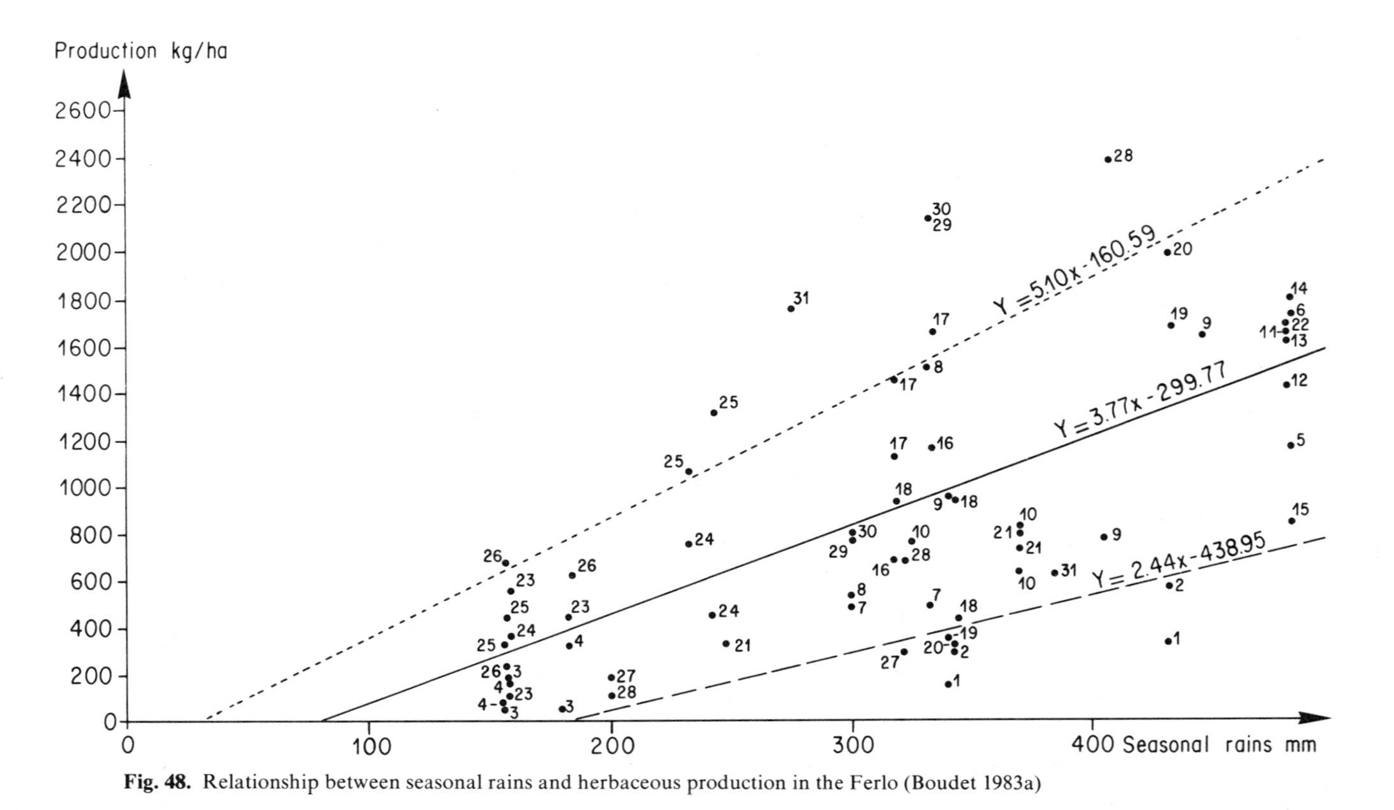

Fig. 48. Relationship between seasonal rains and herbaceous production in the Ferlo (Boudet 1983a)

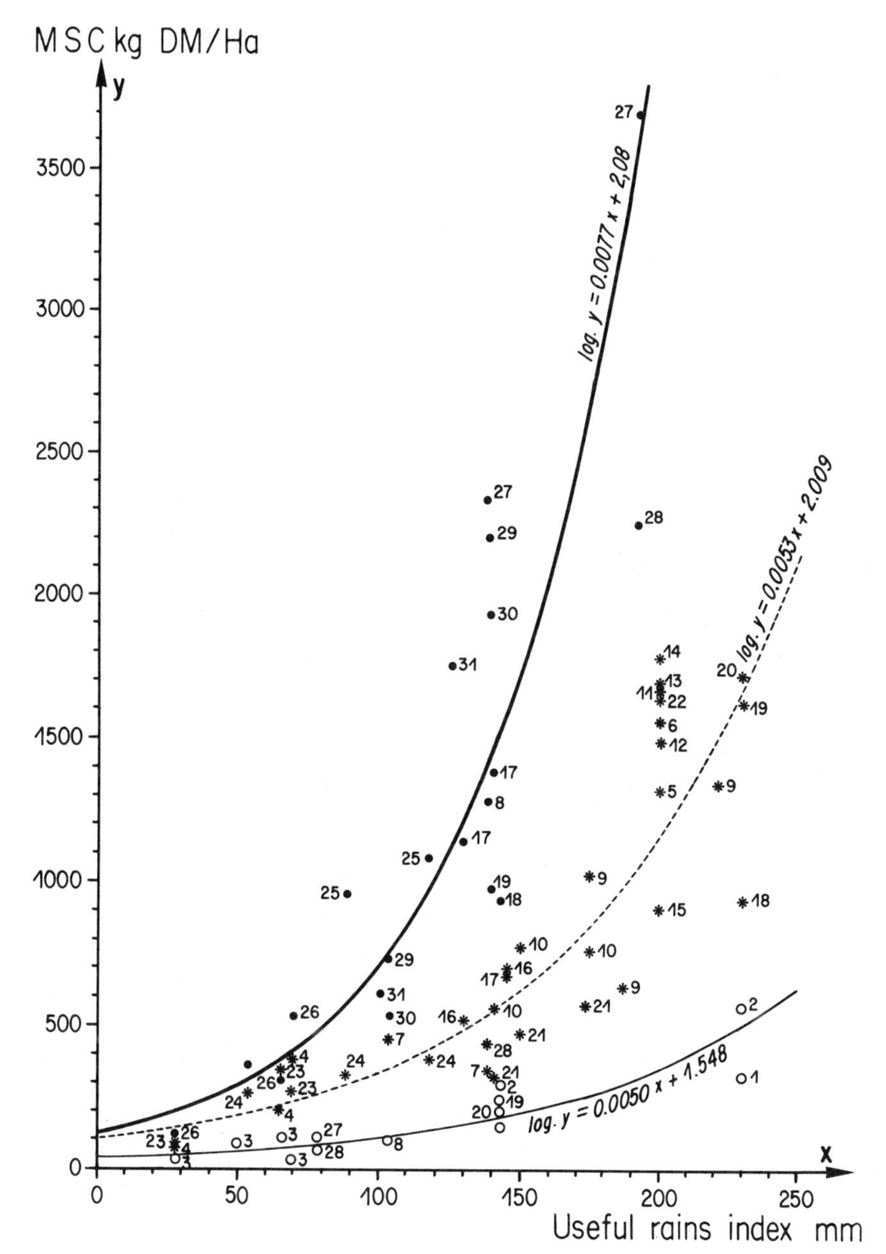

Fig. 49. Correlations between "useful rains" and maximum standing crop of herbage (Boudet 1983). See also Tables 13 and 18 and Figs. 32 and 47

iii) Fair to good yielding grasslands:
$\log Y = 0.0049\ X - 0.6829$
URE = 1.8; $r^2 = 0.90$; n = 25

iv) Low potential grasslands:
$\log Y = 0.00693\ X - 1.829$
URE = 0.6; $r^2 = 0.955$; n = 13

where Y = maximum standing crop;
X = useful rains, as defined above;
URE = useful rains efficiency (kg DM ha^{-1} yr^{-1} mm^{-1} of rain above 0.5 PET);
r^2 = determination coefficient.

The useful rains efficiency factor (URE) thus varies from 0.2 to 9.0 kg DM ha^{-1} yr^{-1} mm^{-1}, averaging 3.0; some 50% of the figures lying in the 1.8–4.5 range. The overall equation was Y (kg ha^{-1}) = 3.8 P mm-300, while Sicot and Grouzis (1981) found Y (kg ha^{-1}) = 2.2 P mm-135 in the Sahelian zone of northern Burkina-Faso. The figures drawn from Le Houérou and Hoste's predictive equation thus fall half-way between those emerging from the two above-mentioned regressions.

Evaluation of green biomass in northern Senegal in 14, combining AVHRR satellite imagery data and ground measurements on a large number of sites, resulted in a good correlation ($r^2 = 0.80$ to 0.83) between herbaceous green biomass, on the one hand, and integrated normalized difference value index (NDVI) and maximum NDVI respectively (Tucker et al. 1985b). Mean RUE varied from 1.07 in 1984 (with a mean intersite precipitation of 120 mm) to 4.28 in 1981 (with a mean intersite precipitation of 321 mm). The mean regional value for the 4-year period was 2.8, i.e. within a 5% range from the figure predicted from Le Houérou and Hoste's equation. This very elegant method is subject, however, to two main constraints, viz. (1) the green biomass must be above 300 kg DM ha^{-1} and (2) the woody vegetation must remain below 10% canopy cover; otherwise the evaluation is markedly biased (Tucker et al. 1985). These constraints do not apply to most of the Sahelian zone. Thus, one may therefore conclude that this technical breakthrough could, and should, become a potent tool towards the rational management of the Sahel grasslands, since it allows for the accurate evaluation of the range resource in space and time at the end of each rainy season for the next 9–11 months to come and therefore allows one to make the appropriate management and administrative decisions in good time at various levels of decision-making: local, provincial, national and international.

2. *Seed Production*

Seed Production in the herbaceous layer is extremely variable, depending on vegetation type, soil and annual rainfall in amount and distribution; in a given year it may fluctuate from 5 kg ha^{-1} to over 800 kg ha^{-1}. Average figures correspond to 10–15% of the maximum standing crop, with extreme values of 6–31%; that is a rain-use efficiency for seed production of 0.10 to 2.0 kg DM ha^{-1}

yr^{-1} mm^{-1} (Gaston and Lemarque 1976; Bille 1977; Grouzis 1979; Hubert et al. 1981; Gillon et al. 1983; Kahlem 1981; De Vries and Djiteye 1982). The higher figures, up to 1200 kg DM ha^{-1} yr^{-1}, were obtained from piospheric communities dominated by *Cassia tora* and *Acacia seyal* in the Sahelo-Sudanian zone at Bandia, Senegal (Kahlem 1981).

Some species are highly productive and their seeds are collected by pastoralists for human consumption, locally known under the collective name of "Fonio". The Fonio-producing species are mainly *Digitaria exilis* and occasionally *Panicum laetum*; they may locally produce in good years several 100 kg of seeds per hectare. These are extremely important in the ecology and distribution of a pest bird, the "millet eater", *Quelea quelea* (Morel 1968b; Gaston and Lemarque 1976). About 10% of the seeds produced are eaten by birds, rodents, ants and other insects (Bourlière 1978). Fonio species are also sometimes sown by women in small fields in the outward reaches of Bambara villages in the Sahel of Central Mali (Wilson et al. 1983).

3. *Root Biomass and Production* (Bille 1977; Legrand 1979, De Vries and Djiteye 1982; Grouzis 1987); Mané 1986)

Root distribution in the soil profile depends on soil type and particularly on texture and structure. Nearly 100% of the herbage root biomass is found in the upper 100 cm of soil depth, in all cases. In the Sahel of Burkina-Faso 80% of the root biomass occurs in the upper 50 cm of the profile; that is 70% in coarse, sandy soils and 90% in clay soils. It was found that the distribution of the root biomass was a logarithmic function of soil depth adjusting to the general equation:

$$R = ae^{bp},$$

where R = root biomass; e = natural logarithm base; p = soil depth in cm and a and b = constants depending on plant communities and soil types. In the investigated examples "a" varied from 365 to 1202 mg per 10^3 cm^3, and "b" from -0.03 to -0.1.

The cumulative distribution of root biomass as a function of depth is shown in Fig. 50.

Root biomass is, on the average, equal to 2.2 above ground MSC; but this ratio varies greatly with plant community and soil type; it is lower (1.3) in coarse, sandy soils and higher (3.5) in heavy soils. Interannual variation of this ratio depends also on the same factors: 1.0 to 2.0 in sandy soils and 2.0 to 5.0 in clay soils (Legrand 1979; Grouzis 1988).

In the Sudano-Sahelian zone of Central Mali, De Vries and Djiteye (1982) found still more shallow herbage root systems, with 80% of the biomass in the upper 20 cm of the soil.

Bille (1977) and Legrand (1979) found that roots represented 50–80% of the total herbage biomass, whereas De Vries and Djiteye reported that annual production of roots was equal to 30–50% of total biomass production. The contradiction between the two sets of data is only apparent since Grouzis

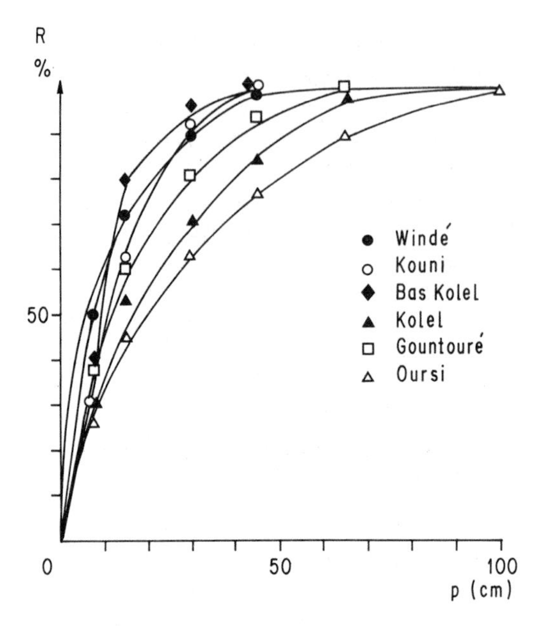

Fig. 50. Variation in the cumulative percentage of roots (R%) as a function of depth in various vegetation units of the Sahel of Burkina-Faso. Each symbol represents the average of three measures taken in the 1978 growing season. (After Grouzis 1988)

(1987) showed that annual root production represents about 50% of root biomass (40–60%). This means that root decay takes about 2 years to complete and therefore the root biomass in any given year is about equal to 150% of annual production.

Mané (1986) found on ferrugineous sandy soils near Niamey in the Sudano-Sahelian zone of the Rep. of Niger an average root to herbage ratio of 0.37; Grouzis (1987) actually found that annual root production averaged 49% of total annual biomass production (43–57%), a figure slightly higher than most references in the literature (Van Keulen 1975, 30%; Floret and Pontanier 1982, 45%; De Vries and Djiteye 1982, 30–50%).

Root turnover was found to average 2.1 ± 0.4 years in the Sahel of Burkina-Faso (Grouzis 1988); it was faster in sandy (1.7) than in clay (2.5) soils.

3.2.7.2 Constraints to Primary Production of the Herbaceous Layer

Contrary to the concept that had been believed for decades, water is not necessarily the major constraint to primary production. De Vries and Djiteye and their team (1982) have shown that nutrients are often the limiting factor, particularly phosphorus and nitrogen. These authors found, from small plot experiments in Central Mali, that joint application of P and N may increase herbage yield three- to five-fold. Large-scale experiments on multihectare plots naturally produce more modest results (Hiernaux et al. 1979); but production may still be doubled or trebled under real, field-scale conditions.

This, however, is only true whenever rainfall is high enough, i.e. above 200–300 mm yr^{-1} (depending on distribution). The nutrient constraint is thus a limiting factor 9 of 10 years in the Sudano-Sahelian transition zone (where De Vries and Djiteye's experiments were carried out), about 6 of 10 years in the Sahel proper and only 1 or 2 of 10 years in the Saharo-Sahelian transition zone.

From the practical range management standpoint, however, range fertilization is not an economically feasible proposition given the respective market values of fertilizers and transport, on the one hand, and of animal products, on the other hand. Local costs of fertilizers at the "farm gate" in 1980 were 0.8–1.0 US \$ kg^{-1} N and 2.0–2.3 US \$ kg^{-1} P, while the price of livestock was 0.6 to 0.8 US \$ kg^{-1} live wt. and the average herd offtake was in the vicinity of 10–12% in cattle and 25–30% in small stock. It would thus take a ten-fold increase in herbage yields to offset the cost of fertilizers; which is obviously impossible, even if the herd offtake were doubled. A four-fold increase in herbage yields would raise the carrying capacity from e.g. 25 kg live wt. ha^{-1} to 100 kg live wt. ha^{-1}; doubling the offtake would raise it to 30%, i.e. to 30 kg ha^{-1}. These 30 kg were worth some 21 US \$ according to 1980 prices, that is, equal to 20 kg N or 10 kg P ha^{-1}, from which the cost of transport to the field and application should be deduced. Increasing herbage yields through the use of fertilizers is therefore not possible, given the present terms of trade between fertilizers and animal products.

3.2.7.3 Productivity of the Woody Vegetation (Figs. 51, 52.)

1. *Browse Production*

From the grassland management and ruminant nutrition viewpoints browse production is of paramount importance in the Sahel; not so much for the amount of energy it provides as for the feed quality and seasonal availability of browse forage. In fact, for the 9–11 months of dry season, the herbaceous layer is extremely poor in protein, carotene and phosphorus (Le Houérou 1980b, c), and quite unable to meet the dietary requirements of herbivores in these nutriments. Conversely, browse species are usually rich in protein, carotene and phosphorus throughout most of the dry season. Browse thus provides a necessary complement to the energy-rich dry herbaceous forage for 9–11 months annually.

Browse productivity studies are rather recent (1970 onwards), for two main reasons:

i) Its nutritive importance has been grossly overlooked for decades;
ii) Measurement of browse production is a difficult, time-consuming, tedious and costly activity; while appropriate methods of measurement have only been established recently.

The following table was set up on the basis of calculations using the data published by M.I. Cissé (1986). They result from a 10-year study in Central Mali.

Eco-climatic zone (according to Le Houérou and Popov 1981)	Mean annual rainfall (mm)	Mean herbaceous MSC (kg DM-ha^{-1} yr^{-1})	Herbaceous RUE (kg DM-ha^{-1} yr^{-1} mm^{-1}	Total wood Production (kg DM-ha^{-1} yr^{-1})	Wood RUE (kg DM-ha^{-1} yr^{-1} mm^{-1}	Browse production (kg DM-ha^{-1} yr^{-1})	Browse RUE (kg DM-ha^{-1} yr^{-1} mm^{-1}	Total forage production (kg DM-ha^{-1} yr^{-1})	Total forage RUE (kg DM-ha^{-1} yr^{-1} mm^{-1}	Total aerial primary productivity RUE (kg DM ha^{-1} yr^{-1} mm^{-1})
Sahero-Sahelian	250	500	2.0	300	1.2	200	0.8	700	2.8	4.0
Sahelian Sensu Stricto	350	800	2.3	1000	2.9	700	2.0	1500	4.3	7.2
Sudano-Sahelian	500	1300	2.6	1300	2.6	800	1.6	2100	4.2	6.8
Northern Sudanian	750	2400	3.2	2350	3.1	1200	1.6	3600	4.8	7.9
Mean	462	1250	2.5	1238	2.5	725	1.5	1975	4.0	6.5

Note: The above browse production values seem to us well above the usually accepted zonal value (Le Houérou 1980b). According to M.I. Cissé (1986), about 50% of browse production is located below 200 cm height and therefore accessible to cattle; perhaps not more than 25% are accessible to small stock (below 120 cm); probably as much as 75% is available to camels (below 3.5 m), but virtually the whole browse production would be available to giraffes (below 5 m). Cissé further estimated that 33% of the browse production is available during the prerainy season (May-June), while 80% of the annual production is available during the rainy season (July–September), 60% during the postrainy season (October-November), 34% during the cool dry season (December-February) and 25% during the hot dry season (March-April). RUE = Rain Use Efficiency (kg DM ha^{-1} yr^{-1} mm^{-1}).

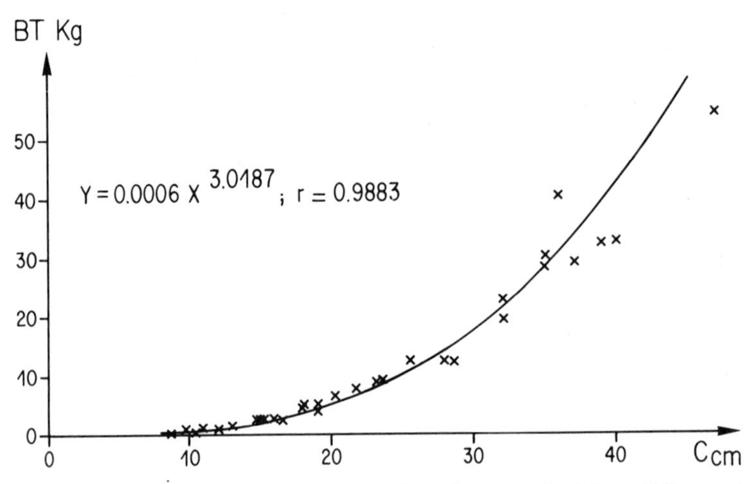

Fig. 51. Allometric relations between stem circumference and total wood biomass *Acacia laeta* Sahel of Burkina-Faso (After Menaut 1983)

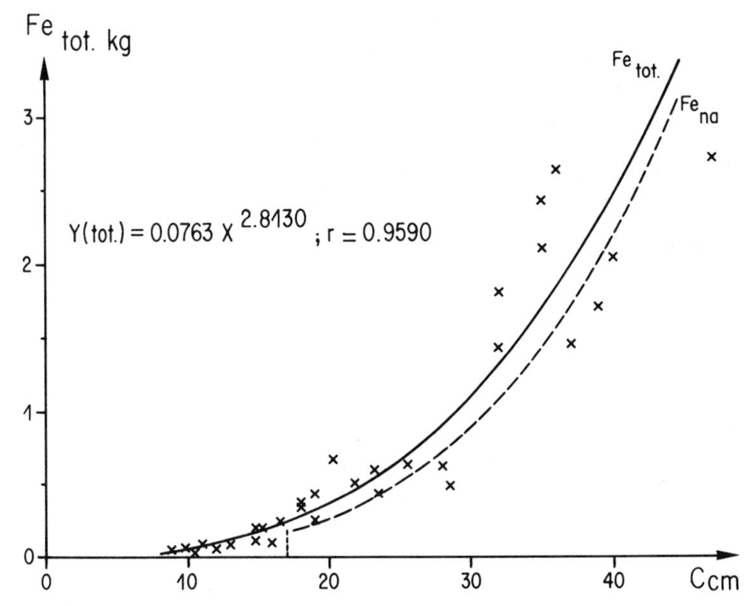

Fig. 52. Allometric relationship between stem circumference and leaf production in *Acacia laeta*, Sahel of Burkina-Faso (After Menaut 1983) (___ = total leaf production kg DM --- = part in accessible to stock)

A synthesis on the subject was published by the present author in 1980, the conclusions of which are as follows (Le Houérou 1980b; Von Maydell 1983):

a) There are about 100 important browse species in the Sahel.
b) Mean crude protein (c.p.) content (N × 6.25) is 12.5%;
   Mean mineral (min.) content is 10.9%;
   Mean phosphorus content is 0.15%.
c) The best quality browse in protein and mineral is provided by:
   i) Capparidaceae 21% c.p. 14% min.
   ii) Leguminoseae 17% c.p. 7% min.
d) Browse production may vary from 0.1 to 1 kg DM per individual "trub" per annum, depending on species and local conditions.
e) Deciduous production is about 45% of total aerial production and about 55% above-ground wood biomass increment.
f) Underground biomass is nearly equal to aerial biomass; the fugacious roots and large root increment are approximately equal.
g) Overall geographical productivity was evaluated as follows in 1980:
   Saharo-Sahelian transition subzone: 150 kg Dm ha^{-1} yr^{-1};
   Sahel proper: 300 kg DM ha^{-1} yr^{-1};
   Sudano-Sahelian transition subzone: 500 kg DM ha^{-1} yr^{-1};
   Northern Sudanian zone: 700 kg DM ha^{-1} yr^{-1};
   Southern Sudanian zone: 1000 kg DM ha^{-1} yr^{-1};
   50% of which, according to the estimations, is accessible to livestock.

Browse RUE is thus about 1.0 kg DM ha^{-1} yr^{-1} mm^{-1} under pristine Sahelian range conditions, i.e. about one-third of herbaceous overall productivity. Overall RUE, i.e. deciduous production plus biomass increment, is about 2.5 kg DM ha^{-1} yr^{-1} mm^{-1} with large variations according to range types and dynamic status: 1 to 10 kg DM ha^{-1} yr^{-1} mm^{-1}. Various studies showed that production can be predicted from allometric relations, particularly from stem circumference or diameter, and various predictive equations for different species have been published (Bille 1977, 1978; Cissé 1980a; Poupon 1980; Menaut 1983).

2. *Fruit Production*

Fruit production from the woody layer averages 10% of the deciduous production (leaves 80%, flowers and bark 10%) (Bille 1977, Poupon 1980). But there are wide variations depending on the site, botanical composition and rainfall distribution. In the agropastoral production system of millet and *Faidherbia albida*, mean annual pod production may reach 200–1000 kg DM ha^{-1} yr^{-1} for densities of 10–50 adult trees per hectare (Le Houérou 1978, 1979c, 1980d; Cissé 1980d). Similar amounts may be produced in dense natural stands of *Acacia seyal* of the Sudano-Sahelian and northern Sudanian zones (Kahlem 1981).

3. *Wood Production*

Wood production in the Sahel is extremely important as it represents 90 to 95% of the energy consumption of the rural population (Le Houérou 1979; Keita 1982). Wood production (or annual increment of aerial woody biomass) is about 40–60% of total annual phytomass production of the woody layer; 40–60% of the production being deciduous (flowers, fruits, leaves, bark).

In the Sahel of Burkina-Faso for instance, under a mean actual rainfall of 400 mm during the study period, Menaut (in Toutain et al. 1983), found for the period 1980–1983 a wood production of 200 kg DM ha^{-1} yr^{-1} in an *Acacia laeta* open savanna, with a woody biomass of 2400 kg DM ha^{-1} and a canopy cover of 11%. Leaf and twig productions were 150 and 100 kg DM ha^{-1} yr^{-1} respectively, 25% of which was directly available to livestock (height below 160 cm). One example of allometric relations between woody biomass and leaf biomass, on the one hand, and stem circumference at ground level are given in Figs. 51 and 52.

The above data are in good agreement with the overall estimates mentioned above, and also with the data from Central Kordofan on *Acacia senegal* published by Olsson (1985) and mentioned earlier in Section 3.2.1. Foresters estimate mean zonal wood production as follows (Bailly et al. 1982; Keita 1982):

Saharo-Sahelian subzone: 0.075 m^3 ha^{-1} yr^{-1} or 50 kg DM ha^{-1} yr^{-1}
Sahel proper subzone: 0.30 m^3 ha^{-1} yr^{-1} or 200 kg DM ha^{-1} yr^{-1}
Sudano-Sahelian subzone: 0.75 m^3 ha^{-1} yr^{-1} or 500 kg DM ha^{-1} yr^{-1}

that is a rain-use efficiency for wood production of 0.3, 0.66 and 1.00 kg DM ha^{-1} yr^{-1} mm^{-1}; these figures are consistent with the rare experimental data available (Bille 1977; Poupon 1980; Le Houérou 1979c, 1980b; Menaut 1983).

3.2.8 The Screen Function, Energy Flow Conversion, Photosynthetic Efficiency

In multistory vegetations, including trees, shrubs and herbs, energy conversion seems twice to three times more efficient than in single-layer pure grassland stands (Bille 1977; Bourlière 1978; Le Houérou 1979c, 1980b).

The net energy conversion factor for the rainy season varied from 0.3 to 1.5% at Fete-Ole under 213 mm of annual rainfall, i.e. under nearly Saharo-Sahelian conditions; and 0.08% on a year-round basis.

At Niono, in Mali, under 500 mm of annual rainfall, i.e. under typical Sudano-Sahelian conditions, the total phytomass produced (aerial and subterranean) is about 660 g m^{-2} yr^{-1} in the open herbaceous vegetation and about 2400 g m^{-2} yr^{-1} in multistory bush (Boudet and Leclerc 1970; Hiernaux et al. 1978; De Vries and Djiteye 1982). Net energy storage is thus 250 and 990 cal cm^{-2} yr^{-1} respectively, estimating an average content of 4500 cal g^{-1} DM in woody species and 3750 cal g^{-1} DM in herbaceous species.

As global radiation averages 180 to 200 kcal cm^{-2} yr^{-1} (Landsberg et al. 1965; Bille 1977; De Vries and Djiteye 1982), net incident and photosynthetically active radiation average about 70–80 kcal cm^{-2} yr^{-1} for both variables, the conversion rate is thus 0.33 and 1.32% of incident energy on a year-round basis or 1.3 and 5.3% for the growing season (Le Houérou 1980b).

The reason for this four- to five-fold increase in photosynthetic efficiency of multistory ligneous/herbaceous vegetation with respect to simple structured monostratal grassland may be ascribed to:

1. Better water budget due to reduced PET (reduced by 50–70%);
2. Lower temperature of soil surface due to shade, hence less evaporation and more transpiration, increased organic matter;
3. Increased water intake and storage due to more stable structure, greater permeability; less runoff;
4. Quicker and more intensive turnover of geobiogene elements (Ca, N, P, K, Mg, S, etc.). The latter fact is particularly well documented (Charreau and Vidal 1965).

3.2.9 Water and Rain-Use Efficiencies (Figs. 48, 49, 53; Tables 14 and 43)

In the southern half of the Sahel, under natural field conditions, about 75% of rainwater infiltrates and 25% runs off; 60% evaporates from the soil surface and only 15% is actually transpired (De Vries and Djiteye 1982; Breman and De Wit 1983). The water-use efficiency (WUE), as measured at Fete-Ole and Dahra in Northern Senegal, ranges from 3 to 9 kg DM produced per mm of water *evaporated and transpired*; 0.88 g DM kg^{-1} H_2O and 1136 kg H_2O kg^{-1} DM (72% transpiration and 28% evaporation in the *infiltrated rains*) (Cornet 1981; Cornet and Rambal 1981).

At Niono, Mali, the coefficient of transpiration in the grass layer is 260–300 kg H_2O kg^{-1} DM on natural grassland and down to 170 kg H_2O kg^{-1} on fertilized grassland (De Vries and Djiteye 1982).

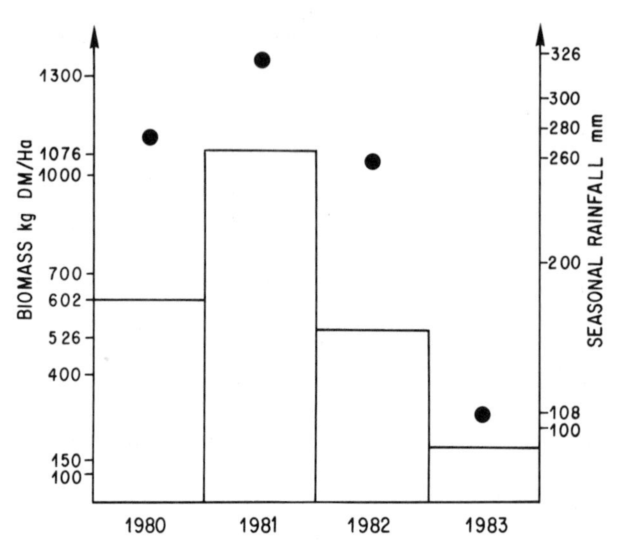

Fig. 53. Relationships between herbage production and seasonal rainfall in the Ferlo area, as measured in 22 weather stations (•). See also Tables 14 and 46 (Vanpraet et al. 1983)

Daily growth during the peak of the rainy season in August and September was about 20 kg DM ha^{-1} day^{-1} at Fete-Ole (Bille 1977) and 35 kg ha^{-1} day^{-1} on natural grasslands at Niono and up to 125–250 kg ha^{-1} day^{-1} on fertilized grasslands at this same latter site (De Vries and Djiteye 1982).

As mentioned in Section 3.2.7.1 under the heading of primary production, rain use efficiency (RUE) is of the order of magnitude of 3 kg DM ha^{-1} yr^{-1} mm^{-1} for the herbaceous layer, a figure 25% below the worldwide mean of 4.0 kg for arid zones (Le Houérou 1982, 1984; Le Houérou et al. 1988). The average RUE is 1 kg DM ha^{-1} yr^{-1} mm^{-1} for the woody layer. But variation in RUE is very large, depending mainly on the dynamic status of the ecosystems under consideration. Depleted herbaceous vegetation and soils may yield RUE's of 1 or less, whereas well-developed vegetations on undisturbed soils may yield RUE's of 6 and more. Variation of RUE in the woody layer is still larger (0.1 to 5.0), depending on present and past management, but also on topographic and edaphic conditions (Le Houérou 1980b, 1987b).

The variability of annual primary production is also high. The quotient of the coefficient of variation of annual production by the coefficient of variation of annual rainfall (Production to Rain Variability Ratio = PRVR) varies from 1.2 to 1.8 as average long-term figures in ecosystems in fairly good condition. In other words, the variability of primary production is 20 to 80% larger than the variability of rainfall, on an annual basis. But in depleted ecosystems PRVR increases steadily to reach values of 2 to 5 (Le Houérou et al. 1988).

3.2.10 Grazing Value (Tables 15a and 15b)

3.2.10.1 Palatability

Most grasses and many forbs are palatable in the early stages of their development. Among common grasses, the most appreciated by livestock are:

Table 15a. Palatability of herbaceous and woody species

Andropogon spp.	*Eragrostis* spp.
Aristida spp.	*Hyparrhenia* spp.
Brachiaria spp.	*Panicum* spp.
Cenchrus biflorus	*Pennisetum* spp.
Cenchrus spp.	*Schoenefeldia gracilis*
Chloris spp.	*Setaria* spp.
Dactyloctenium aegyptium	*Tragus racemosus*
Digitaria spp.	

Among the graminoids:

Cyperus conglomeratus	*Cyperus jeminicus*

Some grasses, however, are little grazed, or completely unpalatable:

Aristida longiflora	*Elytrophorus spicatus*
Aristida pallida	*Loudetia togoensis*
Aristida papposa	*Oryza longistaminata*
Cymbopogon spp.	*Pennisetum pedicellatum*
Elionurus elegans	*Schizachyrium exile*

Some forbs are highly palatable:

Alysicarpus spp.	*Limeum* spp.
Cleome tenella	*Merremia* spp.
Commelina spp.	*Mollugo* spp.
Crotalaria podocarpa	*Rynchosia memnonia*
Gisekia pharnaceoides	*Rynchosia minima*
Hibiscus asper	*Tephrosia nubica*
Indigofera spp.	*Tephrosia purpurea*
Ipomaea pes-tigris	*Zornia glochidiata*

Others are unpalatable:

Abutilon spp.	*Geigera alata*
Achyranthes aspera	*Gynandropsis gynandra*
Aerva javanica	*Heliotropium* spp.
Alcalypha spp.	*Indigofera* spp.
Cassia italica	*Jatropha* spp.
Cassia mimosoides	*Pavonia* spp.
Cassia nigricans	*Rogera adenophylla*
Cassia tora	*Sida* spp.
Chrozophora spp.	*Solanum* spp.
Corchorus olitorius	*Tephrosia* spp.
Crotalaria spp.	*Vernonia pauciflora*
Geigera acaulis	

Among woody species, most are sought for during the dry season, particularly legumes and Capparidaceae. Some, however, are unpalatable or toxic:

Adenium obesum	*Combretum micranthum*
Annona senegalensis	*Combretum nigricans*

Calotropis procera
Cissus spp.
Combretum glutinosum
Diospyros mespiliformis
Euphorbia balsamifera
Pergularia tomentosa
etc.

Among the most palatable and common species are:

Acacia ehrenbergiana
A. laeta
A. mellifera
A. senegal
A. seyal
A. tortilis
Balanites aegyptiaca
Bauhinia rufescens
Boscia salicifolia
Cadaba farinosa
C. glutinosa
Celtis integrifolia
Combretum aculeatum
Commiphora africana
Crataeva adansoni
Dobera glabra
Faidherbia albida
Feretia apodanthera
Grewia bicolor
G. spp.
Maerua crassifolia
M. spp.
Pterocarpus lucens
Salvadora persica
Ziziphus mauritiana

3.2.10.2 Intake

Actual daily forage dry matter intakes by livestock are of the order of:

1.5–2.5% of live weight in cattle; (Dicko 1980, 1981, 1983, 1984, 1985)
2.0–2.2% of live weight in camels;
3.0–4.0% of live weight in sheep;
5.0–6.0% of live weight in goats; (Le Houérou 1980c).

Apparent intake (difference between amount of DM available and left over) may be twice to three times the actual intake (Hiernaux et al. 1978; Dicko 1981), with apparent consumptions up to 10–20 kg DM TLU $^{-1}$ (Tropical Livestock Unit) per day[4]. The rate of consumable forage varies from one grassland type to the other, usually in the range of 30–70% of the maximum standing crop (MSC). The year-long proper use factor is probably in the vicinity of 25–30% MSC, instead of the 40–50% which had been assumed until recently.

3.2.10.3 Feed Value, Chemical Composition

Herbage digestibility varies from 65–70% of the DM in the rainy season to 45–50% at the end of the dry season with a weighted average annual value of 54% (Dicko 1981).

Digestibility in browse is usually 45–55%, with some exceptions up to 64% on the dry matter (Le Houérou 1980c; Rose-Innes and Mabey 1964–66).

Feed value of herbage is good in the rainy season with 8–14% crude protein but very low in the dry season with only 1.8–3.5% crude protein, i.e. virtually no

[4]TLU = conventional dry zebu cow weighing 250 kg kept at maintenance (Boudet and Rivière 1968).

Table 15b. Forage value of the main herbaceous species in the Ferlo (Boudet 1983a)

Good grasses G3: medium size, medium to good tillering, well consumed when green or dry (except at the fructification stage for some species).

Andropogon gayanus
Andropogon penguipes
Brachiaria lata
Brachiaria xantholeuca
Cenchrus biflorus
Cenchrus prieurii
Chloris pilosa
Chloris prieurii
Diheteropogon hagerupii
Echinochloa colona
Eragrostis cilianensis
Eragrostis lingulata
Eragrostis pilosa
Eragrostis tremula
Hackelochloa granularis
Panicum laetum
Pennisetum pedicellatum
Schoenefeldia gracilis
Setaria pallide-fusca

Medium value grasses G2: medium to small size fine but fairly hard stems, well consumed when green but hardly so when dry.

Andropogon pseudapricus
Aristida adscensionis
Aristida funiculata
Aristida mutabilis
Brachiaria distichophylla
Dactyloctenium aegyptium
Digitaria horizontalis
Eragrostis aegyptiaca
Tetrapogon cenchriformis
Trichoneura mollis

Poor quality grasses G1: small size, hard, stem, few leaves.

Aristida sieberana
Aristida stipoides
Ctenium elegans
Elionurus elegans
Eragrostis ciliaris
Loudetia togoensis
Schizachyrium exile
Sporobolus pectinellus
Tragus berteronianus

Medium value legumes L2: medium to small size, palatable when dry and often when green.

Aeschynomene indica
Alysicarpus ovalifolius
Cassia mimosoides
Indigofera aspera
Indigofera pilosa
Tephrosia purpurea
Zornia glochidiata (bloating)

Poor value legumes L1: small size, prostrate habit or hardly consumed.

Cassia obtusifolia
Indigofera astragalina
Indigofera senegalensis

Miscellaneous forbs of medium value M2: fairly well consumed either green or in fruits.

Blepharis lineariifolia
Citrullus lanatus
Commelina forskhalei
Cyperus esculentus
Ipomaea coscinosperma
Merremia pinnata
Merremeia tridentata
Tribulus terrestris

Miscellaneous forbs of poor value M1: small size, hard stem, hardly palatable, except fruits in some.

Achyranthes aspera
Boerhavia repens
Borreria stachydea
Cleome tenella
Corchorus tridens
Cyperus iria
Fimbristylis hispidula
Jacquemontia tamnifolia
Gisekia pharnaceoides
Hibiscus diversifolius
Limeum diffusum
Limeum pterocarpum
Sida cordifolia

Table 15. *Cont.*

Unpalatable species U: no forage value.	
Borreria chaetophylla	*Mollugo nudicaulis*
Borreria radiata	*Monsonia senegalensis*
Ceratotheca sesamoides	*Polycarpea linearifolia*
Cleome viscosa	*Polygala erioptera*
Corchorus depressus	*Portulaca foliosa*
Euphorbia aegyptiaca	*Pycreus macrostachyos*
Heliotropium strigosum	*Tripogon minimus*
Hygrophila senegalensis	*Urginea indica*
Limeum viscosum	*Waltheria indica*
Mollugo cerviana	

digestible protein and a very low content of phosphorus (0.05%) and no carotene (Boudet and Rivière 1968; Boudet 1975a, 1984a and b; Le Houérou 1980c).

The chemical composition of browse is high in crude protein: 12.5%, averaging 17% in the legumes and 21% in Capparidaceae. Minerals are also high, averaging 11%, with 7% in legumes, 14% in Capparidaceae. Phosphorus is in sufficient supply, averaging 0.15%, with 0.25% in legumes and 0.12% in Capparidaceae (Le Houérou 1980c).

There are, however, exceptions of low quality browse particularly in species which are rich in silica, such as:

Celtis integrifolia
Gardenia erubescens
Gmelina arborea
Hyphaene thebaica
Maytenus senegalensis
Oxytenanthera abyssinica
Sterculia setigera
Stereospermum kunthianum
Vitex cuneata
etc.

which are poor in nitrogen ($< 0.4\%$) and rich in silica ($> 10\%$) (Le Houérou 1980c).

4 Wildlife

(Claude J. Grenot and Henry N. Le Houérou)

4.1. General (Tables 2, 16)

The structure and functioning of the Sahelian ecosystems, in their natural state, is not well known. Only the Fete-Ole savanna in the northern Ferlo region of northern Senegal has been the subject of in-depth investigations carried out over more than 10 years by a resident team of the French section of the International Biological Programme, within the ORSTOM/Sahel programme (Bourlière 1978; Poulet 1982).

The Fete-Ole site is little modified by man, except for the fact that the large wildlife species such as ostriches, elephants, giraffes, antelopes, lions and leopards are extinct, replaced by a low and perhaps somewhat equivalent biomass of livestock. It is unlikely that the ungulate biomass could have been above 2000 kg km^{-2} (Coe et al. 1976).

The fact is that in a typical Sahelian zone, with a mean grass cover of 40 and 30% bare ground, the maximum standing crop is of about 1000 kg DM ha^{-1} yr^{-1}; while in the Saharo-Sahelian transition subzone with 5–10% grass cover, the herbage biomass does not exceed 400 kg DM ha^{-1} yr^{-1} (Le Houérou 1980 f). The site under consideration is a typical Sahelian one with a highly variable rainfall averaging 400 mm of the long-term mean (300 mm during the period of the investigations). There is no clear-cut hydrographical network, no permanent water source and a mediocre plant cover composed of various mosaics having a highly seasonal primary and secondary production. Range exploitation is of a nomadic type as a result of the high variability and seasonality of precipitations and primary production. The Sahelian fauna has not been the subject of studies as detailed as for vegetation, so that the overall evaluation of the secondary production is not possible as yet.

The weak rains, narrowly concentrated in a 2–3 month period, result in strong and multiple productivity constraints and a meager overall production. Moreover, soils are poor, and medium-term environment conditions are fairly unpredictable. The temporary abundance of plants and animals during the rainy season avails the coexistence of sedentary, nomadic and migratory consumers, in order to exploit the seasonal excess of energy, while residents are too few to possibly make use of such a-short-lived seasonal abundance. Conversely, at the end of a very long and severe dry season, the rarity of the food resource sets a very low level ceiling to the sedentary populations. The latter have to face the same constraints as plants: lack of free water, very high diurnal temperatures and scantiness of nutrient availability.

Table 16. The status of some Saharan and Sahelian large mammals (after Newby 1984)

Country	Scimitar horned-oryx	Addax	Dama gazelle	Dorcas gazelle	Slender-horned gazelle	Aoudad	References
Ex-Spanish Sahara	Extinct 1973	Very rare	Extinct	Very rare	Very rare	Extinct	Valverde (1968) Gillet (1969)
Mauritania	Extinct early 1960's	Rare	Very rare	Local	Rare	Very rare	Trotignon (1975) Lamarche (pers. communication)
Morocco	Extinct 1932	Extinct	Very rare	Rare	Very rare	Extinct	Cabrera (1932) Chapuis (1961)
Algeria	Extinct	Very rare	Very rare	Rare	Rare	Rare	Grenot (1974) de Smet (pers. communication)
Tunisia	Extinct 1906	Extinct 1885	Extinct	Rare	Rare	Very rare	Lavauden (1924) Schomber and Koch (1961)
Libya	Extinct 1940's	Extinct late 1960's	Extinct	Rare	Rare	Rare	Osborn and Krombein (1969) Hufnagl (1972)
Mali	Extinct	Rare	Local	Local	Very rare	Local	Sayer (1977) Lamarche (1980)
Niger	Very rare	Very rare	Local	Local	Very rare	Local	Newby (1975a) Newby (1982)
Chad	Rare	Very rare	Local	Local	Very rare	Local	Newby (1975b) Newby (1981)
Sudan	Extinct mid-1970's	Very rare	Very rare	Local	Very rare	Very rare	Brocklehurst (1931) Wilson (1980a)
Egypt	Extinct 1850	Extinct ca. 1900	Extinct	Rare	Rare	Extinct	Flower (1932) Jones (pers. communication)

4.2 Invertebrates

The invertebrates of the upper soil layers and in the grass layer include essentially Arthropoda (Tables 17–19). No earth worm or mollusc seems able to survive (Gillon and Gillon 1973, 1974 b).

4.2.1 Non-Social Arthropods

The abundance of spiders remains fairly constant year-round; it is higher at the beginning of the dry season when they represent 11.5% of the non-social Arthropoda (Tables 17–18). Large numbers of a small acarid *Dinothrombidium tinctorium* appear at the time of early rains. *Thysanoura* are present yearlong. Among orthopteroids, only acridids are permanently present; their biomass, however, culminates at the beginning of the dry season, reaching 50% of non-social arthropods while crickets and grasshoppers totally disappear from the grass

Table 17. Numbers and biomass of non-social arthropod populations in the herbaceous layer at Fete-Ole (Gillon and Gillon 1974b)

Non-social arthropods	Density (10^3 individuals ha^{-1})	biomass (g DM ha^{-1})	Annual biomass (%)
Arachnids (spiders)	40.4	90.8	7.2
Thysanoura	7.8	27.9	2.2
Orthopteroids (Acridids)	12.7	281.7	22.4
Hemiptera (Hateroptera)	12.2	32.6	2.5
(Homoptera)	33.7	44.4	3.5
Coleoptera	43.2	559.5	44.5
Lepidoptera	21.7	161.8	12.9
Hymenoptera	6.3	16.2	1.3
Diptera	20.1	39.4	3.1
Miscellaneous	0.4	3.1	0.2
Total	198.5	1257.4	100.0

Table 18. Numbers and biomass of primary and secondary consumer non-social arthropods in the herbaceous layer at Fete-Ole (density: 10^3 individuals ha^{-1}; biomass: g DM ha^{-1}; Gillon and Gillon 1974b)

Non-social arthropods	Season				Annual average	
	Dry		Rainy			
	Density	Biomass	Density	Biomass	Density	Biomass
Primary consumers	39.4	388.2	301.4	1877.8	111.6	917
Secondary consumers	27.2	61.7	141.8	489.7	86.5	338
Total	66.6	444.9	443.2	2367.5	198.1	1255

layer. The annual biomass of *Orthoptera* and *Mantidae* represents 20% of the biomass of non-social *Arthropoda* (Gillon and Gillon 1974 b). Some years of pullulation of grasshoppers may devastate the tree layer (Poupon 1977 a, b). Among the *Heteroptera Hemiptera* the lygeids reach a high density in the early season: 21% of the numbers and 3% of the biomass of the non-social Arthropoda. Pentatomids, Coreids and Reduvids are also well represented groups. In some years *Homoptera Hemiptera* may constitute an important group in the herbaceous layer during the rainy season.

The *Coleoptera* group is well represented, mainly the *tenebrionids* and *carabids*, which are present yearlong. The annual biomass of *Coleoptera* constitutes 45% of non-social *Arthropoda* (Table 3). Of 19 species of *Tenebrionidae* collected at Fete-Ole, five (one *Pimelia* and four *Tentyriiane*) make up 85% of the individuals and 92% of the biomass (Gillon and Gillon 1974 a). *Elateridae* and *Meloidae* are tied to the rainy season. *Lepidoptera*, present yearlong in the grass layer, constitute 14.5% of the annual biomass of non-social *Arthropoda*.

The biomass of secondary consumers in non-social *Arthropoda* is always higher (three- to six- fold) than that of primary consumers (Tables 17–18). The rainy season promotes primary consumers, whereas at the end of the dry season the secondary consumers, less dependent on the annual vegetational cycle, become more numerous.

4.2.2 Social Arthropods

Two groups: ants and termites are abundant (Table 19). Only termites were the subject of elaborate studies at Fete-Ole (Lepage 1972, 1974a, b). Some ant species are termite eaters (*Camponotus, Crematogaster, Pheidole, Brachyponera*). The production of termites is maximal at the beginning of the rainy season (July), at the time of swarming; this is also the time of mass emergence of the small acarid *Dinothrombidium* which predates on termites. Four species of termites of the 23 existing on the Fete-Ole site represent 90% of the numbers and biomass (Table 19). The low density of the colonies is similar to large populations (0.5 to 1 million individuals per colony). Numbers may reach 2 to 3 million termites per hectare. The total biomass (1–3 kg DM) represents 60 to 90% (depending on environment) of all other wild primary consumers, with the exception of ants (Bourlière 1978). The net total production of termites is 9 kg DM ha^{-1} yr^{-1}, of which 30% are winged. Production is maximal in the early rainy season at the time of swarming. Most of the winged termites will be consumed by decomposers. Termites levy 100–150 kg DM ha^{-1} yr^{-1} from the surface ground grass layer and litter, i.e. 10% of the added biomass of the grass layer and deciduous woody stratum together. Termites show highly adaptive traits to long dry spells.

Table 19. Ecological characteristics of social arthropods (ants and termites) at Fete-Ole, Senegal (Lepage 1974b)

Social arthropods	Species (No.)	Food regime[a]	No. of nests ha^{-1}	Density (10^3 individuals ha^{-1})	Biomass (g DM ha^{-1})
Ants	10			100–145	
Monomorium	4	G	20		
Tetramorium	2	G			
Brachyponera		O			
Pheidole		O	20		
Crematogaster		O			
Camponotus		I			
Messor		C	0.04		
Termites	23				
Main species					
Psammotermes hybostoma	1	H,X	2.1	1050	325
Odontotermes smeathmani	1	H,X	1.8	525	351
Bellicositermes bellicosus	1	H,X	0.5	392	958
Trinervitermes trinervius	1	H	4.8	177	115
Total	9	-	-	2286	1845

[a] G = granivorous; 0 = omnivorous; I = insectivorous; H = herbivorous; X = xylophagous.

4.3 Vertebrates (Tables 2, 16, 22–23 for mammals and Tables 20–21 for birds)

The vertebrate fauna is poor in species numbers; mammals, particularly pullulating rodents, produce the strongest impact on primary production (Poulet 1982).

4.3.1 Herpetofauna

At Fete-Ole amphibians include eight species of *Anoura,* three of which are common: *Bufo pentoni, Tomoptera delalandi* and *Discoglossus occipitalis.* Their cycle is adapted to seasonal desiccation of temporary ponds. These amphibians spend the dry season hibernating underground in the burrows of foxes and honey badgers. At the first rains, toads break out of the ground in large numbers and head for ponds where they lay their eggs. Embryonic and larval development are very fast (Forge and Barbault 1977); swarming termites constitute their favorite food.

Seven species of lizards, five snakes and one tortoise have been registered on the Fete-Ole site, but none of those seems to build populations of any importance (Bourlière 1978).

Table 20. Specific richness, density and biomass of avifauna at Fete-Ole (Morel and Morel 1974; Bourliere 1978)

Years	1969–1972	1969–1970	1972–1973
Rainfall (mm)		209	33
Species number	108	81	75
Residents	60	51	48
Doubtful status	4	3	3
Migratory			
Paleotropical	17	8	7
Palaearctiics	27	19	17
Density (ind. ha^{-1})		6.3 ± 1.4	2.9 ± 1.1
Biomass (g ha^{-1}) live wt.		402 ± 41	186 ± 95
DM[a]		121	56

[a] DM = Live wt × 0.3

Table 21. Avifauna at Bandia, Senegal: average numbers (or nests) and average biomass; mean annual consumption (or harvest) in kg DM ha^{-1} by birds, rodents and ants in a Sudanian environment (Gillon et al. 1983)

Biological years	1978–79	1979–80	1980–81
Rainfall (mm)	616	596	350
Birds (No. of species: 22)			
Average population (ind. ha^{-1})	–	49	38
Average biomass kg ha^{-1} live wt.	–	4.4	3.8
kg ha^{-1} DM	–	1.3	1.1
Consumption seeds (kg DM ha^{-1})	–	124.6	100.6
Consumption insects (kg DM ha^{-1})	–	28.2	25.1
Rodents (No. of species 2)			
Average numbers	12.0	22.0	–
Biomass live wt.	0.6	1.2	0.2
DM	0.2	0.4	0.1
Seed consumption (kg DM)	23.1	44.9	5.8
Ants (No. of species: 1)			
Number of nests	0.50	0.50	0.38
kg seeds harvested (DM)	3.4	3.3	2.4

Table 22. Numbers, maximum and minimum biomass of the rodents *Taterillus pygargus* from 1969 to 1977 and *Arvicanthis niloticus* in 1975–76 at Fete-Ole, Senegal (computed from Poulet in Bourliére 1978)

Year	Precipitation (mm)	Numbers		Biomass			
		Minimum	Maximum	Minimum (Live weight)	Maximum	Minimum (Dry matter)	Maximum
Taterillus pygargus							
1969–70	450	4.1	9.0	148	324	44	97
1970–71	209	0.6	7.8	22	281	7	84
1971–72	202	–	0.7	–	25	–	8
1972–73	33	0.5	0.8	18	29	5	9
1973–74	209	30.0	–	–	–	–	–
1974–75	316	20.0	40.0	1080	1440	–	–
1975–76	311	3.2	180.0	720	6480	216	1944
Arvicanthis niloticus							
1976–77	?		35.0	119	1512	36	454

Table 23. Ecological characteristics of the mammalian fauna of northern Senegal (Poulet 1972)

Species	Abundance[a]	Food regime[b]	Period of activity[c]
Soricidae			
Crocidura sericea	+ +	I	N
C. lusitania	+ +	I	N
C. lamottei	+ +	I	N
Chiroptera, Nycteridae			
Nycteris thebaica	+ +	I	N
Tadarida major	+ +	I	N
Lagomorpha			
Lepus crawshayi	+	H	N
Rodentia, Sciuridae			
Heliosciurus gambianus	+	H,F	D
Xerus erythropus	+ +	H,F	D
Muridae			
Arvicanthis niloticus	+	G	N,D
Mastomys sp.	+	G	N,D
Cricetidae			
Taterillus pygargus	+ +	G,I	N
T. gracilis	+	G,I	N
Desmodilliscus braueri	+	G	N
Hystricidae			
Hystrix cristata	+	Rh	N
Carnivora			
Canidae *Canis aureus*	+ +	C,O	N,D
C. adustus	+	C,O	N,D
Vulpes pallida	+ +	C,O	N
Mustelidae			
Mellivora capensis	+	C	N
Zorilla striatus	+	C	N
Viverridae			
Genetta genetta	+ +	C	N
Ichneumia albicauda	+ +	C,I,F	N
Hyaenidae			
Hyaena hyaena	O	O	N
Felidae			
Felis silvestris libyca	+ +	C	N
F. serval	–	C	N
Tubulidentata			
Orycteropus afer	O	I	N
Artiodactyla			
Phacochoerus aethiopicus	+	Rh	N,D
Bovidae			
Gazella rufifrons	+	H	D
G. dorcas	O	H	D
G. dama	O	H	D
Primates			
Cercopithecidae			
Erythrocebus patas	O	O	D

[a] –, Very rare; O, rare; +, infrequent; + +, common.
[b] G, granivorous; F, frugivorous; C, carnivorous; H, herbivorous; I, insectivorous; O, omnivorous; Rh, rhizophagous.
[c] N, nocturnal; D, diurnal.

4.3.2 Birds

Species numbers are also few in this group (Table 21); the number of species to be found is 139 (Morel and Boulière 1962; Morel and Morel 1972, 1980). Migratory birds represent 40% of the avifauna, most of them from Europe and Africa north of the Sahara. All of them, except the quail, are insect eaters, but occasionally they consume berries. Migratory species represent the main fluctuation factor in the population of predatory birds in the western Sahel, since none of the 53 species in the group seems to be strictly sedentary (Thiollay 1978). Of the birds of prey 44% (70% of all birds) are migratory "Ethiopian" (i.e. Afro-tropical) species, while 27% are palaearctic winter visitors (18% of all the migratory birds). Migrations are necessary due to the seasonal variation in production of the Sahel ecosystem and to the complementarity in time of different savannas at various latitudes, when food availability is concerned.

Migratory birds thus ensure an optimal use of resources while complementing the impact of sedentary species. Palaearctic and African migratory birds never coexist during the dry season, thus avoiding a possible competition for food. Granivorous or granivorous/insectivorous sedentary birds are represented by groups which are well adapted to arid regions: *Columbideae* (7 species), *Pteroclididae* (2 species), *Ploceideae* (13 species). The trophic impact estimated for granivorous species is about 10% of the annual seed production, varying from 2.6 to 4.3 kg DM ha^{-1} yr^{-1}, depending on years. The number of species and the mean monthly biomass vary from one year to the next (Table 6; Morel and Morel 1974). Living biomass may vary from 402 g ha^{-1} $month^{-1}$) (i.e. 121 g DM ha^{-1} $month^{-1}$) and 186 g ha^{-1} $month^{-1}$) (i.e. 56 g DM ha^{-1} $month^{-1}$). The average biomass of palaearctic migrants is only 15.5 g ha^{-1} $month^{-1}$, fresh weight.

4.3.3 Mammals

The descriptive aspect of Sahelian mammals is the only one studied so far (Dekeyser 1955; Happold 1973; Delany and Happold 1979). The present mammalian wild fauna in the Fete-Ole region is rather poor (Poulet 1972, 1982). There are 27 terrestrial species and 5 bats (Table 23). The red-fronted gazelle, which has become rare, was probably the main grazer under pristine ecosystem conditions. Elephants and giraffes are the predominant browsers, while the warthog concentrates on the rhizosphere. All large ungulates migrate with season.

Mammals include some 12 vegetarian species: herbivores (3 ungulates and 1 hare), granivores (8 rodents); 12 animalivores; 9 carnivores, 4 insectivores and 1 consumer of social insects, the antbear (*Orycteropus afer*); and 1 omnivore, the Patas monkey.

The influence of the Fulani herds on vegetation is hardly noticeable more than 14 km from permanent water points. Before the Great Drought (1969–1973) there were some 12 000 head of cattle, 12 000 sheep and goats and some camels in the Fete-Ole region, i.e. a stocking rate of 2300 kg km^{-2} (Bourlière 1978).

4.3.4 Rodents

Due to the fact that ungulates have become rare, rodents constitute the major group of primary consumers among wild mammals. The Gerbillinae are the best adapted to the environment; the most common species is *Taterillus pygargus*. Its density varies according to habitat and season from 0 to 9 individuals per hectare, representing a biomass of 0 to 324 g live wt. ha^{-1} (Table 22). The rate of disappearance of adults of 25% per month is compensated by a high productivity of adult females: a litter two to six young every 6 weeks from September to March. Production varies from 40 to 600 g live wt. ha^{-1}. From an estimated production of 40–60 kg ha^{-1} yr^{-1} of seeds, some 10 kg germinate and only 1–2 kg will produce the herbaceous cover (Poulet 1972, 1982). Of these *Taterillus* populations consume 2.1–6.6 kg ha^{-1} yr^{-1}, i.e. 5–10% of the annual production. The amount of grain available does not seem to be a limiting factor to population numbers; but the effects of drought on the reproductive rate is important. The 1969–1973 drought resulted in a severe decrease of the density level of *Taterillus* and an increase of the desert species, *Desmodilliscus braueri* (Poulet 1974, 1982).

The production of seeds was nil in 1972 (rain = 33 mm), but populations of *Taterillus* could survive becasue of the seeds left over from previous years. Populations of other mammals also sharply decreased during that drought. However, after the rather good rains of 1974 and 1975, populations recovered steadily, so that several Sahelian rodents started to pullulate in 1975: *Taterillus pugargus*, 180 individuals per hectare; *Arvicanthis niloticus*, 100 individuals per hectare, which became a pest of catastrophic proportion. Such a demographical explosion may be explained by the extension of the reproductive season and the higher fertility of females (Poulet and Poupon 1978; Poulet 1978, 1982).

The Nile rat, normally restricted to humid zones, acquired a diurnal and arboreal behaviour, eating bark and twigs of *Commiphora africana, Acacia senegal* and *Balanites aegyptiaca,* which resulted in considerable havoc to the ligneous population. A sequence of "rainy years" coming in succession may thus produce conditions favourable to pullulation.

4.3.5 Consumption of Seeds by Granivores

In such regions as the Sahel, with sharply contrasting climatic conditions and long dry seasons, herbaceous vegetation can only develop for a short period each year. Recently a study on the production and consumption of grain by birds, rodents and ants was carried out under the Sudano-Sahelian conditions of Bandia, in Senegal (Table 21). Annual rainfall is much higher than in Fete-Ole (570 mm yr^{-1}) Gillon et al. 1983). Vegetation is an open woodland of *Acacia seyal*, Combretaceae, Tiliaceae and Capparidaceae with a herbaceous layer dominated by a ruderal legume weed, *Cassia tora*. Annual seed production varied from 286 to 1061 kg ha^{-1} yr^{-1}, depending on the year; thus, it was about ten times higher than in Fete-Ole. the 22 main species of birds, all of them at least partly granivorous, consumed 6 to 26% of these seeds (Table 21). The rodent population, *Mastomys erythrolocus* and *Taterillus gracilis*, consumed 1–15%, and the ant *Messor galla* 0.4–2.0%. The high bird stocking rate resulted from the particular site of Bandia next to the Atlantic seashore at the southern margin of the semi-arid zone, the presence of permanent water supply, the existence of extensive millet and groundnut crops and the abundance of trees and nesting facilities. Rodents may experience sharp fluctuations in numbers, and in some years their impact may be more pronounced than that of birds; in 1975–1976, for instance, they consumed 276 kg ha^{-1} of seeds (Hubert 1982a, b; Hubert et al. 1981). In fact, one-third to two-thirds of the seeds produced are actually eaten, if due allocation is made for endophagous insect parasitism (bruchids) which develop in the seeds of *Cassia tora* and *Acacia*. The former, although dominant, are consumed neither by birds nor by rodents, perhaps because of the presence of phytoagglutinins (Gillon et al. 1983). Ants (*Messor galla*) harvest the most abundant seeds, thus allowing a greater diversity of species in the vegetation. They constitute the less selective group among granivores in the Sahelian environment and also the group having the lesser impact (Gillon et al. 1983).

4.3.6 Large Mammals

The density level of the four main species of carnivores in the Fete Ole region, i.e. *Canis aureus* (jackal), *Felis libyca* (Saharan cat), *Genetta genetta* (genet) and *Vulpes pallida* (pale fox), varies from year to year from 0.4 to 3.1 individuals km^{-2} (Poulet 1972, 1974), representing a live weight of 1.4 to 14.0 kg km^{-2} (Table 23).

The hare *Lepus crawshavi* (= *L. whytei*) had a population of 0.2 to 1.0 individuals km^{-2}, i.e. a biomass of 0.4–2.0 kg live wt. km^{-2}.

The stocking rate of wild ungulates in the Sahelian zone is very poorly documented. In the Sahelian zone of northern Darfur (Rep. of Sudan) an aerial survey over a territory of 250 000 km^2 found a biomass of large herbivores: 1.0 kg live wt. km^{-2}, with local concentrations of 8.0 and 11.0 kg km^{-2} in limited areas, and only 18 species present (Wilson 1979a, b). In contrast, the Sudanian zones of southern Darfur yielded some 28 kg km^{-2} over an area of 160 000 km^2 with local concentrations of up to 488 kg km^{-2} and some 49 larger mammals (Watson et al. 1977; Wilson 1979a, b, 1980a). One can only say that the figures found in N. Darfur and N. Senegal are consistent in their mediocrity.

In a region quite similar to the Sahel in terms of climate, soil and vegetation, the Kalahari "Desert", the biomass of large wild ungulates was still 400–440 kg km^{-2} in 1980 (DHV Engineering 1980, Williamson and Williamson 1981). The present Kalahari wild ungulates' stocking is thus only 20% of the theoretical potential in the absence of livestock, according to the predictive equation of Coe et al. (1976) and to Bourlière (1978). The Kalahari stocking seems to be a reflection of what the Sahel could have borne some 100–150 years ago (Mauny 1957) and the Sahara 3000 to 4000 BP (Hugo 1974).

The biomass of wild ungulates in the Sahel has sharply decreased since the 1950's due to overhunting and poaching; until the 1960's it reached 0.3 to 190 kg live wt km^{-2} (Bourlière 1962). The biomass of the four major ungulates in existence in Chad in the late 1950's (*Oryx dammah, Addax nasomaculatus, Gazella dama* and *Gazella rufifrons*) reached concentrations of 80 kg km^{-2} in central-eastern Chad along the 16° N lat. between Largeau and Abéché in the early 1960's (Gillet 1965).

Thus, the scimitar-horned oryx used to be found in herds of several 100 in Chad until the early 1960's; for the country at that time they were estimated at 4000 to 5000. This species of oryx also used to be common in the western Sahara and Mauritania, but became extinct in the 1960's (Valverde 1968). The addax, the best-adapted antelope to desert habitats, one of the rare species which never drinks free water, is virtually extinct at present, while Monod (1958) counted some 12 500 fresh spoors in December 1952 in the SW empty quarter of Mauritania, the Majabat al Kubra. Oryx and addax are now nearly extinct as a result of persecutive hunting for the past 30 years, with increasing mechanical means such as helicopters, also within the so-called game reserves. Large mammals have already become vestigial and the day is not far ahead when the only living specimens will be found in zoos and zoological gardens. Table 2 shows the ecological distribution of mammals throughout the northern/tropical zones of Africa; Table 16 (Newby 1984) shows the present status of some large mammals.

4.4 Conclusions

Wildlife has to face very harsh environmental conditions in the Sahel. Primary consumers, however, enjoy one advantage over their cognates from temperate and humid tropical climates: the dryness of the atmosphere and soils allows for a longer conservation period of plant tissues and therefore their availability over longer periods (Bourlière 1978). This stored energy mitigates, for primary consumers, the effects of the long seasonal drought. Furthermore, a number of shrubs and trees keep their green foliage throughout the dry season or most of it. The pods of *Acacias* and other legume trees and shrubs bring additional stored sources of energy and protein for a number of mammals, including ungulates, particularly livestock. The same applies to the fruits of *Boscia* for birds. During the course of the dry season, phytophagous species reduce the aerial biomass of the grass layer by 30 to 60%. For at least 6 months annually herbaceous vegetation cannot ensure a balanced diet to herbivores; consumption of leaves and fruits from shrubs and trees is then a necessary complement to herbaceous production (Bille 1978; Le Houérou 1976b, 1980; Gillet 1980).

Because of their adaptive and demographical strategies, Sahelian wildlife species are able to cope with drought spells and return to their initial stock in rather short periods, they are resilient, however, only under one condition, i.e. that the destructive interference of human activities is reduced or discontinued (Bourlière 1978; Le Houérou 1980c).

5 Livestock

5.1 General

Holocene rock painting and engraving in the Saharan mountains and in the Sahel show that a so-called 'bovidian' pastoral period extended in the desert and its southern fringe between 5500 and 1000 BC (Hugo 1974; Mori 1965; Lhote 1958). All cattle shown on the rocks belong to the *Bos taurus* type (*Bos africanus, Bos ibericus*). The zebu type, *Bos indicus*, which now represents over 95% of the Sahelian cattle population, came later, around 400 AD from India and Pakistan, through the Strait of Aden to East Africa and then spread over most of the continent, south of the Sahara. Some hamitic zebus, however, were known to Africa from ca. 2000 BC, along the lower Nile Valley. The only non-zebu cattle in the Sahel nowadays is the "Kouri" breed from Lake Chad and its crossbred "Sanga" cattle in the neighbouring regions. Other taurine cattle in Africa are the Ankole breed of the Upper Nile, the West African Short Horn and the Guinean N'Dama breed from the Fouta Djalon and Nimba mountains, renowned for their tolerance to trypanosomiasis (sleeping sickness), and therefore their adaptation to the humid tropics, contrary to the zebus. Along the contact zone, under the 800–1200 mm isohyets, crossbreds between taurines and zebus are common (Mason 1951; Mason and Maule 1960; Doutresoulle 1947; Eppstein 1971; Zeuner 1963; Williamson and Payne 1965; see Figs. 56 and 57).

Livestock husbandry, probably originating in the Near East and along the Lower Nile Valley, has thus been in existence in the Sahel for at least 7500 years.

5.2 Livestock Numbers, Evolution of Numbers, Densities, Stocking Rates

The number of animals in the Sahel countries is shown in Tables 24–29. The grand total is a cattle equivalent of almost 51 million tropical livestock units (TLU), an abstract unit defined as a 250-kg mature dry zebu kept at maintenance. The conversion factors used are somewhat different from those found in many publications since we have used the ratio between metabolic weights, and not between live weights, to determine the equivalences between species and breeds (Le Houérou 1981). Mean population weights are according to Wilson (1982b): cattle, 183 kg; sheep, 25 kg; goats, 21 kg; camels, 307 kg; assess, 107 kg. Thus, the

Table 24. Livestock populations in the Sahel countries in 1983 (FAO production yearbook 1983)

Country	Numbers of heads in 1000				
	Cattle	Sheep	Goats	Camels	Asses
Burkina-Faso	2950	2000	2500	6	200
Chad	3600	2300	2100	421	255
Mali	5400	6450	7500	240	425
Mauritania	1500	5000	3000	750	143
Niger	3521	3448	7470	410	501
Senegal	2250	2100	1050	6	240
Sudan	19550	19500	12900	2500	688
Conversion factor	0.81	0.18	0.16	1.16	0.53
Total TLU	31404	7333	5844	5026	1299
Percent of all stock TLU	62	14	11	10	3

Grand total TLU: 50 906

Table 25. Livestock populations in the Sahel countries in 1950 (FAO production yearbook 1950)

Country	Numbers of heads in 1000				
	Cattle	Sheep	Goats	Camels	Asses
Burkina-Faso	1120	590	934	3	167
Chad	3260	2000	2000	266	370
Mali	3000	4080	3860	74	225
Mauritania	514	1065	1614	160	85
Niger	2818	1324	3494	224	300
Senegal	720	506	747	4	70
Sudan	3957	5660	4440	1550	521
Total	15387	15225	17089	2281	1738
Conversion factor	0.81	0.18	0.16	1.16	0.53
Total TLU	12465	2740	2734	2646	921
Percent of all stock TLU	58	13	13	12	4

Grand total TLU: 21 506

Table 26. Overall increase in stock numbers between 1950 and 1983 (FAO production yearbooks 1950, 1983)

Species of stock	1950	1983	$\frac{1983-1950}{1950}$	$\frac{1983}{1950}$
Cattle	15 389	38 771	152	2.52
Sheep	15 225	40 738	167	2.67
Goats	17 089	36 528	114	2.24
Camels	2281	4333	90	1.90
Asses	1738	2452	41	1.41
Total	51 722	122 822	137	2.37

Table 27. Evolution of livestock numbers by country between 1950 and 1983 (FAO production yearbooks)

		Numbers in 1000				1983–1950 × 100
Country	Stock	1950	1968	1973	1983	1950
Burkina-Faso	Cattle	1120	2600	1600	2950	163
	Sheep	590	1700	1050	2000	239
	Goats	934	2400	2200	2500	168
	Camels	3	4	5	6	100
	Asses	167	250	360	200	20
Chad	Cattle	3260	4500	4100	3600	10
	Sheep	2000	1800	1600	2300	15
	Goats	2000	2000	2100	2100	5
	Camels	266	325	330	421	8
	Asses	370	450	375	355	- 3
Mali	Cattle	3000	5067	3700	5400	80
	Sheep	4080	5200	3900	6450	58
	Goats	3860	5100	3800	7500	94
	Camels	74	231	150	240	224
	Asses	225	702	440	425	89
Mauritania	Cattle	514	2100	1900	1500	192
	Sheep	1065	2600	3000	5000	369
	Goats	1614	2200	2200	3000	86
	Camels	160	500	700	750	369
	Asses	85	231	254	143	68
Niger	Cattle	2818	4100	3000	3521	25
	Sheep	1324	2500	2000	3448	160
	Goats	3494	5870	5000	7478	114
	Camels	224	380	350	410	83
	Asses	300	490	575	501	61
Senegal	Cattle	720	2747	1750	2250	212
	Sheep	506	1347	980	2100	315
	Goats	747	1490	900	1050	40
	Camels	4	5	6	6	- 50
	Asses	70	356	240	240	243
Sudan	Cattle	3957	11 200	15 200	19 550	394
	Sheep	5660	11 000	15 400	19 500	244
	Goats	4440	8400	12 000	12 900	190
	Camels	1550	2500	3400	2500	61
	Asses	521	611	681	688	32

Table 28. Geographical densities of livestock in 1983

Country	TLU (10^3)	Area (km^2)	Density (TLU km^{-2})
Burkina-Faso	3262	274 200	11.90
Chad	4289	1 284 000	3.34
Mali	7238	1 240 000	5.84
Mauritania	3541	1 030 700	3.44
Niger	5410	1 267 000	4.27
Senegal	2503	196 190	12.76
Sudan	24 675	2 505 810	9.85
Total	50 918	7 797 900	6.53

Table 29. Proportions of various species of livestock in the Sahel countries, as of 1983 (FAO production yearbook 1984)

Country	Units[a]	Cattle	(%)	Sheep	(%)	Goats	(%)	Camels	(%)	Asses	(%)	Total
Burkina-Faso	Heads	2950		2000		2500		6		200		
	CF	0.81		0.18		0.16		1.16		0.53		
	TLU	2389	73.2	360	11.0	400	12.3	7	0.2	106	3.2	3262
Chad	Heads	3600		2300		2100		421		255		
	TLU	2916	68.0	414	9.6	336	7.8	488	11.4	135	3.1	4262
Mali	Heads	5400		6450		7500		240		425		
	TLU	4374	60.4	1161	16.0	1200	16.6	278	3.8	225	3.1	7238
Mauritania	Heads	1500		5000		3000		750		143		
	TLU	1215	34.3	900	25.4	480	13.6	870	24.6	76	2.1	3541
Niger	Heads	3521		3448		7470		410		501		
	TLU	2852	52.7	621	11.5	1195	22.1	476	8.8	266	4.9	5410
Senegal	Heads	2250		2100		1050		6		240		
	TLU	1823	72.8	378	15.1	168	6.7	7	0.3	127	5.0	2503
Sudan	Heads	19 550		19 500		12 900		2500		688		
	TLU	15 836	64.1	3510	14.2	2064	8.4	2900	11.8	365	1.5	24 675
Mean percentages			60.8		14.7		12.5		8.6		3.3	
Grand Total												50 918

[a]Heads in 10^3; CF = conversion factor; TLU = tropical livestock unit (250 kg) in 10^3.

metabolic weights ($kg^{0.75}$) of TLU are 62.87; cattle, 51.00; sheep, 11.20; goats, 9.80; camels, 73.00; asses, 33.5 $kg^{0.75}$. Thus, the ratios of mean population weights $(MPW)^{0.75}/TLU^{0.75}$ are:

$$\frac{\text{MPW kg}^{0.75}}{\text{TLU kg}^{0.75}} = \begin{array}{ll} \text{cattle:} & 51/62.9 = 0.81 \\ \text{sheep:} & 11.2/62.9 = 0.18 \\ \text{goats:} & 9.8/62.9 = 0.16 \\ \text{camels:} & 73/62.9 = 1.16 \\ \text{asses:} & 33.5/62.9 = 0.53. \end{array}$$

In many pertinent publications one TLU is equated with ten heads of small stock which is nutritionally incorrect by about 40%. The TLU is supposed to ingest an average 6.25 kg DM day^{-1} (a figure that has been proven correct for pastoral stock) (Dicko and Wilson et al. 1983) or 2.5 kg DM per 100 kg live wt. or 100 g DM per $kg^{0.75}$ per day (Boudet and Rivière 1968).

Thus, converted in TLU, the proportions of the different species is 62% cattle, 14% sheep, 11% goats, 10% camels and 3% assess (Tables 24–29).

The geographical density (sum of the national numbers over the sum of national acreages is 6.53 TLU km^{-2}, thus equivalent to a stocking rate of 16.3 kg ha^{-1}. This figure, however, is of little significance since most of the Sahelian countries have huge expanses of true desert virtually void of vegetation and therefore no stocking. If we only consider the Sahel, the overall zonal density would be 50.9 million TLU ÷ 3 million km^2 = 32.74 TLU km^{-2}, or 3 ha TLU^{-1}, i.e. 83 kg live wt. ha^{-1}. But this figure is now too high in turn as large numbers of animals spend part or all of the dry season outside the Sahel, in the Sudanian zone, to the south (because of the presence of the Tsetse flies, *Glossina* spp., vectors of trypanosomiasis, stocks have to leave the Sudanian zone during the rainy season, and then move to the Sahel). The actual stocking rates in the Sahel probably average 6 to 9 ha TLU^{-1} at present, judging from a number of surveys in various countries, i.e. 30 to 40 kg live wt. ha^{-1}. These stocking rates, however, are most uneven throughout the Sahel: there are areas of high stock concentrations like the Niger River Inland Delta and its neighbourhood (Wilson 1982b; Wilson et al. 1983) or in the Niger Bend in Mali and in Burkina-Faso (formerly Upper-Volta) (Barral 1974, 1977; Lhoste 1977; Grouzis 1984). In these two areas stocking rates reach 3–6 ha TLU^{-1} yr^{-1}, and locally more. Conversely, there are large areas, mainly in the northern half of the Sahel, with no permanent water resources, where stocking rates are of the order of magnitude of 10–20 ha/TLU/12 months (3–6 ha/TLU/3 months).

The overall long-term carrying capacity for the Sahel region, as suggested by a number of experiments and stocking trials on state-owned ranches (Dahra, Doli and Bambylor, in Senegal; Niono in Mali; Ekrafan and Toukounous in Niger; Ouadi Rimé in Chad; Leo in Burkina-Faso, etc.), as well as by many surveys, is of the order of magnitude of one TLU for 12 ha, or one head of cattle (MPW) for 10 ha (20 kg live wt. ha^{-1}). This is a theoretical consumption of 200 kg DM ha^{-1} yr^{-1}, i.e. 20–25% of the maximum standing crop (MSC) of an average Sahelian grassland or 15–20% of it when browse production is accounted for. In grasslands in reasonably fair condition, like most of the Ferlo of northern Senegal (10 000 km^2), under 300/500 mm of long-term mean annual rainfall, the actual stocking rates are 20–30 kg live wt. ha^{-1}, averaging 25 kg (73% of which are cattle); the density of cattle ranges from 7.5 ha $head^{-1}$ on the northern sandy sites, to 17.8 ha $head^{-1}$ in southern sites on shallow soils over iron or lateritic hardpans (Naegelé 1971a, b; Valenza and Diallo 1970, 1972; Bourlière 1978; Barral et al. 1983; Sharman 1982; Vanpraet 1985; Le Houérou 1988b).

In northern Burkina-Faso (15 000 km^2), under similar rainfall, the carrying capacity is estimated at 9 ha TLU^{-1} (28 kg live wt. ha^{-1}) (Barry et al. 1983), whilst the actual stocking rate averages 6 ha TLU^{-1} (42 kg live wt. ha^{-1}) (Barral 1974). The result of such a 33% overstocking is what could have been expected, i.e. a grassland in poor to very poor condition, with RUE factors 30 to 50% below the average Sahel values and very low browse availability (Sicot and Grouzis 1981; Piot et al. 1980; Grouzis 1984). Our estimation is an overall overstocking of some 30% in the Sahel as a whole, at the present time, with local heavy overstocking and some local understocking as well.

Table 12 shows an overall increase in stock numbers of almost 30 million TLU from 1950 to 1983, i.e. a 137% increase and a ratio 1983/1950 of 2.37 for the seven Sahelian countries. Again the increase is very uneven from one country to the next as shown in Table 27. The largest increase, by far, is in the Republic of Sudan both in absolute and in relative terms, while the increase is only slight in Chad. The proportions of stock, expressed in TLU's has substantially increased in favour of cattle from 1950 to 1983 as shown in Tables 24 and 25, since cattle represented 62% of all stock in 1983, against 58% in 1950.

The people to stock ratios in the different countries are shown in Table 9. The increase in livestock has been similar to the increase in human population so that the ratio has remained unchanged from 1950 to 1983, i.e. 0.9 TLU per person in the overall population and about one TLU per inhabitant in the rural population. The increase in both livestock and human population has thus been an exponential growth of 2.7% per annum between 1950 and 1983 (Tables 24–29). While the rate of increase in cultivated land was 1% per annum for the same period, according to official statistics, but closer to 2.5% according a number of surveys using remote-sensing techniques; that is a rate almost equal to the growth of human and stock populations (Figs. 54–57).

5.3 Breeds (Doutresoulle, 1947; Mason and Maule 1960; Williamson and Payne 1965) (Fig. 56)

There are many breeds of cattle in the Sahel usually linked to the ethnic groups that raise them, e.g. Moor cattle, White Fulani, Red Bororo, Gobra, Tuareg, Azawak and Baqqara, which are all zebus (*Bos indicus*). They vary a great deal in colour, size, shape, weight, shape and size of horns, ecological adaptation and performance (Doutresoulle 1947; Mason 1951). Mean weight in adults varies

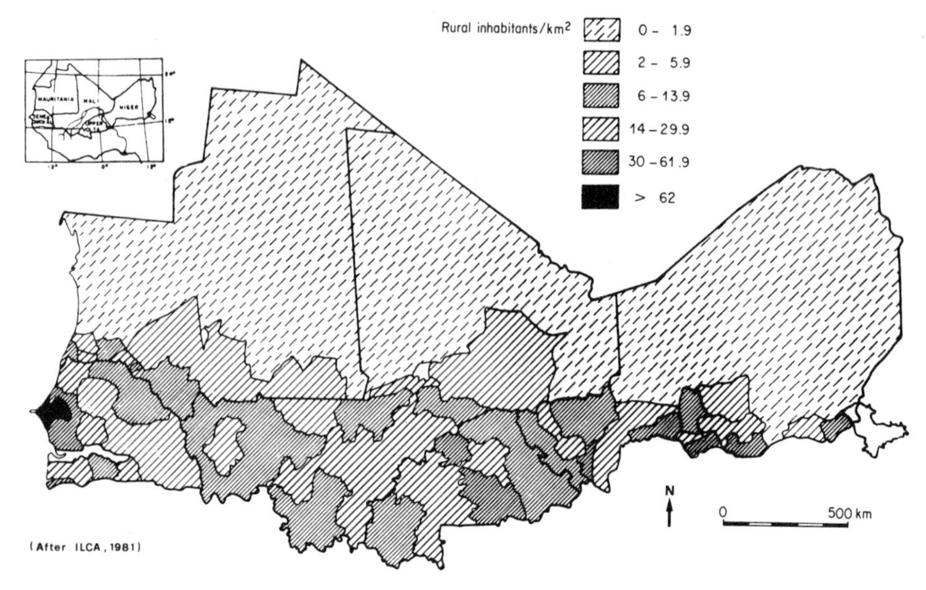

Fig. 54. Density of rural human population in the Sahelian and Sudanian ecoclimatic zones of West Africa (After ILCA 1981)

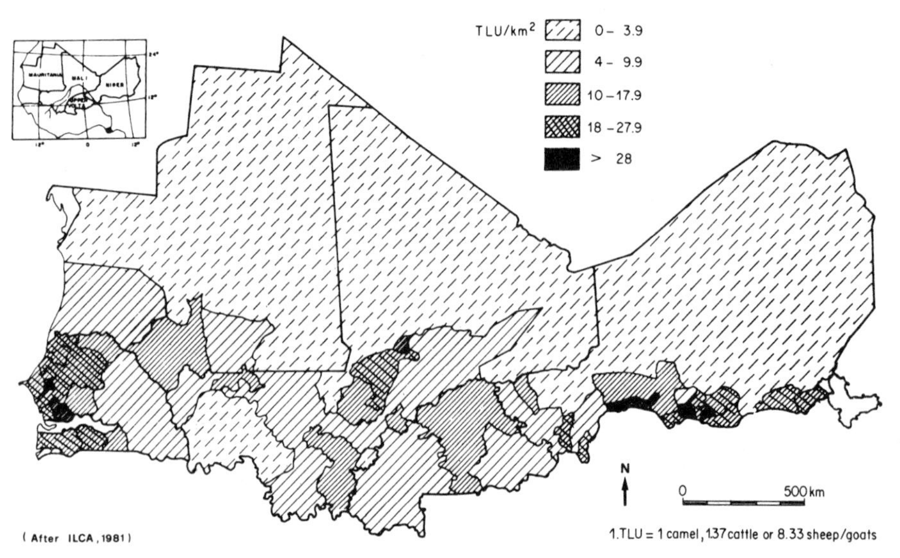

Fig. 55. Density of livestock in the Sahelian and Sudanian ecoclimatic zones of West Africa (After ILCA 1981)

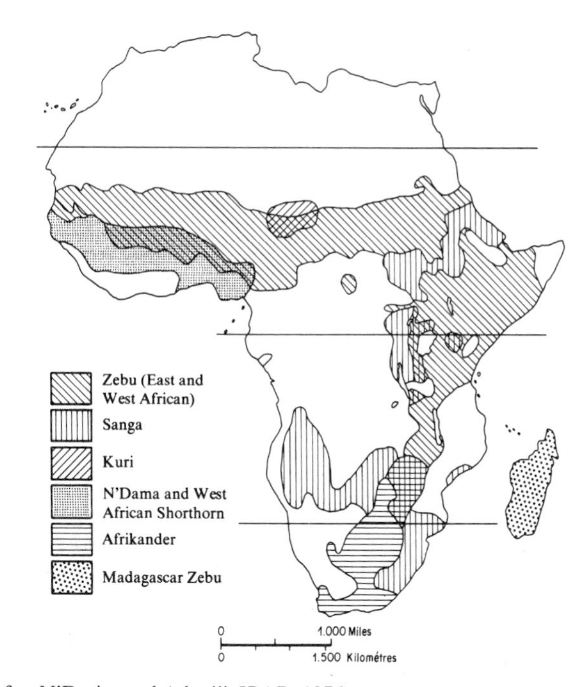

Fig. 56. Cattle breeds in Africa (After N'Derito and Adeniji, IBAR, 1976)

Fig. 57. Cattle distribution in intertropical Africa: one *dot* = 5000 heads (After N'Derito and Adeniji, IBAR, 1976)

from less than 300 kg in small breeds like the Tuareg to more than 500 kg in large breeds such as the Gobra of Senegal, in the males and 200–400 kg in females. Height at withers varies from 125 to 150 cm. There are large numbers of varieties and crossbreds within zebus and between taurines and zebus; the Sanga cattle from the Chad Basin, for instance, is a crossbred between African zebus and the "Kouri" breed belonging to taurines. Sahelian cattle may be separated into four main groups:

1. Short-horn West African zebu (Moor, Asawak, Tuareg);
2. Sanga zebu crossbred with hamitic cattle (*Bos taurus*);
3. Long, lyrate-horned West African zebus (Fulani, Bororo, Gobra);
4. the "Kouri" (*Bos taurus*).

Sheep belong to two main types, the wool sheep and the hair sheep. The Macina wool sheep of northern African origin (Morocco) is restricted to the middle Niger valley: Inland Delta and Big Bend which altogether represent 1 million animals. Most of the Sahel sheep, some 98%, belong to the hairy type (*Ovis aries sudanica*); again there are many subtypes or breeds associated with various ethnic groups of pastoralists. Mean adult heights vary from 65 to 85 cm and weights from 30 to 60 kg.

Sahelian goats (Moor, Tuareg, Fulani types) are also rather tall animals 70–85 cm in height at withers, weighing 25 to 35 kg; colour is usually variegated: black, red and white; hair is thin and short. In the southern Sahel of Niger and Nigeria, the "Maradi" or "Red Sokoto" goat is raised for its skin. It is a small breed

Typical Sahelian zebu cattle: the white fulani "Gobra" breed of Senegal. Average adult weight 300–350 kg (Photo A. Cornet)

ca. 65 cm in height weighing 20–25 kg, reddish in colour. This breed produces a particular quality of skin and leather called "Morocco", subject to international trade.

Camels are like other stock named after the pastoral ethnic groups who breed them and the places where they are bred: Moor, Tuareg, Tubu, Zaghawas, Kabbabish. The desert and northern Sahel Tuareg breed, the "Mehari" from the Air and Iforas are more slender, lighter and faster riding breeds, in contrast to the heavier southern Sahel beast of burden or pack camel. The southern limit of distribution of the camel is also the southern limit of the Sahel: about the 600-mm isohyet of annual rainfall; the controlling factor being disease, in particular pyroplasmosis and trypanosomiasis. Height at withers is 180–215 cm, and weight of adults 350 to 480 kg.

There are also many local breeds of asses, grey in colour, with a narrow black stripe on the shoulder. In general, they are rather small: 90–110 cm in height, weighing 100–130 kg, carrying loads of 50–100 kg.

Horses and mules are of no importance numerically; as in many pastoral societies horses are a matter of prestige, carefully nurtured.

5.4 Herd and Flock Structure and Demography (Wilson 1976–1984)

From six surveys by different teams in various Sahel countries the following average figures emerge (computed from the data published by Wilson and Wagenaar 1983).

Cattle

Type of animals	Percent in the herd	Mean weight	Weighted average
Breeding cows	42.7	220	94
Heifers	17.1	150	26
Female calves	11.5	50	6
Total females	71.3	—	—
Adult males (> 4 yrs)	8.7	300	26
Young bulls (1–3 yrs)	8.6	150	13
Bull calves (0–1 yr)	11.4	50	6
Total males	28.7	—	—
Total	100.0		171

Hence, 1 TLU = 250/171 = 1.46 heads;
one head = 171/250 = 0.68 TLU.

Sheep

Averages of seven surveys: Tuaregs, Moors, Baqqara, Fulani, Arab, Masai, Afar; males, 22%; females, 78%; breeding females, 58% (Wilson 1982b). In the Tuareg flocks in Niger (Wilson and Wagenaar 1983):

Age (months)	Males (%)	Females (%)		Total
0–6	7.4		12.0	19.4
6–15	4.7		11.8	16.5
15–21	3.9		16.0	19.9
21–27	1.7	58.0	8.2	9.9
27–35	0.5	Breeding	6.5	6.9
35–60	0.1		20.7	20.8
> 60	0.0		6.6	6.6
Total	18.2		81.8	100.0

Goats

Averages of the same seven surveys mentioned above: males, 21%; females, 79%; breeding females, 56%. In Tuareg flocks in Niger (Wilson and Wagenaar 1983):

Ages (months)	Males (%)	Females (%)		Total
0–6	10.3		13.9	24.2
6–14	2.6		10.5	13.2
14–19	1.0		4.8	5.7
19–24	0.2		8.4	8.6
24–30	0.0	61.6	9.1	9.1
30–60	0.0	Breeding	25.4	25.4
> 60	0.0		13.9	13.9
Total	14.1		85.9	100.0

Camels

Ages (years)	Males (%)	Females (%)	Total
< 2	10.3	11.3	21.6
2–5	10.6	14.6	25.2
5–8	11.2	13.7	24.9

Camels (continued)

Ages (years)	Males (%)	Females (%)	Total
8–10	4.5	8.3	12.7
10–13	3.2	4.9	8.1
13–15	0.2	3.9	4.1
15–17	0.2	1.5	1.7
> 17	0.0	1.7	1.7
Total	40.2	59.8	100.0

Asses

Average of four surveys (Niger, Sudan, Mali)
(Wilson and Wagenaar 1983).
Males total, 59%; working, 50%. Females total, 41%; breeding, 28.5%.

5.5 Average Production Parameters

The main production parameters that permit the calculation of stock productivity are shown in Table 30. These figures originate from a number of recent surveys (Coulomb 1971, 1972; Dumas 1977, 1980; Wilson 1976–82; Wilson 1983; Wilson and Wagenaar 1983; Wilson et al. 1983; Wilson 1984).

The conclusions are shown in the four last lines of Table 30 showing offtakes and productivity indices. Production is very low according to modern husbandry standards; but small stock are far more productive than cattle, although they have received, until now, far less attention from scientists, particularly from veterinarians.

5.6 Productivity (Table 30)

Overall livestock productivity in the Sahel is difficult to appreciate in a precise and quantitative manner because:

1. Unlike in the arid zones of developed countries meat production is not the first and foremost goal and output of African pastoralists and farmers. There are many others, e.g. dung and manure, draught power, transport, hides and skins, wool, hair, milk, occasionally blood and, most important, social status.
2. No records of performance or outputs are furnished by stockowners thus productivity is computed from production parameters gathered from surveys. These parameters cannot take into consideration all aspects of productivity

Table 30. Mean production parameters in Sahel stock (Wilson 1982b, 1984; Wilson and Wagenaar 1983; Wilson et al. 1983; ILCA 1981.)

Parameters	Cattle	Sheep	Goats	Camels	Asses
Age of females at first parturition (months)	54	17	15	60	29
Parturition interval (months)	19	10	9	26	29
Annual reproductive rate (%) (Litter size × 365/parturition interval)	65	94	108	45	42
Breeding females, percent of herd/flock	43	58	56	35	29
Number of young per parturition	1.0	1.08	1.40	1.0	1.0
Number of young/female/yr	0.65	1.4	1.8	0.5	0.6?
Percentage of multiple births	1.0	4.5	13.5	1.0	?
Number of young per female in lifetime	2.1	5.5	6.5	2.7	4
Maximum young production in lifetime	8	10	12	9	8
Longevity (years)	11	5	6	17	16
Number of animals/herd/flock	40	103	60	21	–
Weaning age (months)	11	5	5	12	?
Weight at birth (kg)	18	2.6	1.8	30	12
Weight at weaning	80	16	11	150	?
Weight of adult males	300	35	30	400	120
Weight of adult females	250	30	25	300	100
Mortality in young (to 1 year)	30	25	35	50	?
Mortality in adults	7	10	10	5	?
Offtake percent of herd/flock	12	24	28	8	?
Productivity index I[a]	45	30	20	–	–
Productivity index II	21	89	57	–	–
Productivity index III	1270	231	147	–	–

[a] I = weight of young weaned per breeding·female/yr; II = weight weaned + milk per 100 kg of breeding female; III = g weaned per kg of metabolic weight.

since, at the present state of research, production indices are still imperfect as they are only based on biomass production (milk and live weight).

3. Some of the production traits are difficult to quantify, e.g. the considerable impact of dung and manure on soil fertility and grain yields, the social value of having herds and privileges derived thereof that may be highly significant economically (parentalism, clientelism, etc.).
4. Land is generally not privately owned and therefore high levels of production from individual animals are impossible to attain. Management is thus aimed at obtaining smaller returns from a larger number of animals, a strategy that also helps to minimize the effects of climatic vagaries, disease outbreaks and other constraints. The complementary and additive effects of mixed species stocking contribute greatly to increasing the total overall productivity and still more importantly, to ensure a continuous supply of the major products which the husbandmen need (Wilson 1984).

5.6.1 Transport

Transport is an important factor in the daily life of pastoralists; pack animals are donkeys, camels and bullocks and sometimes cows. Donkeys perform duties that are often overlooked by surveyors and scientists: daily transport of water, lifting water, riding and carting. Transport performances of oxen, donkeys and camels are shown in Table 31.

Table 31. Transport performance of Sahel stock (Wilson 1984)

Species and type	Country	Average weight	Load	Load/weight ratio	Type of work
Ox	Sudan	350	110	0.33	Pack and riding
	Mali	300×2	500	0.83	Carting
Donkey	Sudan	110	80	0.73	Pack
	Sudan	160	80	0.50	Riding
	Mali	105	380	3.62	Carting
	Mali	105	400	4.00	Carting
Camel	Sudan	450	400	0.89	Pack
	Mali	400	210	0.53	Pack

5.6.2 Draught

Draught power is of rather recent use in the Sahel: oxen, cows and donkeys are mainly used to plow the land for grain cultivation and occasionally for weeding. Camels and donkeys are also used for lifting and hauling water; working hours are 5–6 per day for cattle and 3–4 for donkeys. The daily area cultivated is around 1600 m^2 for a pair of oxen, with an annual output of 4–5 ha (Wilson 1984). Draught performance of oxen, camels and donkeys are shown in Table 32.

The power developed in lifting water from wells is higher than in plowing as it is produced in short bursts; a single ox develops some 60 kg m^{-1} s^{-1} and lifts 700 l of water per hour from a depth of 50 m, while a donkey develops 23 kg m^{-1} s^{-1}, lifting 200 l h^{-1}, i.e. 2.0 and 1.6 kg per kg body wt. per hour respectively.

Table 32. Draught power performance of Sahel stock (Wilson 1984)

Type of animals	Average weight (kg)	Draught (kg)	Speed (m s^{-1})	Power developed (kg m s^{-1})
Camel (Sudan)	475	75	1.0	75
Oxen (Mali)	400	55	0.8	45
Donkey (Mali)	120	35	0.7	26

5.6.3 Hides, Skins, Wool, Hair

Those commodities are important in the international trade and currency earnings of Sahelian countries, but also at the family and village levels, in housing, clothing, bedding, saddlery, containers, craftwork. Many Sahelian hides and skins are of poor quality due to brand marks, poor curing and insect damage. An exception to this is the "Red Sokoto" or "Rousse de Maradi" goat skin producing the so-called Morocco leather of outstanding quality, well known in international trade (gloves, patent leather, suede). These skins are worth 20% of the animal's market value as compared to 10% for a sheep skin and 5% for a cattle skin. Wool is of little importance. The overall wool production of the Macina sheep is about 700 g h^{-1} yr^{-1} (Wilson 1984), mainly used for carpetry and blanket making. The Moor and Zaghawa hair sheep produce some 200 g $head^{-1}$ yr^{-1} per adult male, mainly used for tent making.

5.6.4 Dung and Manure

Unlike in East Africa, dung is not yet used as fuel in the Sahel; its main use is fertilization of cropland and occasionally as mortar (mixed with straw) for building. Agreements exist between pastoralists and agriculturists for the use of millet stubbles by stock. The terms of trade vary from one area to the next; but manuring by pastoral herds is highly valued by farmers in the southern Sahel. Manured village fields under continuous cropping of millet in Central Mali outyield unmanured, periodically fallowed fields by a factor of 3 to 5 (Wilson 1984).

Blood is usually not consumed by Sahel pastoralists, contrary to East African pastoralists.

5.6.5 Meat

As suggested above, meat is not the main production goal of Sahelian pastoralists; most of the meat comes from culled animals and is therefore of poor quality, with the exception of young male sheep and goats, including the "Homestead Sheep", the "Mouton de Case" nurtured for ritual and ceremonial purposes.

Dressing percentage in carcasses is about 45% of live weight in cattle and camels, 43% in sheep and 48% in goats. Carcass weights are around 215 kg for camels, 130 kg for cattle, 18 kg for sheep and 12 kg for goats. Selected animals, fattened for export to the Gulf (of Guinea) states might weigh 20% more (Wilson 1984).

Given the fact that the price of meat grows by a factor of 2.0 to 2.5 from the northern Sahel to the Guinean coast, as shown on the isoprice of meat maps in West Africa (Sarniguet et al. 1975), a number of small-scale attempts towards the stratification of the livestock industry were made in the 1970's. This stratification consists essentially of fattening-finishing operations using cattle raised in the Sahel. The fattening has been sometimes achieved either on sown pastures

(*Stylosanthes guyanensis, Panicum maximum, Brachiaria ruziziensis*) in the Sudanian-Guinean ecological zones (under strict health control), or, more often, in the Sahel itself and in the Sudanian zone in feedlots using agroindustrial by-products such as cotton seeds, rice bran, groundnut cake, molasses and sugarcane tops. These operations have usually been technically successful in various countries (Senegal, Mali, Burkina-Faso, Ivory Coast, Nigeria), since live weight gains of 400 to 1000 g $head^{-1}$ day^{-1} have been recorded. A limited number of commercial operations have been set up. However, these operations have suffered from sharp competition with imported meat from New Zealand, Australia and Argentina and have thus not made a significant development.

5.6.6 Milk

Milk is the main production goal of Sahelian pastoralists. It has been estimated that 75% of the cattle in tropical Africa are maintained for milk production; and the same applies to goats and camels as well (Wilson 1984). The amount of milk produced, however, is very low and there is an acute competition between the pastoralist and the calf. The share between the pastoralist and calf/kid is usually made on a 50% basis, on average; this proportion is, however, subject to variations with season, age of the calf/kid, etc. The average lactation length in cows is about 7 months with a total yield of 500–800 liters (2.4–3.8 l cow^{-1} day^{-1}), about half of which is levied by the pastoralist. This is one of the main reasons for the high mortality rates witnessed in young (Table 30).

In goats, lactation lasts 5–6 months with a total yield of 75–100 liters, i.e. 0.5–0.8 l $goat^{-1}$ day^{-1}; as there is 1.5 lactations per annum (3 lactations in 2 years) the total annual output is 110–150 liters per breeding goat per year.

Milk production by sheep is much less: 120–150 days of lactation yielding 40–50 kg i.e. some 0.6 kg ewe^{-1} day^{-1}; milking ewes is practised by only a few ethnic groups: e.g. Tuaregs, Kabbabish.

Camel lactation lasts about 12 months yielding 1500 to 2000 kg (4–6 l cow^{-1} day^{-1}) (Wilson 1983). Under good nutrition conditions yields may increase above 2500 kg yr^{-1} (Knoess 1977).

5.7 Disease, Parasitism, Predation, Theft (IEMVT 1971, 1977; Hall 1985; Perreau 1973)

The poor performance and high mortality of Sahel stock is due to the conjunction of poor nutrition with poor health conditions.

1. The main diseases are the following:
   - Bovine rinderpest, virus, notifiable disease;
   - Foot-and-mouth disease, virus, notifiable disease;

- Small-ruminants rinderpest, virus;
- African horse sickness, virus;
- Contagious bovine pleuropneumonia, mycoplasm, notifiable disease;
- Anthrax, bacterium, notifiable disease;
- Pasteurellosis, bacterium;
- Brucellosis, bacterium;
- *Clostridium* disease or black quarter disease, bacterium;
- Heart-water, rickettsia (tick-borne);
- Haemorragic septicaemia, bacterium;
- Streptotrichosis, bacterium;
- Sheep pox, virus; caprine pleuropneumonia.

2. Internal parasitism is caused by:

- Cestodes
  Toenia saginata, cysticercosis (measles) in cattle;
  Moniezia expansa, sheep toeniasis;
- Nematodes
  Respiratory and digestive strongylosis;
  Bovine ascaridiosis;
- Trematodes
  Distomatosis: *Fasciola hepatica* and *F. gigantea,* liver fluke;
  Shistozomiasis or bilharziosis: *Schistosoma* spp.;
- Protozoa
  Bovine coccidiosis;
  Piroplasmosis: theleriosis and babesiosis (tick-borne diseases);
  Trypanosomiasis, sleeping sickness, transmitted by biting flies, *Glossina* spp.: *Trypanosoma vivax, T. congolense.*

Disease outbreaks, particularly rinderpest and contagious bovine pleuropneumonia, have been curbed by large-scale vaccination campaigns in the 1960's and 1970's, which resulted in a steady increase in the cattle population (Tables 24–28). Unlike in East Africa tick-borne diseases are not of great significance, although some heart-water disease has been periodically reported. Internal parasitism is high and kept flourishing by permanent and temporary ponds in which animals "walk in" to drink and water is thus continuously reinfested by excrement. The main parasite is worms in the respiratory and digestive tracts and in the bloodstream; and protozoa in the bloodstream and digestive tract.

Predation is not significant, although small stock and calves may be taken by jackals and hyenas. Theft may be a problem around major boreholes where large numbers of animals are watered in great confusion (Bernus 1981).

5.8 Production Systems and Herd Management (ILCA 1978a, 1981; Dicko 1980–1985; Dicko and Sangaré 1984; Dicko and Sayers 1980; Wilson 1982b)

There are four main groups of animal production systems in the Sahel:

1. Sedentary agropastoral;
2. Short-range transhumant agropastoral;
3. Medium- and long-range transhumant and nomadic;
4. Agro-sylvo-pastoral systems.

There are many graduations within these systems and many transitions between them.

5.8.1 The Sedentary Agropastoral Systems of Central Mali

These systems are located either in the southern half of the Sahel, around irrigated areas, along main rivers such as Senegal, Niger, Chari, Niles or in flooded depressions. There are, of course, large differences according to whether one is dealing with rain-fed or irrigated cropping, or retreat-flooding cultivation.

Rain-fed farming is essentially millet cropping (*Pennisetum typhoides*) with, in addition, greater or smaller acreages of groundnuts (*Arachis hypogeia, Voandzeia subterranea*), cowpea (*Vigna sinensis*) and sorghum (*Sorghum bicolor*). Short cycle (90–100 days) "Souna" millet is farmed in manured fields around villages, whereas longer cycle (130–150 days) "Sanyo" millet is cultivated in outward, unmanured, fallowed fields. Further away, a third outer reach of shifting bush fields is devoted to groundnut, cowpea, voandzou (Bambara groundnut), sorrel (*Hibiscus sabdariffa*) and "Fonio" rings around villages with decreasing cropping intensity as one leaves the villages. The outer reaches beyond the cultivated zone are rangelands. However, there may be no rangelands, depending on the size of villages and the distance between them. Millet yields vary from 600 kg ha^{-1} yr^{-1} in the inner ring to 200 kg ha^{-1} yr^{-1} in the outer ring. Field size averages about 4 ha in the intense cultivation ring and 24 ha in bush fields in Central Mali, i.e. 0.6 and 3 ha per worker respectively. The "Sanyo" outer fields thus require some 78% of the labour force and the "Souna" 13%, the balance being used for groundnuts and minor crops, while "Fonio" fields occupy about 0.25 ha and are the responsibility of the women. The outer ring extends 5 km from the villages; the cultivated area per village is thus about 785 ha. Each family owns some 50 head of small stock (80% goats) and 0.5 head of cattle; as the size of households averages 12 persons, the livestock to human ratio is thus 0.75 TLU per person among the farmers.

In this system livestock is grazed on rangelands during the rainy season and stubbles and crop residues for most of the dry season. Work oxen represent about one-third of the cattle population, all males together composing around 47% of the herd. Calving percentage is some 60% and the calving interval 600 days.

Weight at calving is about 18 kg for males and 16 kg for females; average weight gain to 2 years is 160 g day^{-1}, i.e. a body weight of 140 kg at 2 years of age against a mature weight of 240 kg for females and 300 kg for males at the age of 5 years.

Weight loss during the dry season is 20% of live weight (50–60 kg in adult cattle); in immature animals about half the weight gained in the rainy season is lost during the dry season, hence, the overall poor performance recorded. The overall mortality is about 6% per annum (21% in calves) and the offtake is very low, around 7.5%.

During the rainy season herds and flocks are communally grazed at distances of 10–50 km from the villages and sometimes up to 100 km by waged shepherds (often Fulani). Males and females are kept together yearlong; therefore, calving occurs year-round with a strong peak at the end of the dry season (April-June), i.e. conception occurs for cattle mainly during the second half of the rainy season or early in the dry season when body condition is at its best.

While cattle constitute a kind of capital or savings deposit account, small stock represents the current account because of the rather high turnover in the latter. All females represent approximately 75% of the flock, while breeding females represent about 55%. Sheep are kept for meat home consumption and market sale, while goats are primarily kept for milk and secondarily for meat (young males and culled females). Sheep have litters of 1.05 and goats of 1.24 young; first parturition occurs around 470 days and the parturition interval thereafter is 260 days. The annual reproduction rate is 1.6 kids and 1.4 lambs per breeding female. Preweaning mortality is 39% for goats and 35% for sheep. Mortality is very significantly affected by the type of birth (single or multiple) and by the parturition interval.

Live weight loss can obviously be avoided by good management, i.e. light stocking rates and rotational grazing; net gains of 80 kg TLU^{-1} yr^{-1} may thus be achieved in cattle, i.e. adult age is then reached at 3 years instead of 5, as accomplished at the Ekrafan ranch (300 mm) in NW Republic of Niger. Mortality loss could also be drastically curbed by appropriate prophylaxis and care.

5.8.2 The Short-Range Transhuman Agropastoral Systems (Barral 1974, 1977; Milleville 1980; Barry et al. 1983; Grouzis 1984; Lhoste 1986; Dicko and Sangaré 1984; Dicko and Sayers 1988; Kieckens 1984)

This type of husbandry is exemplified by the "endodromic" system of the inner part of the Big Bend of the Niger River, the Liptako, Oudalan and Gourma districts of the Rep. of Niger, Mali and Burkina-Faso. The system includes shifting and fallowed millet cultivation associated with livestock husbandry. Herds and flocks follow yearlong circuit itineraries between temporary ponds in the rainy season and permanent ponds in the dry season. The annual circuit takes place over distances of 50–200 km within an area of 20000 km^2, harbouring, yearlong, about 180000 TLU of cattle (75%), small stock and a few camels. This represents an average density rate of 9 ha TLU^{-1} with extreme figures of 3 and 12 ha TLU^{-1}. The human population is 70000 inhabitants: 52000 nomads (34000 Tuaregs; 18000 Fulani; and 18000 settled Songhai and Rimaibé) with livestock to human ratios of 3.0 to 3.5 TLU per person, for the nomads, with extreme values of 1 TLU for 1.5 inhabitants and 4 TLU/inhabitant. Production parameters and productivity in this system are not substantially different from those analyzed above for the millet agropastoral system.

5.8.3 Long-Range Pastoral Systems (Bernus 1981; Wilson 1982b; Le Houérou 1985a)

These systems differ from the agropastoral systems described above inasmuch as livestock production constitutes the major, if not the only, source of livelihood; some more or less occasional cultivation of millet may or may not occur. These systems are very diverse throughout the Sahel, according to the major type of stock which is owned: camels (Reguibat Moors, northern Tuaregs, Tubbus, Zaghawa, Kabbabish) or cattle (southern Tuaregs, southern Moors, Fulani, Baqqara); both categories having various proportions of small stock. One may distinguish (1) very long-range nomads, who move over distances of 1000 km or more, such as the Reguibat Moors, some Saharan Tuaregs who may visit the northern Sahel, the Kabbabish, from (2) the medium- to short-range transhumant or nomadic pastoralists, who move over distances of 150–300 km, such as the Fulani of the Inland Delta of the Niger River in Mali, the Illabaken Tuareg of central-northern Rep. of Niger, the Woodabe (Bororo) of Niger, Nigeria and Cameroons, or the Baqqars of Darfur and Kordofan, in the Rep. of Sudan. The movements are governed by the seasons; the dry season is spent in the northern part of the Sahel where temporary ponds are available, whilst the dry season sees the return to the Sudano-Sahelian and Sudanian zones to the south where permanent water is available, particularly along permanent rivers: Senegal, Niger, Inland Delta and lakes of the mid-Niger Valley, Lake Chad, Chari, Logone, Bahr el Azum, Bahr el Arab, Bahr el Ghazal, etc. The northward rainy season move is also justified by various health-care considerations such as the "salt cure" where stock is taken to areas of saline land or saline springs in order to balance their annual mineral supply. Most pastoralists consider this "salt cure" necessary for the health of their stock (Bernus 1981; Leprun 1978b). The respective roles of major and trace elements in the salt cure have not been investigated to our knowledge. Some browse species such as the *Capparidaceae* and *Salvadora persica* containing 25–35% of non-silica minerals may play a similar role (Le Houérou 1980b, c). Transhumance also permits herds and herders to escape parasitism and harassment from flies and mosquitoes that plague the Sudanian lowlands in the rainy season.

Herd and flock structure production parameters and productivity were discussed in Sections 5.4–5.6 and shown in Table 30. The importance of the nomadic livestock production in the Sahel is probably over 50% of the total of the overall livestock sector, the balance coming from sedentary agropastoral and short-range transhumant systems (Wilson 1982).

Among the Delta Fulani herds of Central Mali, studied by ILCA, herd size is of the order of 50–500 head and stock performances are 20–30% higher than in the two other types of systems: 65% calving rate, growth rate to 12 months of 180 g head^{-1} day^{-1}, calf mortality 28%, milk offtake 0.1 1 head^{-1} day^{-1}, total milk production being 235 kg/11.3 months (1.44 1 cow^{-1} day^{-1}).

5.8.4 Agro Sylvo-Pastoral Systems (Figs. 58, 59)

Two very original agro-sylvo-pastoral systems occur in the Sahel:

1. *The Acacia Senegal fallow-cropping system* of Kordofan and Darfur in the Republic of Sudan between the isohyets of 300 and 800 mm of annual rainfall

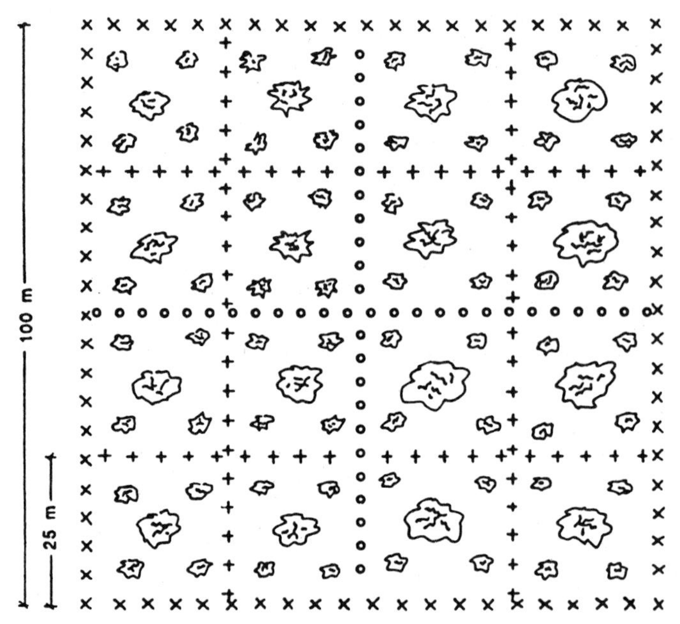

Fig. 58. Sketch of an agro-forestry model for the Peanut Basin of Senegal (main crops peanut and millet associated with livestock production, draught power and fuel wood production) (Le Houérou 1987a)

× × × × × × Primary windbreak: $4 < h < 8$ m
Acacia senegal or *Faidherbia albida* or *Prosopis juliflora*

o o o o o o o Secondary windbreak: $2 < h < 4$ m:
Acacia holosericea

+ + + + + + Tertiary windbreak: $1 < h < 1.5$ m:
Combretum aculeatum + *Ziziphus mauritiana,* Sahelian Zone
Bauhinia rufescens + *Ziziphus mauritiana,* Sahelo-Sudanian Zone
Feretia apodanthera + *Ziziphus mauritiana,* Sudanian Zone

Groves of: *Acadia tortilis* 10 • 10m:Sahelian Zone
Faidherbia albida 10 • 10m: Sudano-Sahelian and Sudanian Zones
16 × 5 = 80ha^{-1}

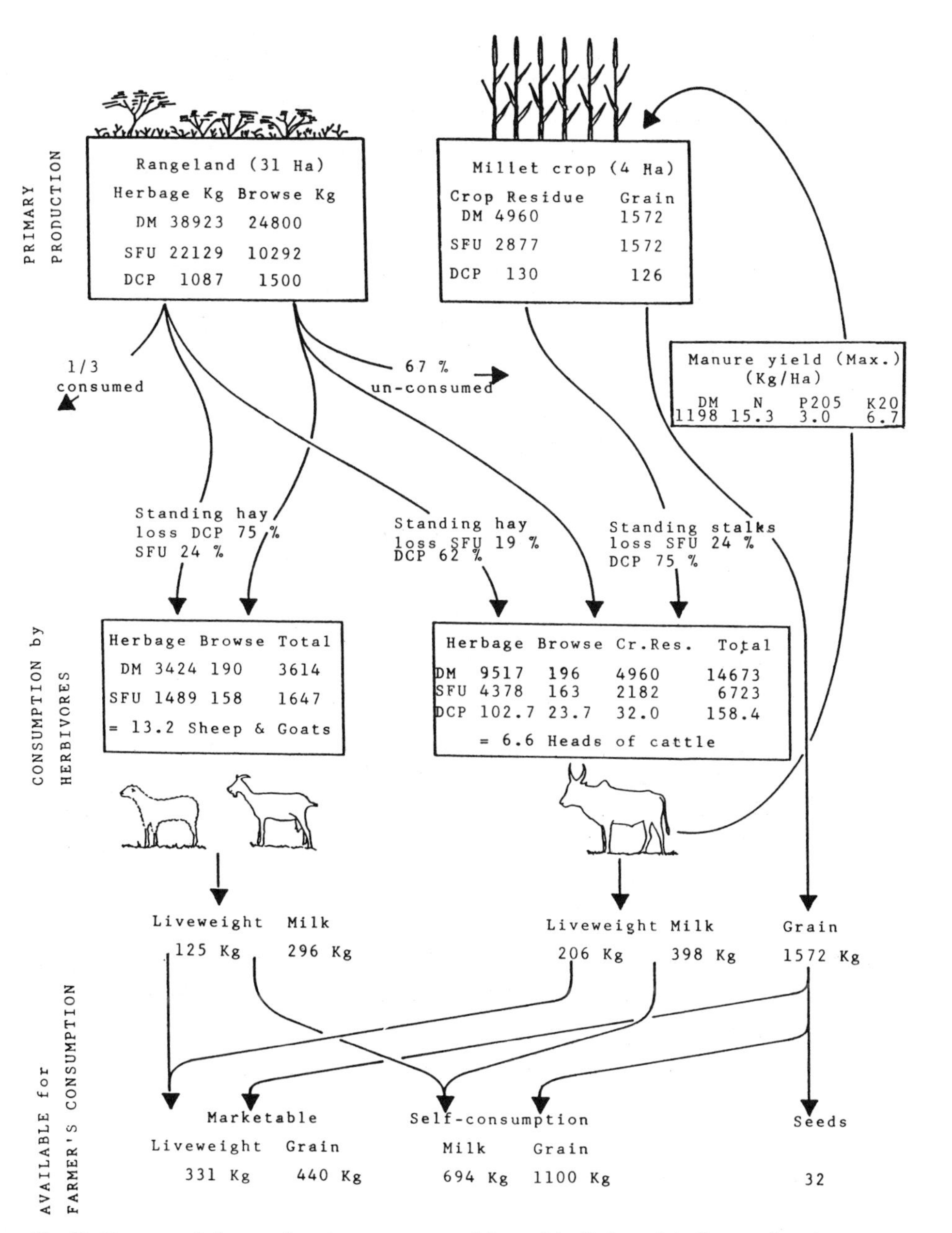

Fig. 59. Energy and nitrogen flows in an agropastoral farm of the Sudano-Sahelian ecoclimatic zone of northern Burkina-Faso (After Kiekens 1984)

(Kassas 1968, 1970; Seif el Din 1965). This system is responsible for large proportions of the production of gum arabic in the Rep. of Sudan (44 000 metric tons annually worth some 40 million US $). The whole cycle lasts 20 to 30 years: 4–10 years of cropping (millet and sorghum), 8–10 years of fallow and colonization by *A. senegal* and 6–10 years of gum arabic production.

2. *The Faidherbia albida (= Acacia albida) agro-sylvo-pastoral system* is found in the southern half of the Sahel and in the northern Sudanian zone between the isohyets of 500–1000 mm of annual rainfall. It combines the cultivation of millet (*Pennisetum typhoides*) and groundnuts (*Arachis hypogeia*) with orchards of *Faidherbia* (10–50 trees ha^{-1}) which allows a millet production 2.0 to 2.5 times higher than in unfertilized open-land millet fields in the same areas, due to the recycling of geobiogene elements (Charreau and Vidal 1965; Giffard 1964, 1971, 1972; Radwanski and Wickens 1967; Wickens 1969; Dancette and Poulain 1969; Jung 1970; Felker 1978; CTFT 1988). This system permits permanent cropping with rather high yields by local standards and an additional one to two head of small stock per hectare or one TLU for 5–10 ha, i.e. equivalent to ranges in good condition. The system is thus about three times more productive than open-land millet cultivation, besides bringing year-round cash flow to the farmer/husbandman (Montgolfier-Kouevi and Le Houérou 1980), producing food crops, meat, milk, wood and fencing material. This system has been one of the bases of old agrarian civilizations which have been in existence for centuries both in West and East Africa (Rift Valley of Ethiopia) permitting high densities of population: 30–70 persons km^{-2} (Pelissier 1967; Le Houérou 1978, 1980d).

5.9 Nutrition, Plant-Animal Relationships, Grazing Patterns, Grazing and Herding Management Stock Behaviour, Stocking Rates and Carrying Capacities (Dicko 1980, 1983, 1984, 1985; Dicko and Sangaré 1988; Dicko and Sayers 1988)

From a detailed study that lasted 4 years in Central Mali, the following conclusions can be made:

1. *Grazing time* varied from 6 h day^{-1} at the peak of the rainy season to 10 h day^{-1} at the end of the dry season; the weighted average being 8 h day^{-1}. Night grazing took place particularly during the second half of the dry season when mid-day temperatures are very high (40°C and above). *Daily water consumption* in cattle varied from 25 kg day^{-1} in January to over 40 kg in May, i.e. 10–16% of live weight. Walking distance from water may vary from 5–10 km

in the rainy season to 18–20 km in the dry season. The watering interval may vary from once a day in the rainy season to once every second or third day in the dry season; but only mature, dry animals may be watered every third day in the second half of the dry season.

2. *Browsing* varied largely depending on species, opportunities and season. It may vary from 4 to 35% of the grazing time for cattle, but averages 34% for sheep and 87% for goats. In cattle it may be 5% in the dry season and up to 45% in the second half of the dry season, averaging 25% over the dry season as a whole in the southern Sahel of Senegal (Blancou et al. 1977). Similar figures were found in Niger with Fulani herds (Granier 1977) and Tuareg herds (Wilson 1982b).

3. *Digestibility* of herbage measured by five different methods averaged 54% DM year-round, varying from 66–72% in August-September to 44–52% in April-June; digestibility of browse is 45–55% DM.

4. *Daily intakes* averaged 2.6 kg DM per 100 kg live wt. (1.8–3.8%), with daily fecal output of DM of 1.23% live wt. Apparent intake, i.e. the difference between herbage available and left over was 150–200% of actual intake: 10–20 kg DM TLU^{-1} day^{-1}.

5. *Live-weight change* varied from –520 g $head^{-1}$ day^{-1} in November to + 1 150 g $head^{-1}$ day^{-1} in July during the compensatory gain period. The average loss throughout the dry season was 20% of body weight as from the end of the previous rainy season. Live-weight change is highly correlated with the amount of digestible protein in the diet ($r = 0.89$) and, to a lesser extent, to the amount of dry matter ingested ($r = 0.68$).

6. *Behaviour: Grazing and walking* activities require an amount of energy equal to 42% of the minimum maintenance needs:

   Time spent in various activities for cattle were:

Walking	2–4 h day^{-1}
Resting	6–10
Ruminating	4–6
Grazing	6–10

7. *Grazing patterns* depend essentially on water availability; areas with temporary ponds are used in the rainy season; while dry season grazing occurs within a radius of 20 km from permanent water. Camels do not depend on water availability in the rainy season but must be watered once to twice a week in the dry season; they may thus travel very far from any water source in the rainy season and up to 80–100 km in the dry season. Small stock behave like cattle or are a little less dependent on free water. From the vegetation dynamics and range condition viewpoints, areas farther than 15 km from permanent water are usually in fair to good condition.

8. *Stocking rates* vary with range types naturally, but also to a large extent with distance from water; stocking rates may reach 3 ha TLU^{-1} yearlong some 10 km

around permanent water sources (and up to 3 TLU ha^{-1} yr^{-1} in flooded "bourgou" pastures (*Echinochloa stagnina*). In areas further than 15–20 km away from permanent water, stocking rates are of the order of 10–15 ha TLU^{-1} yr^{-1}, which approximates the carrying capacity. High stocking rates inevitably lead to pasture deterioration and desertization, and therefore to carrying capacities decreasing far below the initial potential under pristine conditions (one-third to one-fifth). The case has been particularly well documented around recent boreholes (Bernus 1971, 1981) gathering 15 000 to 25 000 TLU's in the dry season (i.e. 12.5 and 5.0 ha TLU^{-1} for 8–9 months respectively, that is two to four times that which had been initially, correctly, planned).

6 Development Outlook; Constraints and Limits

6.1 Range Improvement and Development

6.1.1 Range Seeding

Many experiments attempted in various areas of the Sahel have proved that reseeding of depleted pastures is not feasible with the presently available technology and the commercially available seed pool below the 600-mm isohyet of annual rainfall (Le Houérou 1977, 1979b; Wilson 1982b). Of the 80 herbaceous tropical arid zone species of forage which have been tried at Niono (550 mm) in Mali in 1977–1980, not one single species became established and amenable to produce a grazing impact. Only a few perennial and annual legumes survived the first dry season in small numbers:

Alysicarpus ovalifolius
Alysicarpus vaginatus
Macroptilium atropurpureum (Siratro)
Macroptilium lathyroïdes
Stylosanthes fruticosa
Stylosanthes hamata
Stylosanthes subsericea
Stylosanthes viscosa

The main reasons for the failure of introduced (and local) species in reseeding programmes are:

1. Irregular and unpredictable rainfall in space and time;
2. Very long and severe dry season with very high evapotranspiration rates (among the highest on earth);
3. Strong competition from annual grasses and forbs.

No long-term success has been recorded either in establishing browse species below the 400-mm isohyet of annual rainfall, even with species belonging to the native flora, in spite of some encouraging results with

Bauhinia rufescens
Combretum aculeatum
Piliostigma reticulatum
Ziziphus mauritiana

One noticeable exception to this is the plantation of *Acacia tortilis* and *Acacia senegal* at M'Bidi in Senegal under about 400 mm of the long-term mean rainfall (Hamel 1980; Le Houérou 1979b, 1980b, 1987a; Von Maydell 1983). We do not mean that reseeding or planting of browse species is impossible in the Sahel; it is, however, very difficult and this difficulty should not be underestimated by Sahelian neophytes.

6.1.2 Range Management

The main constraints to improved range management practices in the Sahel are of a various nature: sociopolitical, economic and technical.

1. *Sociopolitical Constraints.*
   These are due to the fact that land and water are communally owned, whereas the stock is privately owned. Such a situation inevitably leads to a "looting strategy" (Le Houérou 1976b) or "the tragedy of the commons" (Hardin 1968), whereby it is in every individual's interest to take immediate and full advantage of whatever resource is available irrespective to what may happen in the long term and on the whole. There is no responsibility for resource management and therefore no possible private investment of any sort; furthermore, public investment is jeopardized by the generalized lack of responsibility. Until this sociopolitical situation is cleared there does not seem to exist any hope for any improvement of the Sahel grassland in the foreseeable future.

2. *Economic Constraints*
   The producer is not being given a fair price for his animal products. Most of the profit resulting from the 2.0–2.5-fold difference in meat prices between the Sahel and the Coast States to the south is taken by stock traders, with whom the real power remains. All attempts to break this monopolistic situation have failed so far, regardless of political regime

3. *Technical Constraints*

a) Fencing. Supposing it would be socially acceptable, fencing is not economically feasible: 1 km of simple, five-barbed wire fences would cost in excess of 3000 US $, while the market value of livestock is less than 1 US $ kg^{-1}. Given these terms of trade, it would take 5 to 10 years of production to offset the cost of fencing each hectare, supposing paddock sizes of 100 to 1000 ha, and supposing also that fencing would raise the productivity of the range by 50% above the traditional system of free-ranging husbandry.
b) Firebreaks are necessary in ranching or in settled operations. The cost in 1980 was about 40–50 US $ km^{-1} for a 6-m-wide firebreak using draught power, and a little less with mechanical power. That is a cost of 0.6–0.8 US $ ha^{-1} against a gross production of 4–6 US $ ha^{-1}, i.e. about 15% of the gross output; firebreaking is thus feasible and viable (besides being necessary).
c) Rotational and deferred grazing have been successfully practiced in various state-owned ranches in virtually all Sahelian countries. But these have inevitably been financial disasters because of the high cost of fencing and water development, with respect to the market value of meat, since there is no quality pricing.
d) Water development. Borehole drilling costed 100 to 200 US $ m^{-1} and an additional 50 000 to 100 000 US $ for the equipment (engine, pump, tank, trough).

6.2 Development Outlook

6.2.1 Range Development

In the Sahel proper, under typical range conditions with little or no farming, range development is technically feasible using the classical tools of range management, i.e.:

1. Adjustment of stocking rate to long-term carrying capacity;
2. Applying deferred and rest-rotation grazing management;
3. Establishing some type of economical fencing, firebreak networks and avoiding trespassing of alien herds and flocks;
4. Water development ensuring a permanent supply throughout the dry season;
5. Application of sensible livestock management strategies and techniques with regards to prophylaxis, reproduction, weaning, culling, supplementary feeding and stratification, including fattening and finishing operations.

A number of small-scale development projects have been successfully carried out using these simple techniques over limited surfaces of a few 100- to a few 1000-ha ranches.

One of the latest such projects was the West Germany–Senegal project in the Ferlo of northern Senegal, near the borehole of Viendou-Thiengoly (P = 350 mm) (Tluczykont and Gningue 1986). In that particular project, which lasted almost 10 years, some 14 000 ha natural rangeland were cut off the traditional pastoral management system, fenced and stocked with a stocking rate of 12 ha TLU^{-1}. The stocking applied resulted from preliminary grazing trials. This stocking rate was only 60% of the current rate in surrounding traditional systems. The stock used was the usual local white Fulani zebu of the "Gobra" breed plus an equal number of local small ruminants. Locally made metallic fences of the Australian "Cyclone" type divided 14 × 1000-ha paddocks, each provided with permanent water piped from the nearby borehole of Viendou-Thiengoly. Each paddock was allocated to two families owning 35 cattle and 35 small stock females. The capital investment cost was reimbursed by the recipient families over a 10-year period, with no interest charges, the running cost being entirely borne by the beneficiaries.

Cattle and small stock were managed in a rational way including vaccination, eradication of parasitism, appropriate weaning, destocking and culling practices and supplementary feeding as appropriate.

The net result was a steady improvement of the range both in the quantity and the quality of forage available, including browse and the regeneration of the herbaceous and woody layers, in spite of two extremely dry seasons in 1983 and 1984.

The overall productivity per animal unit and per unit of surface area were more than doubled as compared to the local traditional herds under otherwise identical environmental conditions. The increase of productivity came essentially from better feeding and care which resulted in:

1. Sharp increase in breeding rate (75 vs 50% in cattle);
2. Considerable decrease in mortality, particularly in the young (2 vs 25%);
3. Higher percentage of calves weaned per breeding cow (62 vs 45%);

4. Shorter calving intervals (350 vs 560 days);
5. Earlier weaning (9 vs 12 months);
6. Heavier weaners (150 vs 90 kg).

Overall the system was found economically feasible; there was actually a pressing demand from neighbouring pastoralists to extend the system with a quite significant financial participation of their own.

The main constraint in the system was that the destocked calves and culled animals, instead of being sold to fattening operations, were transferred to the traditional herds, thus aggravating stock pressure on the range. The project thus did not fulfill its overall goal of increasing productivity of both rangeland and livestock. There is thus the general problem of transferring small-scale results to regional development. This problem has not found any viable solution to date.

It is a fact that such projects, although technically successful, can hardly be extended to substantial geographical areas. This failure in extension comes much more from a lack of managerial know-how than from the want of appropriate scientific and technical knowledge. There have been many successful small-scale projects in the past, but none of them, to my knowledge, gave way to any large-scale project that could be regarded as successful according to economic criteria, neither in the Sahel nor in East Africa.

The main constraints are thus of an organizational, administrative and legal nature, including:

1. Land tenure and ownership;
2. Credit and marketing;
3. Effective extension services;
4. Price policies.

6.2.2 Development of Livestock Production (SODESP 1980, 1984)

A large-scale project of a different nature from the above mentioned one was developed in the Ferlo of northern Senegal for some 12 years (1975–1987), aimed at improving livestock production without any particular action geared to rangeland. The project was based on health control, supplementary feeding, calf husbandry, destocking, postweaning husbandry and fattening operations (both commercial and at the small farm level in the "Peanut Basin").

Stock performance was thus increased by 60 to 80% with respect to traditional, local Fulani husbandry; but the productivity per surface area of grazing land was not improved.

The project encompassed 1300 pastoralists husbanding 30 000 cow-calves and 9000 ewe-lambs. In spite of encouraging results from the purely zootechnical viewpoint, the project collapsed due to organizational and managerial reasons after some 12 years of existence and an annual expenditure of 10 million US $ over the last 5 years.

Cattle fattening and finishing schemes were developed in several Sahelian and Sudanian countries of West Africa (Senegal, Mali, Nigeria, Ivory Coast, etc.)

in the 1970's and the early 1980's. These were based on health control and pen feeding on agroindustrial by-products such as molasses, rice and wheat bran, cotton seeds and cake, peanut cake, etc. and crop residues such as maize stalks and sugarcane tops. These were usually semi-large-scale operations handling hundreds or thousands of animals per annum.

Although these operations were technically successful and economically sound, with daily live weight gains of 500–1000 g in young stock they were all discontinued after various lengths of periods of functioning (5–10 years and sometimes less). No large-scale, lasting commercial operation ensued.

In fact, about 1 billion US $ have been spent in range-livestock development in the African dry lands over the last 30 years, with virtually no positive large-scale development outcome.

6.2.3 Agropastoralism (Figs. 58, 59)

In the Sudano-Sahelian and northern Sudanian ecoclimatic zones, where farming and animal husbandry compete, the development prospects look less gloomy, when based on mixed farming including the introduction of annual and short-lived perennial fodder species as sown pastures in the crop sequence. This has been shown to be technically feasible in the Sahelo-Sudanian and Sudanian zones, but not in the Sahel sensu stricto (FAO-CILLS 1980–86; Milleville 1980; Kiekens 1984).

The main succesful species used in multilocal trials throughout the Sudano-Sahelian and Sudanian zones of Mali, Niger and Burkina-Faso were: *Andropoyon gayanus* var. *bisquamulatus, Pennisetum pedicellatum, Cenchrus ciliaris* cvs *Biloela* and *Gayndah, Alysicarpus glumaceus, Stylosanthes hamata* cv *verano, Macroptilium atropurpureum* (siratro), *Macroptilium lathyroides, Lablab purpureus* cvs *Highworth* and *Rongai, Cajanus cajan, Vigna unguiculata* (cowpea = niébé).

The development scope of the introduction of these fodder plants is four-fold:

1. To replace the traditional fallow by a fodder legume which will enrich the soil in nitrogen.
2. Ensure the integration of crop and livestock husbandry in order to increase animal production including draught power via improved feeding.
3. Increase food production via higher soil fertility resulting from manuring.
4. Combat soil erosion in various ways including contour benches sown to *Andropogon gayanus* along field edges.

It has been shown that these species are able to produce 200 to 10 000 kg DM ha^{-1} yr^{-1} depending on sites and soil characteristics. The legume species may be cultivated either in pure stands or associated with Millet (*Pennisetum glaucum = P. typhoides*) or sorghum (*Sorghum bicolor*).

The main constraints to the extension of these techniques are:

1. Fencing of forage fields in order to prevent livestock vagaries and thus to ensure successful cropping.

2. Local seed production in order to ensure the perpetuation of the crop and free it from the import of costly foreign seed material.
3. Low efficiency of the extension services.

The annual energy flow, in an average Sudano-Sahelian agropastoral farm of 35 ha (31 ha grazing land and 4 ha millet crop), sustaining a typical family of 5.5 persons in Northern Burkina-Faso (450 mm) under optimal management conditions, is shown in Fig. 59 (Milleville 1980; Kiekens 1984).

The total net annual production of such a farm is:

1. 694 kg milk, used for self-consumption;
2. 1572 kg millet grain, of which 32 kg are kept as seeds, 1100 kg are consumed on the farm and 400 kg are marketed.
3. 331 kg meat, partly marketed, partly self-consumed.

That is a production, at the human consumption level, of 6.33 million cal yr^{-1} or 181×10^3 cal ha^{-1} yr^{-1}, a global consumption of 4.78×10^6 cal $farm^{-1}$ yr^{-1} and a daily consumption of 2250 cal $person^{-1}$, i.e. about equal to the nutritional needs of the family. Protein production is 224 kg or 6.4 kg ha^{-1} yr^{-1}; consumption is 157 kg or 78 g $person^{-1}$ day^{-1}, 30% of which is animal protein.

In other words, under optimal management conditions, the South-Sahelian agropastoralist can barely meet his nutritional needs, with a slight energy deficit and a slight protein excess. Again, in the Sudano-Sahelian and northern Sudanian ecoclimatic zones, although agropastoral development has been shown to be technically feasible and economically viable, in principle (overall production could be trebled under appropriate management), little lasting large-scale development action actually occurred for the same reasons as in the Sahel proper, that is:

1. Lack of managerial know-how;
2. Unsolved organizational and administrative problems;
3. Inefficiency of extension service.

6.2.4 Agroforestry

Agroforestry has been successfully practiced in several Sahel and Sudan zones in the past, particularly the use of *Acacia senegal* in a "bush-fallow" rotation and the millet *Faidherbia albida*-millet system, which is common as well in West and in East Africa.

But, unfortunately, most of these systems are on the decline due to excessive human pressure on the land. In the Peanut Basin of Senegal, for instance, with a rural population density of 60 inhabitants km^{-2}, under rainfalls of 400–700 mm, there is a continuing decline in soil fertility and the *Faidherbia albida* groves are being progressively cleared (Le Houérou 1987a) (Fig. 58).

The recommended action is an increased integration of agroforestry and animal production in order to increase draught power and restore fertility without reducing the area cropped. This includes the plating of forage hedges of well-adapted local shrub species in a network of windbreaks along field edges, together with the development of the traditional *Faidherbia albida/Acacia tortilis* groves, additional hedges of fast-growing fuel species (*Acacia holosericea*) and service wood-producing species (*Eucalyptus camaldulensis*) and fruit trees (mango, guava, citrus) which complemented the scheme and made it attractive to farmers by meeting their most acute needs and wishes.

The models proposed (Fig. 58) were composed of: *Acacia senegal* or *Prosopis juliflora*, as primary windbreaks; *Acacia holosericea*, as secondary fuel-producing windbreaks, and tertiary windbreaks composed of fodder hedges of *Combretum aculeatum* and *Bauhinia rufescens* mixed with *Ziziphus mauritiana* (the spiny *Ziziphus* serving to protect the two other species from overbrowsing). A sketch of this basic model is shown in Fig. 58. These species are easy to establish at a very low cost using the technique of pregerminated seeds planted when the rainy season is well established with an actual water storage of 50 mm or more in the soil.

This production model would result in a stabilization of soil and the suppression of wind and water erosion over a period of 2–3 years and in the steady increase in soil fertility due to the manuring, hence in grain and animal production that could increase more than three-fold the present overall net productivity in less than 10 years, with an additional net income of 200 to 500 US $\$ ha^{-1}$, depending on management and local climatic and edaphic conditions.

Such models, inspired from actual real-life conditions of geographically limited extension (for ethnic reasons) have not been popularized for the same reasons mentioned above.

6.2.5 Livestock Health and Water Development

The only persistent and large-scale livestock development actions which took place in the Sahel over the past 40 to 50 years are concerned with only two types of activities: animal health and borehole development.

Large-scale vaccination campaigns have been launched in all Sahel countries since the 1950's; veterinary services have been set up to carry out large-scale prophylactic activities. As a result, the main epizootic diseases, particularly rinderpest and bovine peripneumonia, have been curbed; parasitism has also been reduced, but to a much lesser extent. These actions have been quite successful in the Sahel; the net result has been a 2.5-fold increase in livestock between 1950 and 1986.

As range production declined and stock increased, the final consequence of health development schemes has been an acute shortage of feed on the range, hence, an ever-faster depletion of the rangelands leading to desertization, starvation and heavy losses of stock in drought years. The main constraint to livestock development, which used to be animal health during the first half of the present century, has shifted to poor range conditions and feed shortages during the second half of this century. Paradoxically, the cure has been worse than the disease in the long run.

The second development action usually taken by governments has been water development, in particular the drilling of boreholes for discharging large quantities of water. As no range management policy has been enforced at the same time, the result has been a large concentration of animals during the dry, and even the rainy season; 20 000 to 40 000 head of stock have been routinely counted around such boreholes. Naturally, what one could expect under such circumstances took place: rangelands were destroyed over a radius of 15–20 km around those boreholes. It is quite clear that during the very severe droughts of 1971–73 and 1983–85 most animals starved to death despite large supplies of water existing nearby.

7 Monitoring: A Case Study, the Ferlo Region of Northern Senegal

7.1 Introduction

A specialized project was established in Senegal in 1980, following an agreement between the government of this country, on the one hand, and FAO, UNEP, with the collaboration of NASA, on the other hand. The scope of the project was to test, on a real scale in a typical Sahel zone, the practical feasibility of a large-scale rangeland monitoring project that would integrate various techniques of remote sensing with ground control, i.e. orbital, aerial and ground remote sensing.

The project lasted from 1980 through 1985. It produced very interesting results that have been hailed as a scientific breakthrough by the world scientific community, giving way, as it did, to a large number of publications (Tucker et al. 1983, 1985a, b; Justice 1986).

7.2 The Pilot Zone of the Ferlo

7.2.1 Geographical Zoning and Administrative Setup

The pilot area may be subdivided into three to four subzones corresponding to the local Fulani terminology (Santoir 1977; Barral 1982).

1. The Senegal River Valley or "Walo"
2. The seasonal dry season transhumant zone along the river valley, locally known as the "Dieri" in Fulani language, which is a strip some 50 km wide along the river. Dieri, in turn, is subdivided into lower Dieri or "Diedegol" at a distance of 0–25 km from the valley and "upper Dieri" located some 25–50 km from the valley. The upper Dieri used to be grazed during the early dry season (Nov.-Jan.) and the lower Dieri for the second half of the dry season (Feb.-June).
3. The Koya or "sandy Ferlo" in the central and western part of the project area is a traditional zone of rainy season transhumance.
4. The "ferrugineous Ferlo" to the SE is a traditional dry season grazing land because of the permanent water supply in wells dug in the Ferlo valley and some of its tributaries.

The region is divided, for administrative purposes, into two regions and six departments.

River Region:	Louga Region:
Dagana Dept.	Linguère Dept.
Podor Dept.	Kebewem Dept.
Matam Dept.	Louga Dept.

7.2.2 Climate

7.2.2.1 Rainfall

Mean annual rainfall for the 1920–1969 period varied from 320 mm to the north[5] (Richard-Toll 330, Dagana 318, Podor 318) up to 520 mm to the south (Linguère 517, Matam 520, Dahra 526). See Tables 33–34 and Figs. 13, 16, 20–28, 36–38.

Table 33. Variability of annual rainfall in the Ferlo (Barral et al. 1983)

Periods	1920–1969			1970–1981			1920–1981		
Stations	No. of years	x̄ (mm)	C.V.[a] (%)	No. of years	x̄	C.V.	No. of years	x̄	C.V.
Podor	44	318	39	12	191	33	56	287	42
Dagana	48	318	35	12	218	34	60	298	37
Matam	42	520	29	11	302	42	53	474	35
Linguére	36	517	24	12	339	18	48	473	29
Dahra	33	526	24	12	342	35	45	477	30

[a]C.V. = Coefficient of variation.

Table 34. Rainfall probabilities in the Ferlo (Le Houérou 1986)

Mean annual rainfall (mm)	Probabilities 0.1	0.2	0.3	0.4	0.5	0.6	0.7	0.8	0.9
300	430	385	350	315	287	260	232	205	170
400	570	500	460	420	390	360	325	290	250
500	665	610	567	525	492	460	422	385	335
600	780	720	675	630	595	560	520	480	420

[5]Dagana, 1920–69; Podor, 1923–69; Matam, 1923–69; Dahra, 1934–69; Linguère, 1934–69.

However, rainfall in the years from 1970 to 1984 was consistently less than 60% of these amounts. Over this period of 15 years not one single year attained the 1920–69 mean. There were three extremely dry years (1972, 1983 and 1984) in which rainfall was only 30 to 40% of the 1920–69 mean. At individual stations in some of these years, the amount of annual rain was only 10 to 20% of the 1920–69 mean: for example, 33 mm at Fété-Olé in 1972, 79 mm at Mbeuleukhé in 1983, 84 mm at Mbidi in 1984 and 77 mm at Mbar Toulab in 1984. For 6 years (1970, 1971, 1973, 1977, 1980, 1982) in this same period of 15 years, the mean varied from 40 to 60% of the 1920–69 average, for 5 years (1974, 1975, 1976, 1978, 1979) the mean was between 60 and 80% of the 1920–69 figures and only 1 year (1981) was approximately 90% of that value.

Variability in annual rainfall is high and increases with aridity (Tables 33–34, and Figs. 33–38). The annual median is slightly below the mean and this difference also increases with aridity from 2% at 500 mm to 6% at 300 mm, e.g.

Mean (mm)	Median (mm)
300	282
400	390
500	490

The rate of reliability of annual rains [ratio of 0.8 probability to the mean: $f(0,8)/\bar{X}$] is 0.68 ± 0.02 between the 300- and 400-mm isohyets and 0.77 ± 0.03 between the 400- and 500-mm isohyets (Le Houérou 1986). In other words, in 4 years of 5, rainfall will reach 205 mm or more at the 300-mm isohyet and 385 mm or more at the 500-mm isohyet (Dancette 1979).

Rainfall probabilities for a given mean are shown in Table 34 (Le Houérou 1986) and Figs 33–35 (Rodier 1982).

7.2.2.2 Evaporation and Evapotranspiration

Class A pan evaporation shows little variation throughout the area: 3550 mm yr^{-1} at Podor, 3360 mm yr^{-1} at Linguère. PET (Penman) reaches 1724 mm yr^{-1} at St. Louis, 1859 mm at Podor, 2116 mm at Guédé, 1690 mm at Linguère and 1675 mm at Matam (Frère and Popov 1984). The mean monthly values of 0.35 PET are shown in Figs. 13, 16, 20–28. These figures show that there is little difference in the length of dry/rainy seasons regardless of whether one selects the threshold criterion of $P = 2t$ or $P = 0.35$ PET (Le Houérou and Popov 1981). The rationale behind the selection of the 0.35 PET threshold has been discussed by Le Houérou and Popov (1981). It stems from the fact that evaporation from bare, unsaturated ground is usually between 0.3 and 0.5 PET. Whatever amount of rain exists above 0.35 PET is available for plant growth. This value is therefore a good criterion for the discrimination of rest and growing season. According to this criterion, the growing season lasts from an average of 75–80 days in the north to 100–110 days in the south (Figs. 13, 16, 20–28).

7.2.2.3 Air Humidity

The influence of the Atlantic Ocean is scarcely felt beyond 30 km from the shore, to the east of a line joining Richard-Toll – Louga – Thiès (Giffard 1974; Bille 1977). However, the mean annual air humidity, 45% at Podor and 49% at Linguère, is clearly higher than the mean for the Sahelian zone of 35% (Le Houérou 1980f).

The mean minimal monthly values are usually below 20% and never reach 50%. The mean monthly maxima vary from 25 to 50% in the dry season and 50 to 90% in the rainy season.

7.2.2.4 Temperature

The mean annual maximum temperature varies from 35° to 37°C and the mean annual minimum from 19° to 21°C. The mean maximum of the hottest month may reach 41–42°C and the mean minimum of the coldest month (January) does not drop below 13–15°C.

7.2.2.5 Incident Radiation, Insolation, Photoperiodism

Global radiation varies from 400 Langleys (Ly) day^{-1} = cal cm^{-2} day^{-1} in December to 600 in June with an annual average of 520, i.e. nearly 190 kLy yr^{-1} which corresponds to a global radiative evaporation potential (GREP)[6] of 1615 mm (Le Houérou 1972a, 1984). The annual number of sunshine hours is 2730 at Linguére and the daily photoperiod is 11 h day^{-1} at the winter solstice and 13 h day^{-1} at the summer solstice, i.e. a difference of 2 hs. The average annual cloudiness is four octets, ranging from six-eighths in August to two-eighths in March.

7.2.2.6 Wind

The continental trade winds and the Harmattan occur from the E and NE in the dry season and the Gulf of Guinea monsoon blows from the SW in the rainy season. Wind speed averages 2.4 m s^{-1} at 2 m above ground; it is higher in February-March (2.9 m s^{-1}) and lower at the end of the rainy season in September-November (1.8 m s^{-1}).

---

[6]GREP: 0.5 Rg/59 = 95000 Ly/59 = 1615 mm;
Rg denotes global radiation and 59 cal the latent heat of vaporization.

7.2.2.7 Seasons

The Fulani calendar year recognizes five seasons based on temperatures, rains, water availability and grazing land phenology, all of which rule the daily life of pastoralists. They are:

1. "Dabundé", the cool dry season from December to February;
2. "Tchedio", the hot dry season from March to April;
3. "Setsellé", the prerainy season in May-June;
4. "N'duggu", the rainy season in July-September;
5. "Kaulé", the post rainy season in October-November.

7.2.3 The Substratum

7.2.3.1 Geology

During the Mesozoic and Cenozoic eras, a sedimentary deposition accumulated in the Senegalese-Mauritanian Basin several 100-m thick. These sediments are the Maestrichtian sands, clays and limestones, Paleocene, Lower Eocene and Lutetian marls and limestones. The Tertiary marine series grows thicker from N to S from about 20 m W of Dagana to some 50 m near Yaré-Lao to the SE. After the Lutetian, the central and eastern part of the Ferlo rose above sea level and continental sediments of clayey sandstones, the "Continental Terminal", were deposited over a depth of 10–50 m from the NW to the SE. The Continental Terminal is topped by a thick and compact iron hardpan 1–2 m thick, overlain by 5–40 m of fluvio-lacustrine deposits more or less weathered into laterite/ferralite. Local sediments contain gravels of limonite coming from the dismantled iron duricrust during the Pleistocene erosion cycles (see Figs. 60–62)

7.2.3.2 Deep Water Resources (Fig. 42)

Groundwaters include three aquifers. The first is the Maestrichtian water level, discovered in 1938 and exploited by some 70 deep boreholes drilled since 1949. These boreholes yield 20 to 100 $m^3 h^{-1}$. The Maestrichtian aquifer spreads over an area of some 100 000 km^2 and therefore extends far beyond the limits of the project area; the amount of water stored is estimated to be about 5×10^{12} m^3 (see Figs. 63, 66–70).

There is a discontinuous aquifer of little importance, 30–100 m deep, at the bottom of the Continental Terminal deposits. This aquifer is reached by ordinary, stone-built wells having small discharges of a few m^3 per day.

A phreatic, shallow aquifer is also sometimes present at the bottom of the Pleistocene sands at the contact of the latter with fluvio-lacustrine deposits. But this is a sporadic and unreliable aquifer of little importance.

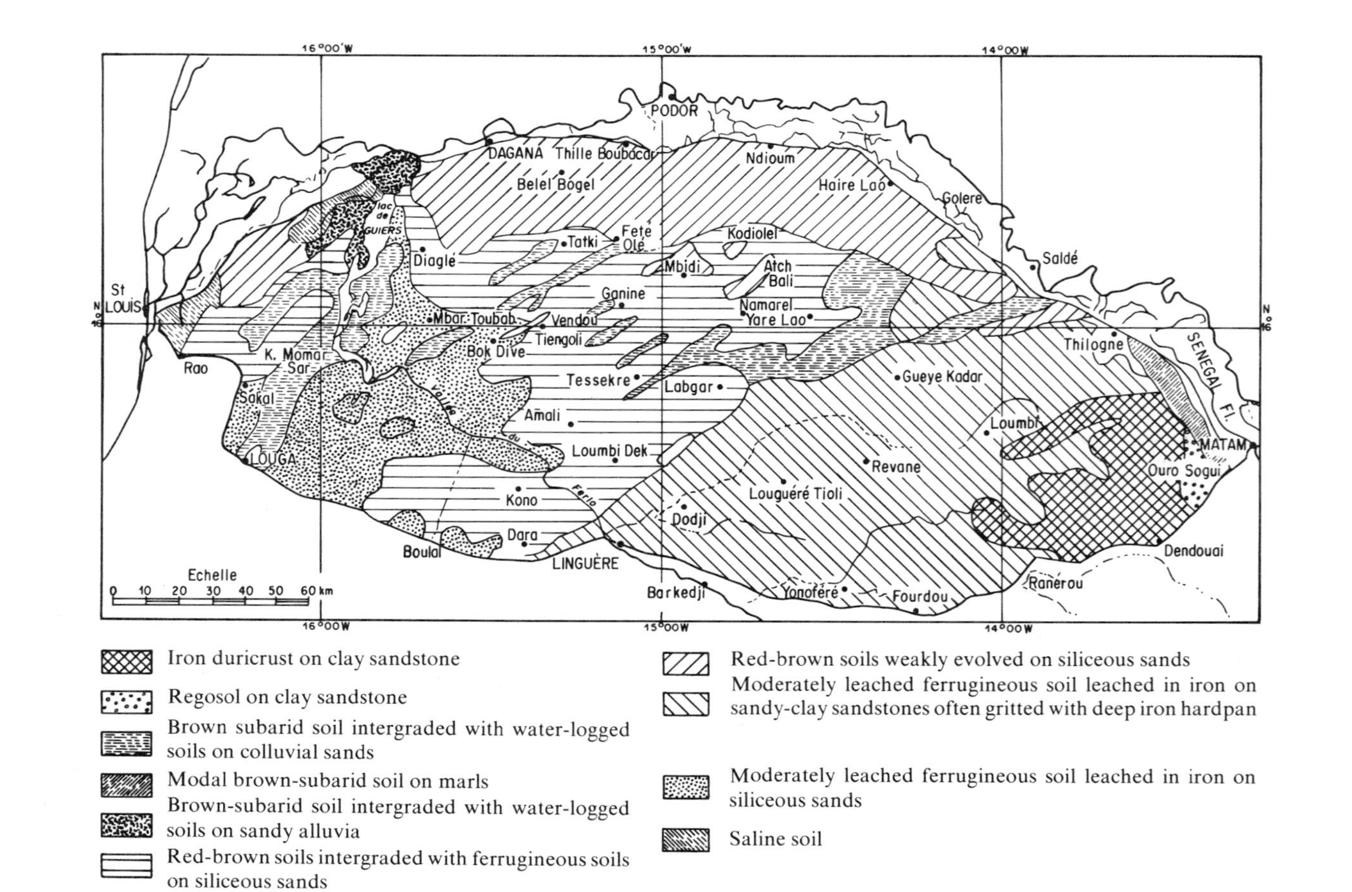

Fig. 60. Sketch of a soil map of northern Senegal, excluding water-logged soils (Maignien, ORSTOM, 1965)

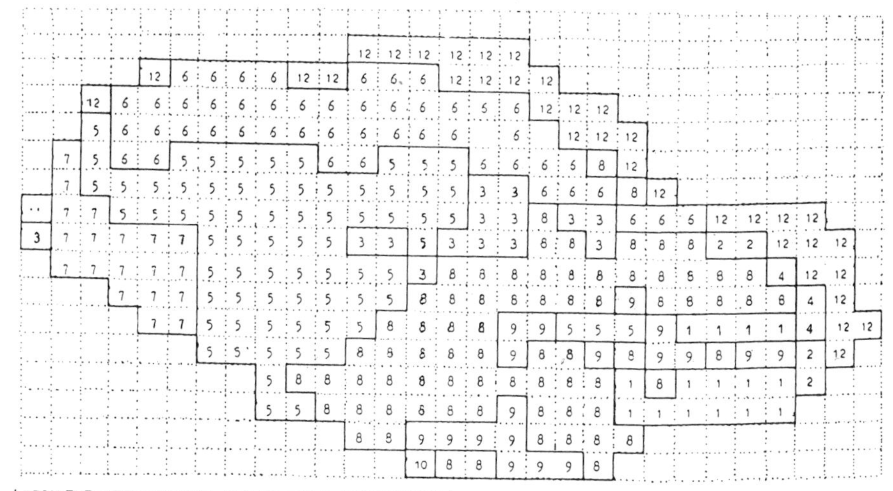

Fig. 61. Digital soil map of the northern Ferlo, scale 1:1 000 000 (Vanpraet 1985). *1* Eroded, lithic, raw mineral soil, iron hardpan (duricrust) on clay sandstone; *2* eroded, regolithic, raw material soil on clay sandstone; *3* isohumic brown-subarid soil, intergraded with water-logged soils on colluvial sands; *4* modal isohumic brown-subarid soil on marls; *5* isohumic brown-red intergraded with ferrugineous soil on siliceous sands; *6* isohumic brown-red, slightly leached soil on siliceous sands; *7* slightly leached ferrugineous soils on siliceous sands; *8* tropical ferrugineous soil, slightly leached in iron, on siliceous sands; *9* tropical ferrugineous soil, leached in iron, on sandy-clay alluvia; *10* leached tropical ferrugineous soil, water-logged, with iron grit (pisolithic limonite); *11* hydromorphic, fairly organic soil; water-logged with gley in the lower layers; *12* hydromorphuc mineral soil, water-logged with gley from the surface

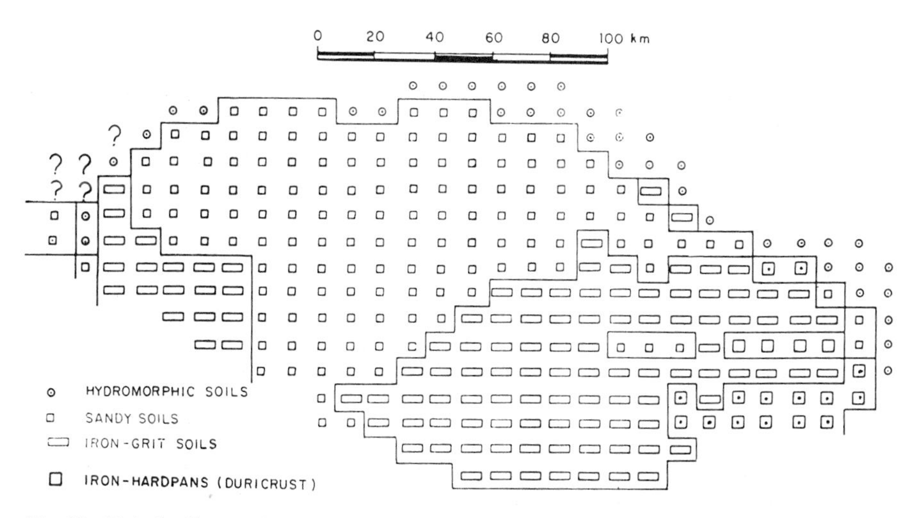

Fig. 62. Digital soil map of northern Senegal (Vanpraet 1985)

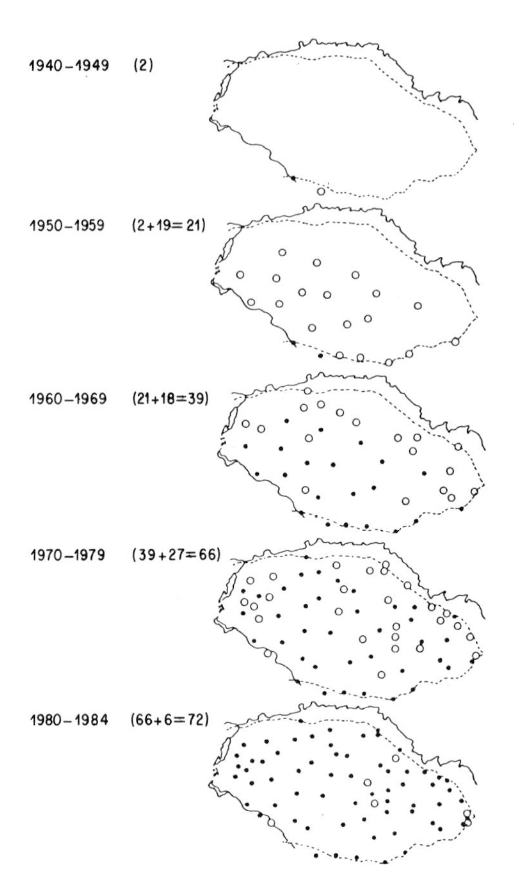

Fig. 63. Changes in numbers of boreholes in the Ferlo from 1940 to 1984 (Vanpraet 1985)

The project area has some 600 wells, 488 of which were in service in 1984; 250 of them discharge 0 to 2 m^3 day^{-1}, 185 yield 2–5 m^3 day^{-1}, 55 produce 10–20 m^3 day^{-1} and 112 are no longer used.

7.2.3.3 Geomorphology

The sand-dune formations or "Seno" (pl. Tchéwé) in Fulani include three geomorphic units. The older one "ante Inchirian" is probably aged 50 000 to 60 000 years BP. It is an old, smoothed, levelled erg 1–5 m thick that geologists usually call the "covering sands".

The Ogolian system of "red dunes", aged 15 000 to 20 000 years BP, is made up of asymmetric ridges ("Tullé", in Fulani) 10–30 m high, 20–50 km long and 0.5–5.0 km wide, spaced 1–5 km apart. These sand ridges are oriented in a NE-SW direction, the direction of the continental trade winds and the Harmattan. This orientation is of primary importance when planning systematic reconnaissance flights (SRF).

The Ogolian dunes have been locally reshaped into smaller parabolic barkhanoid dunes in a NNE-SSW direction, also called transverse dunes, approximately aged 7500 years, of a lighter colour than the Ogolian sands.

The sand cover may be up to 50–100 m thick. Ridges are separated by longitudinal depressions with greyish, more or less temporarily water-logged clay sands that may occasionally be calcareous, locally known as "Baldiol" in Fulani. These Baldiols are the beds of rainy season ponds.

7.2.3.4 Soils (Figs. 60–62, Table 35)

The sandy soils of the dunal systems (the subarid red and brown red soils of the French taxonomy) are neutral to slightly acidic $5.8 < pH < 7.2$. They contain 90–95% coarse sand and 3–5% clay in the upper layers and 80–90% sand and 8–10% clay in the lower layers. Their Fe_2O_3 content is 1.5 to 3.0%, and porosity is high, at 18–26%. Water content at pF 4.2 is 1.1 to 2.5 and at pF (water potential) 3.0, 2.0 to 5.0. Organic matter content is 1–4%. These soils are saturated in Ca^{2+} and Mg^{2+} cations, or nearly so: 2–4 mEq/100 g. They are deficient in N, with 0.1–0.4% and in P with 0.05–0.15%. Clay, sandy brown, subarid soils of the Baldiol depressions develop in the fluvio-lacustrine deposits and may contain 0.3–35% CO_3Ca.

The ferrugineous tropical soils (oxysols), locally known as "Dior soils", are sandy or sandy-clayey, of a strong red colour. They are found in the so-called ferrugineous Ferlo or iron hardpan Ferlo located in the SE of the project area, in contrast to the sandy Ferlo of the central part of the project area. Colloids (clay and iron) are slightly bleached off in these soils and the content of organic matter is low (0.2–0.5%).

The Senegal River valley has mainly water-logged soils of either the superficial hydromorphic psuedo-gley, locally known as "Fondé", which is of a fair quality for irrigation or the heavy-clay vertisol, locally called "walo", which is much more difficult to utilize and develop.

Table 35. Main soil categories in the Ferlo (Maignien, ORSTOM, 1965)

Item No. in soil map caption	Soil categories (French taxonomy)	Area (km²)	Area (%)
1, 2	Raw mineral soils and iron-pan duricrust	1 900	6.0
3, 4, 5, 6, 14	Isohumic brown-red soils on sand or sandstone	14 300	45.1
7, 8, 9, 10	Tropical ferrugineous soils little leached on sand or sandstone	11 200	35.4
11, 12	Hydromorphic soils	3 900	12.4
13	Halomorphic soils	200	0.6

The early Pleistocene iron duricrust soils cover some 5% of the project area, in the SE. The duricrust is subhorizontal and 0.5–1.0 m thick overlaying Continental Terminal sediments; it is locally dissected by linear or gully erosion.

The acreages occupied by various soil categories are shown in Table 35 and Figs. 60–62.

7.2.4 Surface Water

The area of the sandy Ferlo is virtually arheic; there is no organized hydrological network. Runoff is always localized in small watersheds of a few hectares feeding temporary rainy season ponds. Water may remain from 1 to 5 months in these ponds, which, at the peak of the rainy season in August, may cover up to 5% of the landscape. Ponds begin to fill up in July and to dry out in September-October. They usually provide most of the drinking water for man and animals during the 3 months of the rainy season.

The "gravelly Ferlo", on the contrary, has an organized hydrological network with periodically functional main streams and side channels (Loumbol, Tiangol, Louggère, Tiangolkossas, Mboune, etc.) whose outlet is the subfossil Ferlo valley and its lower reaches the Bounoum, leading to the Lake of Guiers. However, this watercourse no longer functions. Many wells and "Ceanes" (funnel-shaped, unmasoned water holes) are dug in these valleys having small outputs which provide water for local transhumance towards the southern border of the project area.

7.2.5 Vegetation and Rangelands

In the Ferlo, as in the Sahel as a whole, the natural vegetation is a *Mimosaceae* scrub to the north, and a *Combretaceae* savanna to the south. This vegetation has been studied and described with great detail in its various aspects in more than a dozen main publications: Raynal (1964); Adam (1966); Audru and Lemarque (1966); Fotius and Valenza (1966); Naegelé (1967, 1971a, b); Diallo (1968); Valenza and Diallo (1972); Bille (1977); Poupon (1980); Cornet (1981); Dieye (1981); Boudet (1980, 1983a); Barral et al. (1983); Valenza (1984). A good synthetic description to which the reader could refer was made by Naegelé and published by FAO in 1971.

When rangelands are concerned, we shall follow the local Fulani terminology, which is accurate and realistic, based as it is on physiographical features which in turn command the nature, composition and productivity of the range. They are the "Seno" range type on sandy, more or less dunal soils, the "Baldiol" in depressions, the "Tiangol" range type in the dry valleys and the "Sangaré" range type on duricrust or iron grit in the pediplains. These main range types are modified by various types of browse and particular vegetation types of limited extent around termitaria, ponds, fallows, etc.

1. The Seno range types may be further subdivided into upslope, downslope and shelf, according to the topography which influences the water budget and

hence the botanical composition and productivity (Bille 1977; Boudet 1983; Barral et al. 1983).

Among the commonest dominant annuals, there are some 30 main annual grass species which altogether represent over 70% of the herbaceous biomass at any site:

Aristida mutabilis	*Cenchrus biflorus*
Aristida stipoides	*Ctenium elegans*
Brachiaria hagerupii	*Elionurus elegans*
Eragrostis tremula	*Digitaria gayana*

There are also a small number of rare perennials:

Aristida longiflora *Andropogon gayanus*

In addition to these grasses, there are some 30 common species of forbs belonging to various botanical families, among which are found a few legumes: *Alysicarpus ovalifolius, Zornia glochidiata, Indigofera* spp., *Tephrosia* spp., *Crotalaria* spp. (see Tables 14–15).

The main woody and browse species associated with the "Seno" range are:

Mimosaceae
- *Acacia tortilis*
- *A. senegal*
- *A. laeta*
- *A. macrostachya*

Anacardiaceae
- *Selerocarya birrea*

Burseraceae
- *Comiphora africana*

Simaroubaceae
- *Balanites aegyptiaca*

Capparidaceae
- *Maerua crassifolia*
- *Capparis decidua*
- *Cadaba farinosa*

Combretaceae
- *Combretum aculeatum*
- *C. glutinosum*
- *Guiera senegalensis*

Tiliaceae
- *Grewia bicolor*

Asclepiadaceae
- *Leptadenia pyrotechnica*
- *Calotropis procera*

2. The "Baldiol" range is dominated by two annual grasses:

   Aristida funiculata and *Schoenefeldia gracilis*

   The woody layer includes:

Acacia seyal	*Adansonia digitata*
Boscia senegalensis	*Combretum micranthum*
Capparis corymbosa	*Grewia bicolor*
Ziziphus mauritiana	*Cordia sinensis*
Adenium obesum	*Dichrostachys cinerea*

3. The "Tiangol" range develops in dry valleys of the ferrugineous Ferlo with a more or less dense, woody layer of thickets under which are found remnants of *Andropogon gayanus* which is probably the pristine perennial grass.

4. The "Sangaré" range is linked to iron grit and iron hardpan soils and dominated by annual grasses:

   Loudetia togoensis | *Diheteropogon hagerupii*
   Andropogon pseudapricus

   The woody layer is dominated by:

   Pterocarpus lucens | *Acacia ataxacantha*
   Combretum nigricans | *Bauhinia rufescens*

7.2.6 Wildlife

Game animals seem to have been abundant until the turn of the century, including elephant, giraffe, roan antelope, hartebeest, topi, greater koudou, oribi, waterbuck, kob, gazelle (red-fronted, dama, dorcas), addax, scimitar-horned oryx, aardvark, warthog, lion, leopard, cheetah, spotted hyena, striped hyena, hunting dog and the ostrich. In the past, the biomass may have reached 2000 kg km^{-2} (Boulière 1978). Under the similar ecological conditions of the Kalahari, the biomass of the large ungulates was still 4 kg ha^{-1} in 1984. In the Ferlo the destruction of predators with strychnine has been encouraged by the Livestock Service since 1950 and further enforced by the settlement of the nomads.

Most of the large mammals are now extinct. Only a few red-fronted gazelles and still fewer dama, some striped hyenas, warthogs, jackals and small mammals (hare and rodents) may still be occasionally found (Poulet 1972, 1974, 1982). The mammalian biomass is now less than 20 kg km^{-2}, which less than 1% of the theoretical potential.

7.2.7 Livestock

According to the estimates of the Livestock Health and Production Service (DSPA), based on vaccination campaigns and some field surveys, the 30 000 km^2 of the project area had in 1981 the stock numbers in Table 36. Since the project area and administrative units have different limits, the number of animals within the project area was calculated by multiplying the number of animals in each administrative unit by the proportion of this unit within the project area, a procedure which, to be valid, postulates a regular distribution of numbers within territorial units. The coefficient was: Dept. of Linguère, 0.4; Podor, 1.0; Matam, 0.3; Dagana, 0.3. The figures in Tables 36–38 are rounded to the nearest hundredth. The conversion rate of various species into TLUs (Tropical Livestock Units: mature zebu of 250 kg kept at maintenance): cattle, 0.81; small stock, 0.18; horses, 0.8; asses, 0.53; camels, 1.16. (Table 37). The proportions between species expressed in TLU's are quite typical of the Sahel as a whole except for camels for which they are smaller than the zonal mean, and horses for which they are larger. The average density in the project area is thus 3 000 000 ha: 407 000 TLU = 7 ha

Table 36. Estimates of livestock numbers in the Ferlo (DSPA 1982)

Administrative units	Cattle	Small stock	Equines	Camels	TLU[a]
Region of Louga					
Dept. of Linguère (× 0.4)	90 000	120 000	18 000	400	105 800
Region of the River					
Dept. of Podor (× 1.0)	171 000	263 000	9 000	200	191 000
Dept. of Matam (× 0.3)	53 000	119 000	14 000	40	71 000
Dept. of Dagana (× 0.3)	36.000	53 000	500	60	39 000
Total	350 000	555 000	41 500	700	406 800
Conversion rates	0.81	0.18	0.53	1.16	------
TLU's	284 000	100 000	22 000	800	406 800
Percent	69.8	24.6	5.4	0.2	100.0

[a] Conversion factors used are based on the ratio between mean population metabolic weight and metabolic weight of TLU.

Table 37. Livestock conversion rates into TLU as used by various authors in the Sahel

	IEMVT 1968/80	Boudet 1983	Meyer 1980	Planche--nault 1983	Barral et al. 1983	Sharman 1983	Le Houérou 1986
Cattle	0.70 to 0.73	0.75	0.80	0.80	0.80	0.80	0.81
Sheep	0.10 to 0.12	0.15	0.15	0.15	0.15	0.15	0.18
Goats	0.08 to 0.10	0.15	0.15	0.15	0.15	0.15	0.15
Horses	1.00	1.00	1.00	1.00	1.00	1.00	0.80
Donkeys	0.50	0.50	0.50	0.50	0.50	0.53	1.53
Camels	1.00	1.00	1.00	1.00	1.00	1.00	1.16

Table 38. Theoretical water requirements in the pilot area[a] (mean figures, outside the area of dependence of the Senegal River and Lake of Guiers)

Consumers	Numbers	Daily requirements (kg)	Annual needs (m^3) for 275 days
People	92 000	25	840 000
Cattle	360 000	40	3 900 000
Small stock	750 000	7	1 420 000
Horses	7 500	20	41 000
Donkeys	20 000	15	81 000
Camels	2 500	10	7 000
Total	------	--	6 289 000

[a] Under the assumption that water consumption is ensured by ponds for an average of 90 days of the rainy season annually.

TLU^{-1}. The average density of cattle found from counting at three main boreholes (Tatki, Tessekré and Belel-Boguel) was 27 730 over 183 000 ha, i.e. 6.6 ha TLU^{-1} (Meyer 1980; Planchenault 1981, 1983; Barral 1982; Barral et al. 1983). It should be kept in mind here that the conversion rates used by various authors may differ by substantial amounts according to whether live weight or metabolic weight rates are applied to convert between head of livestock and TLU. The degree of departure may be as high as 15% as shown in Table 37.

Naegelé (1971b), using data derived from vaccination campaigns following a series of above normal rainy seasons in the 1950's and 1960's, found overall densities of 7.2 ha $head^{-1}$ of cattle in 1966/67; 8.3 ha $head^{-1}$ in 1967/68 and 7.2 ha $head^{-1}$ in 1968/69. This corresponds to an overall density of 6.5 ha TLU^{-1}, all species pooled together for that period in the Koya area. This area is similar to that considered by the counts at boreholes. Assuming that one head of cattle = 0.8 TLU and that cattle make up 70% of all livestock (69.4 in Table 38) there is a good correspondence between the figures (less than 15% difference) in spite of a 20-year interval between the two studies.

7.2.8 Evolution of Land Use and History of Development (Barral 1982)

7.2.8.1 The Ferlo Before the Boreholes (Figs. 63–65)

Up to the completion of the first borehole in 1950, the Ferlo area was considered a "desert" because of the scarcity of permanent water resources. The Ferlo Desert was only visited 4 to 5 months annually by Fulani pastoralists and their transhumant herds during the rainy season (Bonnet-Dupeyron 1952; see Figs. 64–65). Transhumance took place from the Senegal River valley in the neighbourhood of which herds spent the dry season and from the Ferlo valley where wells and ceanes (water holes) allowed for small permanent or subpermanent settlements. Grazing was plentiful and perennial grasses such as *Andropogon gayanus* were found in dense stands sheltering and feeding an abundant wildlife of elephants, giraffes, roan antelopes, topi, kobs, waterbucks, reedbucks, addax, scimitar-horned oryx, aardvarks, ostriches, lions, leopards, hyenas, hunting dogs, caracals, etc.

Until the First World War, the area had some 50 000 pastoralists for 6 months over 30 000 km^2 with an estimated 30 000 head of cattle and 100 000 head of small stock. This represents a total of some 45 000 TLU, i.e. a density of 1.50 TLU km^{-2} or less than 4 kg live weight ha^{-1}.

The Colonial Administration soon tried to fill this "emptiness" by setting up in 1901 the "Well Brigade", belonging to the Corps of Engineers of the Marines. In 12 years 675 wells were dug in the whole country by 40 teams working mainly along the railway from Dakar to the Djoloff, west of Linguère (Brasseur 1952; Barral 1982). The Wells Brigade vanished with the First World War and it was only in 1925 that a policy of pastoral water development was put into action in the Ferlo. But lack of finances meant that its development was slow and since only shallow aquifers with small outputs were being exploited, the number of animals served in the dry season remained small.

It was only after the Second World War, after the accidental discovery of the Maestrichtian aquifer at Kaolack in 1938 (Audibert 1966), that an active pastoral water development policy began to develop rapidly in the 1950's. In 1950, livestock density was still less than 4 head of cattle per km^2 (8 kg ha^{-1}) in the Djoloff; it trebled from 1950 to 1975 (Santoir 1980; Barral 1982) to reach, as we have seen, over 30 kg live wt. ha^{-1} in 1980. At the same time, in order to safeguard the pastoral identity of the region, the Colonial Administration set up "Forest Grazing Reserves" (Reserves Sylvo-Pastorales) covering some 15 000 km^2 of which some 10 000 km^2 are located in the project area: Koya (4300 km^2), Louggéré, Dodji, Sogobé, Khaddar. These reserves seem to have been fairly efficient in protecting the environment by saving the area from the expansion of agriculture which occurred in most regions of the Sahel after World War II. Thus, to a large extent they fulfilled their role of safeguarding the grazing resources of the Ferlo.

7.2.8.2 The Ferlo with Boreholes (Figs. 66–70)

Thirty-five deep boreholes (80–322 m deep) and 33 "deep wells" (Figs. 66–70) reaching the artesian Maestrichtian aquifer were dug from 1950 to 1980. The static level of the aquifer is subartesian, close to ground level. One single borehole exploits the Paleocene aquifer, at Belel-Boguel in the NW whose use implies pumping from a moderate depth. The total discharge is 7×10^6 m^3 yr^{-1} of which 4.8×10^6 m^3 yr^{-1} comes from boreholes and 2.5×10^6 m^3 yr^{-1} from boreholed wells. Individual discharge varies from 10 to 100 m^3 h^{-1}; 80% of the installations yield approximately 20–50 m^3. The quality of the water is good with about 0.15–0.75 g l^{-1} (150–750 ppm) of mineral residue.

Deficiency in phosphorus provoked the spread of the "borehole disease" a hydric botulism resulting from aphosphorosis. This condition encourages the consumption of bones by livestock and the pollution of water by small mammals, particularly rodents. Aphosphorosis is associated with the changes in husbandry practices brought about by boreholes, particularly the abandonment of the "salt cure" (Calvet et al. 1965). The establishment of boreholes and their equipment deeply upset the traditional pastoral practices by settling, or rather partial settling, of the Fulani pastoralists. Each borehole thus created an area dependency where more or less permanent camping grounds and villages have been set up (Figs. 68–69). The older boreholes exert a stronger influence because their discharge is usually larger; their area of dependency reaches a radius of 15–20 km (70 000–125 000 ha). But the average area of dependency is currently only 13 000 to 73 000 ha for the 68 boreholes.

The theoretical water requirements in the project area, besides the area of dependency of the Senegal River valley, is shown in Table 6. These figures were calculated on the basis of 270 days yr^{-1}, as the stock are watered in ponds for 3 months annually on average. The demand thus calculated is 6.3 million m^3 yr^{-1}, 15% below the volume theoretically pumped of 7.3 million m^3 yr^{-1}. To this output must be added the variable and low outputs of mason work, surface wells and water holes. The presently available water resources represent therefore an ample supply from the quantitative standpoint.

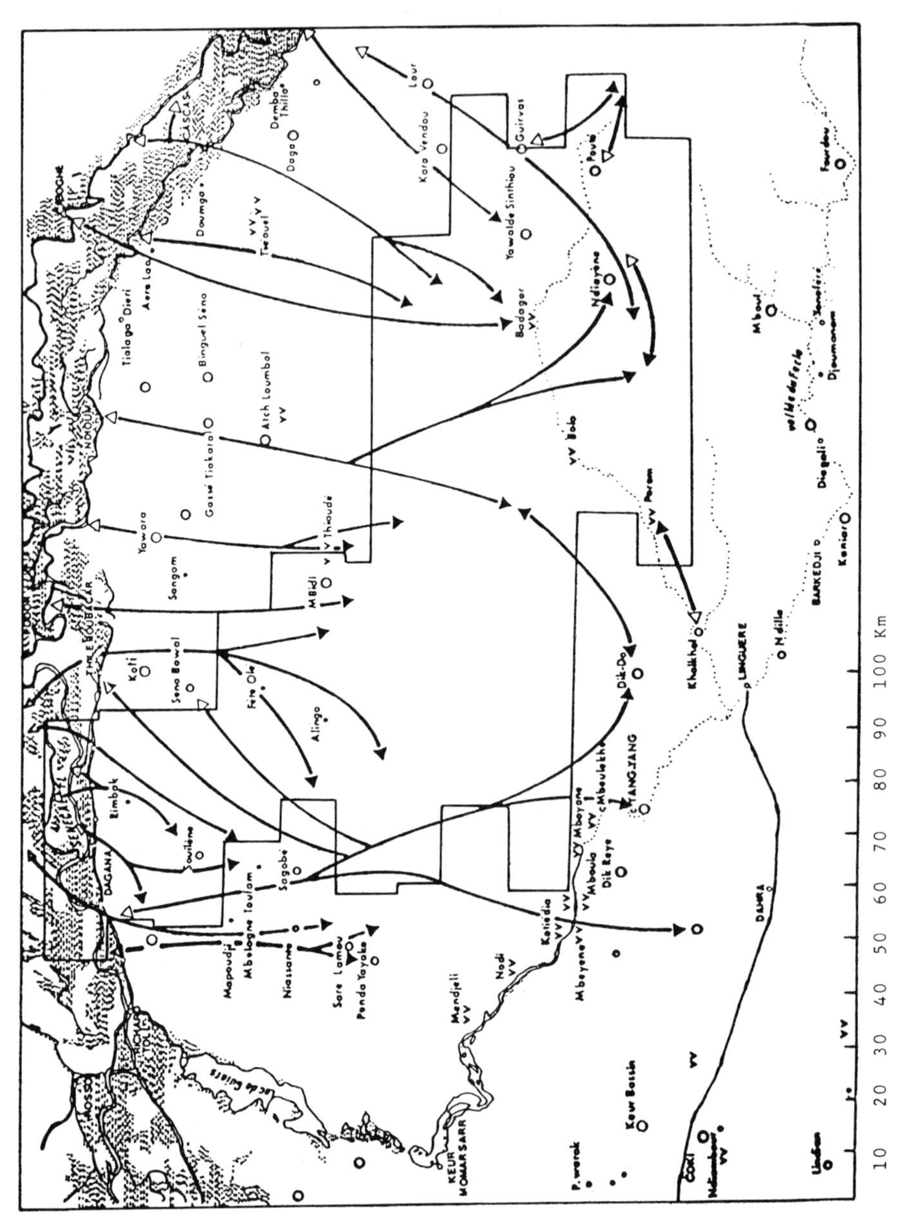

Fig. 64. Transhumance routes before the boreholes (Bonnet-Dupeyron 1952; taken from Barral 1982)

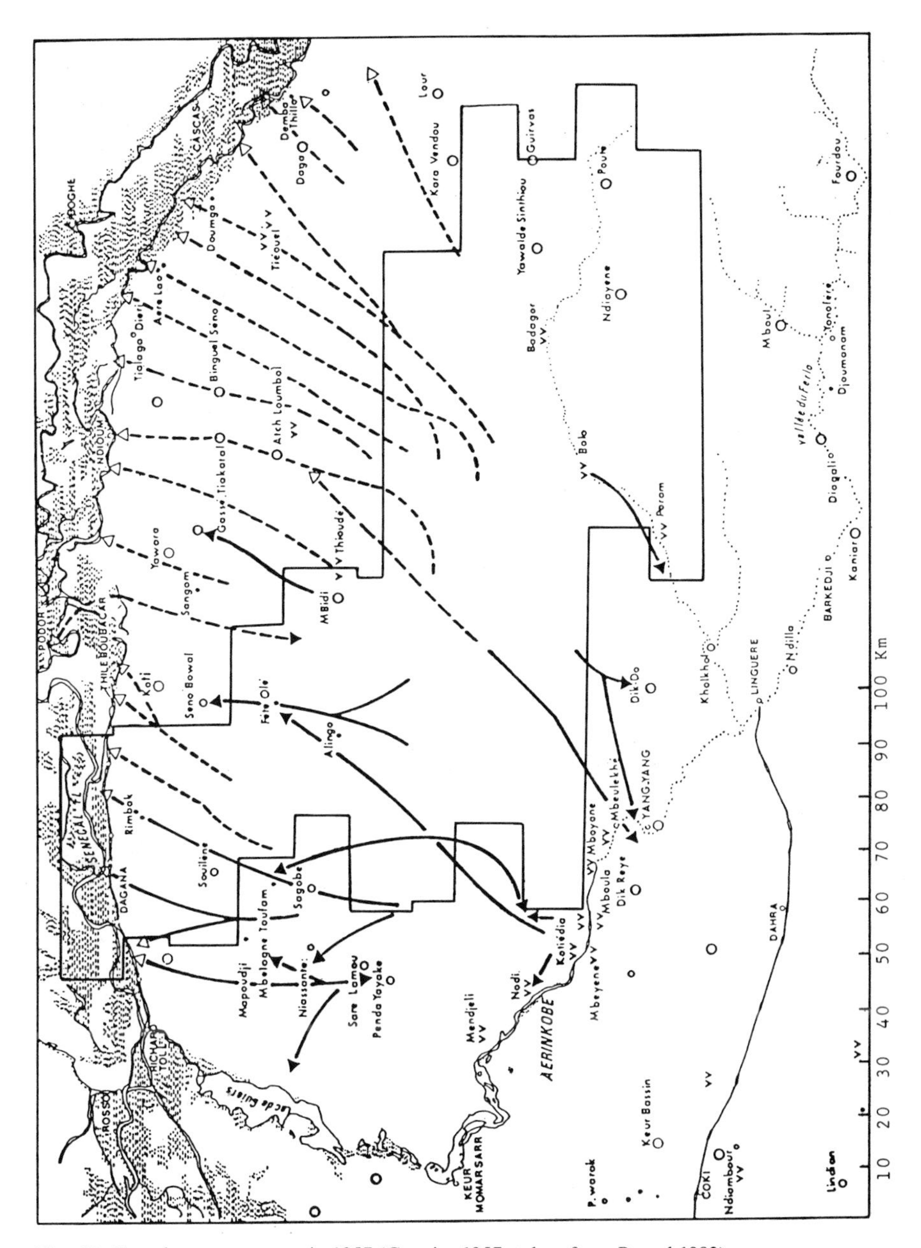

Fig. 65. Transhumance routes in 1957 (Granier 1957; taken from Barral 1982)

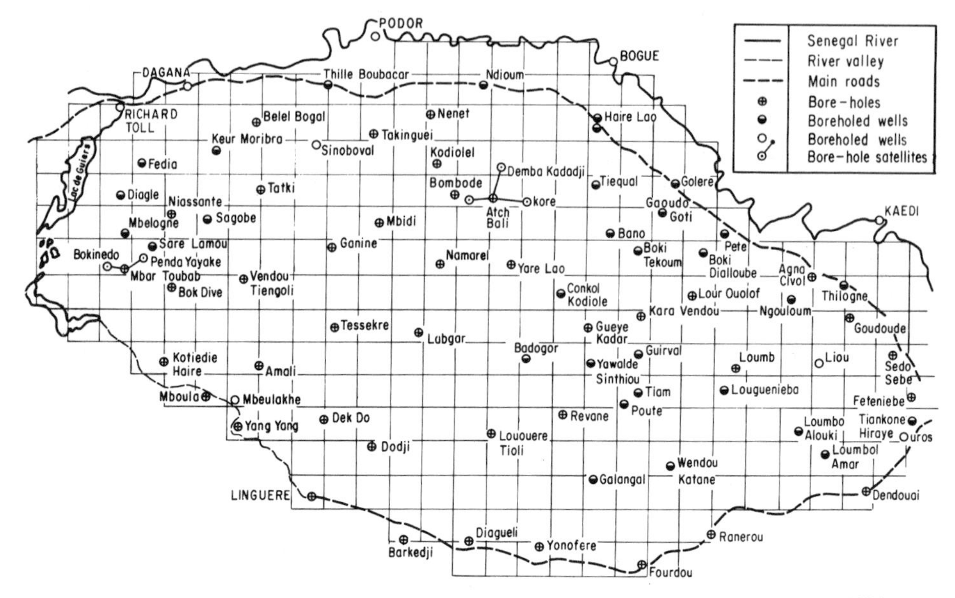

Fig. 66. Distributon of boreholes and other water resources in the Ferlo (1983) (Vanpraet 1985)

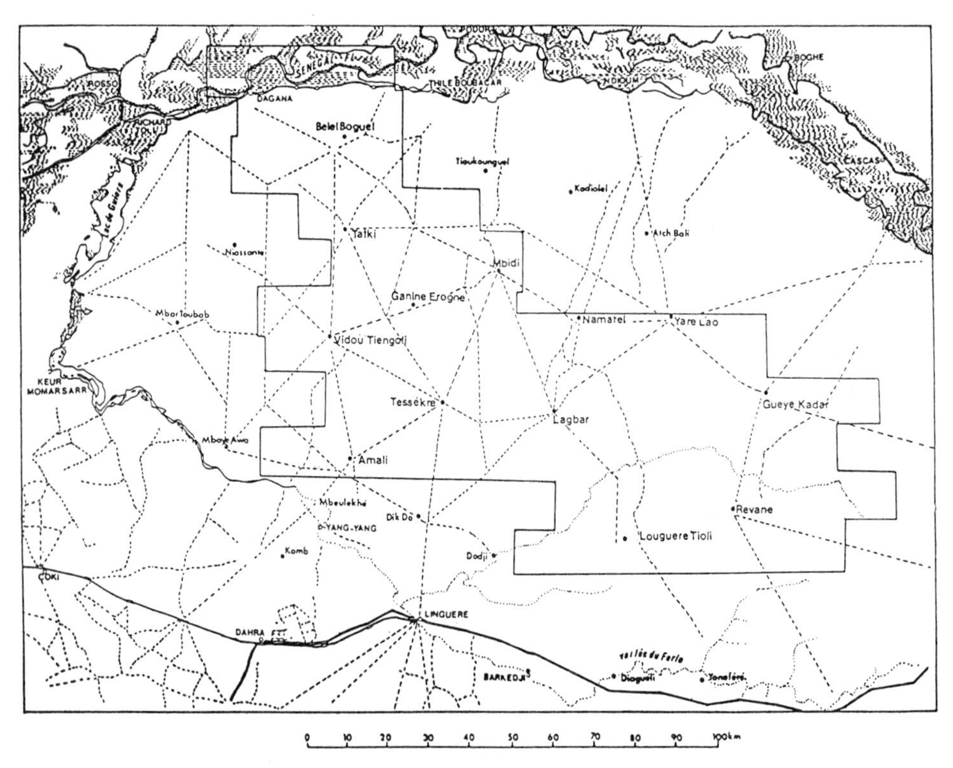

Fig. 67. DGRST/GRIZA/LAT study area showing the distribution of main boreholes and tracks (Barral 1982)

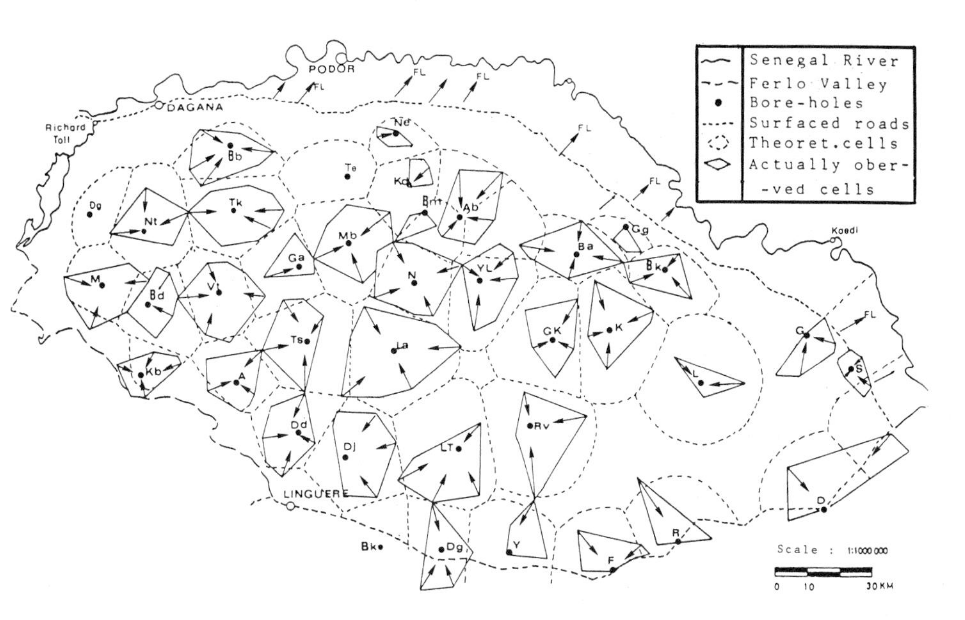

Fig. 68. Boreholes in northern Ferlo and their area of dependency

o BOREHOLES

• VILLAGE OR CAMP RECORDED

o- - - -• VILLAGE OR CAMP WHICH, ACCORDING TO BARRAL, DOES NOT WATER AT THE NEAREST BOREHOLE

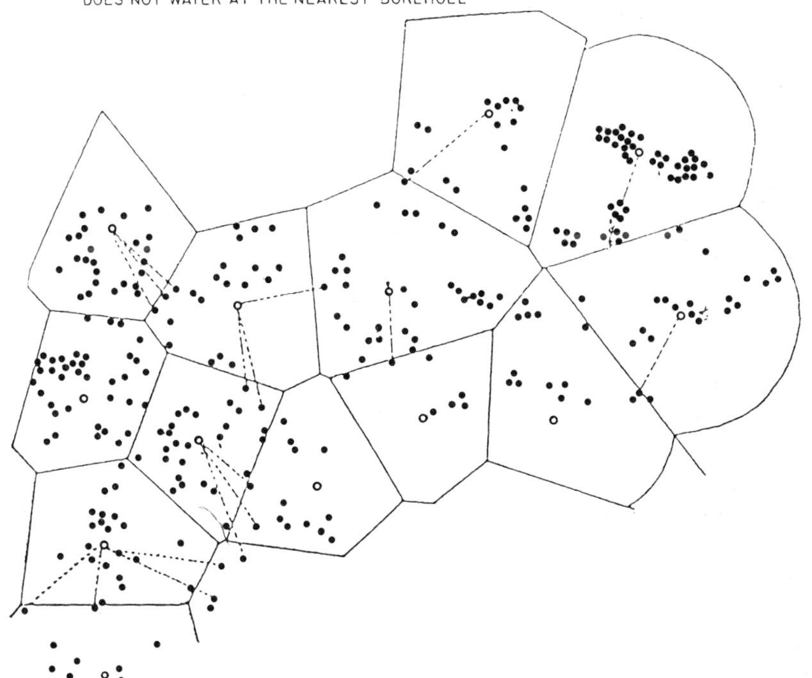

Fig. 69. Villages and Camps and their water dependency in the DGRST/GRIZA/LAT study area (Barral 1982)

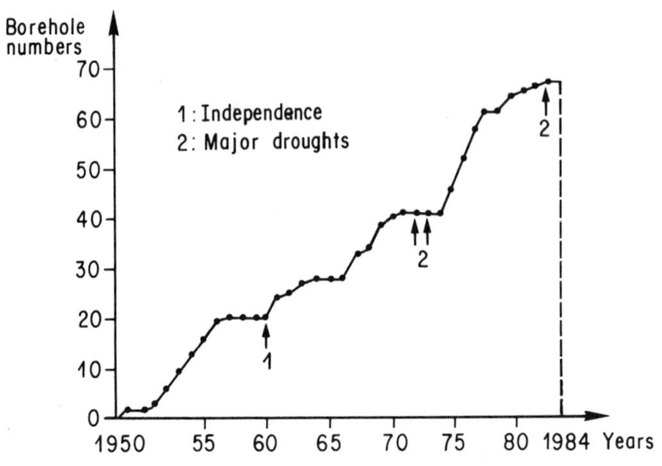

Fig. 70. Evolution of borehole numbers over time as related to major events (Vanpraet 1985)

7.2.8.3 Consequence of Borehole Developments on Range Utilization

In the traditional management system pastoralists and herds used to remain strongly aggregated because of the scarcity of water resources and because of the need for protection against predators (mainly lions and spotted hyenas). In the dry season the "Walo" Fulani used to cultivate retreat-flooding sorghum in the valley of the Senegal River using flooded pastures of the valley (*Echinochloa stagnina, Oryza barthii, Sporobolus helvolus* and *Cyperaceae,* while small stock made use of browse: *Acacia nilotica, A. seyal, Ziziphus mauritiana.* For the rainy season, they used to congregate for transhumance to the "Seno" and "Baldiol" grazing lands of the central and northern sandy Ferlo. Each group had its own pond or group of ponds to which they came every rainy season. The group camping grounds or "Rumano" was made up of a variable number of "Gallé" (household camping grounds). The Walo Fulani used to own relatively few cattle, their herds being mainly small stock. Watering took place daily at the ponds while grazing was unattended and took place in daytime only. Stocks were confined at night in thorn enclosures ("Gallé") at the Rumano. Each camp site Rumano was surrounded by a reserved zone (the "Houroum") within which some millet rain-fed cropping used to take place. In the Houroum, grazing was prohibited to animals belonging to other groups. In this system, stock had yearlong green feed and daily watering.

In contrast, the "Dieri" Fulani used to crop rain-fed millet in the sandy Ferlo (but had no retreat-flooding sorghum cropping in the valley). They moved between the "Diedegol", a 25-km-wide strip on the southern bank of the Senegal River valley, and the Ferlo. Then from the Ferlo, during the rainy season, they moved again to the Djoloff in the SW (region of Linguère, Yang-Yang, Dahra along the lower Ferlo valley, also called Bounoum). The latter transhumance included the "salt cure" in the Bounoum valley between Yang-Yang and the lake of Guiers. In the dry season the stock were watered in the Senegal River and its side channels (Marigots), but they did not graze in the valley (where grazing rights were restricted to the Walo Fulani). The animals were watered every second day. On the day of watering the animals would graze between the camp site and the valley in a strip 10–15 km wide. On the day without watering, they would graze further south in the Diedegol, some 10–15 km south of the camping grounds. The width of land used for grazing during the dry season was thus 25 km along the W and S bank of the Senegal River.

At the onset of the first reasonable rains, both the Walo and Dieri Fulani would start moving to their rainy season quarters in the sandy Ferlo. But in contrast to the Walo, the Dieri would send most of their cattle to a second stage transhumance for the salt cure to the Djoloff. These herds were only attended by a shepherd, while the household with some milking cows and most of the small stock would stay at the Rumano in the Ferlo tending the millet crop. The return from the salt cure transhumance would occur in September at the time of harvesting of the millet crop. After the harvest the whole group would move back to the dry season quarters in the Diedegol in October-November depending on the drying out of ponds along the trail.

Thus, the whole area was used for 4 to 6 months in the annual cycle. The grazing space between the Rumano was subdivided into two areas: the "Diei", reserved for certain groups and the "Laddé" which was free range available to anyone. There were no boundary markers but everyone knew the divisions and would tacitly abide by them. Landscapes and territories were thus organized around a pond or a group of ponds and divided into Houroums, Diei and Laddé. The Houroum thus constituted an organization somewhat similar to the old system of "finage" in medieval west Europe. However, once the millet crop had been harvested, the Houroum reversed to Laddé, so that after the harvest all was "commons"!

The establishment of mechanized boreholes progressively reduced transhumance. From 1950 onwards, the boreholes have acted as attractive forces and poles of settlement for the dry season. As a consequence the rainy season Rumanos have tended to become permanent, but paradoxically, at least in the beginning, mobility was not reduced but rather strengthened. It also became a sort of anarchic "Brownian movement". The Fulani remained pastoralists and did not really settle down (Dupire 1957; Grenier 1957). A study carried out in 1980 over 13 boreholes serving an area of 952 000 ha, i.e. one-third of the project area concluded that there were 89 000 sedentary and 12 000 transhumant cattle for a population of 32 000 people. This represents 9.4 ha per head of cattle, or 8.3 ha TLU^{-1} all stock together, and a stock to human ratio of 3.3. TLU per person (Meyer 1980; Barral 1982). The proportion of transhumant stock thus decreased from 60% in 1950 to 13% in 1980. The aggregation around ponds in the rainy season has remained unchanged; thus movement has been reduced to short journeys within the service area of each borehole. At the same time, groups have split into smaller units and the average number of Gallé per Rumano decreased from 12 in 1950 to 5 in 1980. There are several reasons for such splitting:

1. The removal of predators;
2. Sharper competition for grazing since stock numbers increased by 120% between 1955 and 1980;
3. Reduced primary production due to drought and overstocking.

To these physical and biological causes, can be added sociopolitical reasons: the loosening of solidarity by vaccination campaigns and a decline in the authority of the elders and traditional leaders to the benefit of the representatives of state power. The result has been the progressive disappearance of the Houroum since the 1960's, with the support of the administration. Free ranging has become commonplace at the expense of the old Houroum-Diei structure, thus creating the present anarchic situation of communal grazing.

There is some degree of movement between the areas of polarization of neighbouring boreholes during the dry season as a result of grazing availability, or when the pumping station breaks down for some time, or when grazing is destroyed by bushfire, etc. A new social structure and solidarity has thus developed around the boreholes. This new solidarity expresses itself, for instance, in communal purchase of spare parts, of fuel and the maintenance of pumping stations. The present range used in the rainy season depends on short-distance grazing (less than 5 km) around ponds and Rumano, with the livestock watering every day. The dry season sees the use of pastures further away from the Rumano, which may be temporarily abandoned for short-range transhumance within a radius of 10–20 km of the nearest borehole; watering then taking place every second day. These temporary dry season camps, called "Sedano", may be moved two to three times in the course of a dry season, depending on grazing availability. The system is thus characterized by short-range micro-nomadism, not settlement, i.e. a kind of "borehole endodromy" reminiscent of the system in the Gourma region of Mali and Burkina-Faso described by Barral (1977).

There are occasionally exceptional migrations of larger amplitude in years of extreme drought such as 1972, 1983 and 1984. In 1972 and 1984, most of the stock moved out of the project area to the south. This contrasts with the short-range migrations which occur outside drought conditions, for example, the range has been burned or a borehole has broken down. The mean density of human populations in the Ferlo is 3 inhabitants km^{-2}. The annual growth rate for the past 60 years has been 0.9%. These two figures for density and growth rate are among the lowest, if not the lowest, in the Sahel, which accounts for the relatively unspoiled conditions of the Ferlo as compared to most areas in the Sahel.

7.2.8.4 Range Development in the Ferlo

Grazing improvement in the Sahel has been achieved by the State of Senegal through the technical services of the Ministry of Agriculture. Specific programmes include pastoral water development, establishment of the forest grazing reserve, the creation and maintenance of firebreaks, extension of veterinary services (prophylaxis) and stratification and marketing facilities.

7.2.8.4.1 Pastoral Water Development

1. *Deep Waters.* As mentioned above, there are some 74 boreholes and boreholed wells in which the static level of the aquifer varies from −3 to −60 m. Most of them are equipped with a diesel pumping station. They were drilled and put into service between 1950 and 1980. These installations are fairly well distributed over the project area (Figs. 63, 66, 68, 69). The area serviced by each borehole averages 810 km^2 and the overall density is one installation per 405 km^2. The theoretical discharge is 4.8×10^6 m^3 yr^{-1} for the boreholes (13 150 m^3 day^{-1} = 292 000 TLU day^{-1}) and 2.5×10^6 m^3 yr^{-1} for the boreholed wells (6850 m^3 day^{-1} = 150 000 TLU day^{-1}). A total of 410 000 TLU can therefore theoretically be watered over the whole project area (Table 38). All of these installations exploit the Maestrichtian aquifer, with the exception of the Belel-Bogue borehole set up on the Lutetian aquifer.

2. *Shallow Aquifers.* Some 600 surface wells have been dug, 488 of which were in service in 1982. Most of them exploit the Continental Terminal aquifer at depths varying from 30 to 100 m. Water is usually raised manually or by draught animals. The average discharge is as follows:

Discharge (m^3 day^{-1})	No. of wells	(%)
$D < 1$	137	28
$1 < D < 2$	115	23.5
$2 < D < 10$	183	37.5
$10 < D < 20$	46	9.4
$20 < D$	7	1.4

The "ceanes" are non-lined, funnel-shaped water holes 3–10 m in diameter at the ground surface exploiting shallow superficial aquifers in the valleys, particularly the Ferlo. The number, yield and permanence of these ceanes are poorly known.

3. *Surficial Waters.* Apart from the Senegal River and its side channels cut off as the flood retreats, there are some 305 ponds in the project area used for watering stock. This is equivalent to about one pond for 10 000 ha rangeland. Those ponds consequently water an average of 1350 TLU during the 3 months of the rainy season, i.e. a consumption of 60 m^3 per pond per day or 5400 m^3 per season. The periods for which each pond can be used, i.e. its capacity, would be useful data for range management in the Ferlo and for improving the hygiene of these ponds [parasitism, access, storage capacity (over-excavating)]. Unfortunately, neither the annual dates of filling up and drying out, nor the number of stock they serve, have been recorded. Such records could be a valuable goal for the monitoring of the Ferlo; perhaps high resolution SPOT images could be used.

7.2.8.4.2 *Forest-Grazing Reserves, Firebreaks*

As mentioned above, some 10 000 km² of the project area have been demarcated by the Waters, Forest and Game Service of the Ministry of Agriculture. The areas concerned are the reserves of Koya, Sogobé, Khadar, Louggéré-Tioli, Barkhedji-Dodji, Ndiayène and Lérabé. A network of 2700 km of firebreaks was established by the Water, Forest and Game Service, and fire-fighting committees were organized in each region and department. Mobile fire brigades to combat bushfires were envisaged but do not seem to have been set up except at Linguère. As early as 1971 Naegelé stated that these firebreaks were not very efficient since they are not wide enough and not properly maintained. He made suggestions for their expansion and improved maintenance. Judging from the SRF data on traces of fire, the situation has not much changed in 15 years in this respect.

7.2.8.4.3 *Hay-Making Operations*

A hay-making operation was launched in 1961 and lasted until 1970, with the assistance of FAO (Naegelé 1971b). This operation was aimed at convincing pastoralists of the usefulness and feasibility of hay-making and conservation. The first demonstrations organized by the DSPA took place near Labgar in 1964 with a group of progressive pastoralists who volunteered for the experiment. In 1965 and 1966, the experiment was expanded to the areas served by the boreholes at Tatki (Dept. of Podor) and Yayaté (Dept. of Dagana). After 3 years of demonstration of mowing, harvesting, hay-making, stockpiling and conservation and ox taming, the experiment was extended between 1967 and 1969, with the assistance of FAO and the World Food Programme. The operation was so successful that priority action was signalled in the third 4-year Economic and Social Development Plan (1969–73).

Unfortunately, the programme then collapsed, after the departure of the FAO expert, Mr. Naegelé, due to maintenance problems of the mowers.

Later attempts did not succeed, in spite of the cooperation of SISCOMA on the maintenance problems (Le Houérou 1974). One positive point has been learned, hay-making is technically feasible and socially acceptable to the pastoralists, but the maintenance and management problems remain unsolved.

7.2.8.4.4 *Assistance to the Pastoralists*

Apart from vaccination campaigns against rinderpest and contagious bovine pleuropneumonia, which were developed in the 1950's (and gave rise to a sharp increase in cattle numbers in Senegal, as it did in the Sahel as a whole), the Senegalese administration has set up a para-state organization for helping the pastoralists in animal production. This organization is called the Company for Livestock Development in the Forest-Grazing Zone (Société de Développement de l'Elevage dans la zone sylvo-pastorale: SODESP).

The SODESP, a public, industrial and commercial institution was set up by law on June 2, 1975, in order "to promote livestock production". Its strategy consisted of:

1. Intensifying stock production through calving husbandry (until weaning);
2. Rearing of young in such a way as to obtain faster growth;
3. Fattening and finishing operations for young males and culled animals in order to improve the quantity and quality of the output.

The calving husbandry takes place in the Ferlo, the rearing operation in the Senegal River valley, on the ranch of Doli, and in the so-called groundnut basin. The fattening-finishing operations are carried out by contract with farmers in the Groundnut Belt (peasant-managed fattening) and in feedlots around the main cities (industrial fattening). Both utilize crop residues (groundnut stalk/hay and sugarcane tops) and agro-industrial by-products (cotton seeds, rice bran, groundnut meal and molasses).

Two independent extension areas (called channels) were in operation in 1984 in the areas serviced by the boreholes of Labgar and Mbar Toubab, covering an area of 3000 km^2 (10% of the project area). The calving husbandry operations concerned some 30 000 breeding cows and their calves and 9000 ewes and their lambs, belonging to 1300 pastoralists. The purchase and marketing of stock represented 213 and 279 million FCFA (0.7 and 0.9 million $) in 1984. The SODESP employed 165 people, 36 of whom were field extension technicians. It received assistance and funds from FAC, FED and USAID; its annual budget was 3 billion FCFA (10 million US $).

The performance of the stock husbanded through the SODESP channels are 60 to 80% superior to the traditional practices in terms of total output. The programme is intented to expand to four more areas: Tatki, Gaye Khadra, Ranerou and Lindé. Such an expansion would raise the stock numbers to some 120 000 cattle units and 120 000 sheep units, i.e. some 50% of the cattle and 30% of the small stock population of the project area. Unfortunately the whole organization collapsed in 1983 as a result of poor management and the SODESP was then dismantled.

7.3 Monitoring

7.3.1 Principles, Problems and Methods

The principles of ecological monitoring of natural resources combining various terrestrial, aerial and space remote-sensing methods were described in the early 1970's in East Africa (Gwynne and Groze 1975). None of these methods alone can provide all the complex information required for the ecological monitoring of ecosystems and natural resources, including data on geology, soils, water, vegetation, animals and human activities, their seasonal changes and long-term evolution. Decisive progress has been made over the past 20 years in the field of remote sensing by the improvement in various types of satellite imagery, the development of low altitude systematic reconnaissance flights and a considerable (often underrated) improvement of conventional air photography, which now includes techniques using infrared, false-colour, colour, high sensitivity, high resolution panchromatic emulsions and improved optics.

Each of these techniques permits information to be obtained at various scales, at various detail with various degrees of precision and at widely differing costs. There is almost inevitably a conflict to be resolved in a monitoring programme, between cost and detail. For instance, one must decide if the programme needs numerous data of medium quality at a low cost or high quality data that are necessarily more costly and therefore less numerous. A sometimes difficult choice must be made to fulfill the objectives of the programme with the available means at the most advantageous cost. The diversification of data sources allows for such a selection.

7.3.1.1 Satellite Imagery (Fig. 71)

Satellite images permit a global overview of large scenes (from 8500 to 2 million km²). As the spatial resolution of sensors became more refined from RBV, MSS, AVHRR, TM to SPOT,[7] they have allowed increasingly large-scale utilization. The characteristics of various sensors and types of images are shown in Tables 39–40.

Because satellite images of the same area can be obtained at regular intervals, they can be used to study just those ephemeral phenomena which are so often

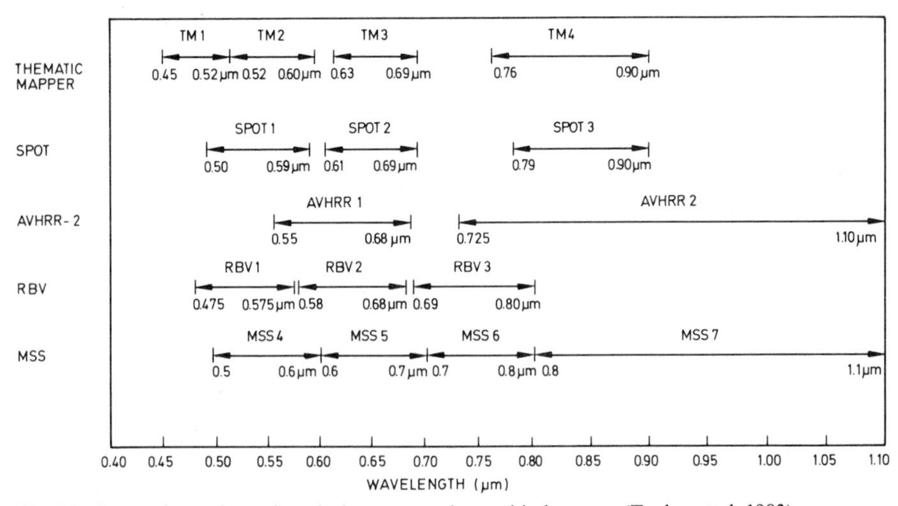

Fig. 71. Comparison of wavelengths between various orbital sensors (Tucker et al. 1983)

[7]RBV = Return Beam Vidicon (ERTS 1 and 2); MSS = multispectral scanner (LANDSAT); TM = thematic mapper (LANDSAT 7); SPOT = Systéme Probatoire d'Observation de la Terre; AVHRR = advanced very high resolution radiometry (NOAA 6 and 7, Meteosat).

Table 39. Main characteristics of satellite imagery used in the inventory of terrestrial resources which were operational in 1985 (Le Houérou 1985; Tucker et al. 1985c)

Types of satellites	Scene (km)	Resolution (m)	Practical scale of enlargement and use	Minimum land surface identified (ha)	Orbital period	Approximate cost of acquisition of colour composites (1987)	Channel spectra
Meteosat	12 500	5000	1/5 000 000	15 625	30′	?	4 Channels: VIS-TIR
NOAA 6/7 AVHRR	2400	1100	1/1 000 000	6250	LAC 9 days GAC 3 days	?	5 Channels: VIS, NIR, MIR TIR
LANDSAT/MSS	185	80	1/250 000	40	16 days	0.02–0.03 US $ per km²	4 Channels: VIS PT
LANDSAT/TM	185	30	1/100 000	6.25	16 days	500–800 $ per 1/4 Scene = 0.06–0.09 $ per km²	7 Channels: VIS, NIR, MIR TIR
SPOT sensor a	60	20	1/75 000	3.5	9 days	0.33 $/km² (1200 $/Scene)	3 Channels: VIS, NIR
sensor b	60	10	1/50 000	1.6	9 days		1 Channel: VIS

MSS Channels	4	500– 600 nm	Green	Thematic Mapper	1	450– 520 nm	Blue-green
	5	600– 700 nm	Orange-red		2	520– 600 nm	Green
	6	700– 800 nm	Red, near-infrared		3	630– 690 nm	Red
	7	800–1100 nm	Near-infrared		4	760– 900 nm	Near-infrared
					5	1550–1750 nm	Medium infrared
SPOT a	1	500– 590 nm	Green		6	2080–2350 nm	Medium infrared
	2	610– 680 nm	Red		7	10.3 –12.5 μm	Thermic infrared
	3	790– 890 nm	Near-infrared				
SPOT b	1	Panchromatic Visible					

Table 40. Wavelength used in the radiometer sensors for the evaluation of green herbaceous phytomass

C1	=	0.55 – 0.68 μm	=	Visible red
C2	=	0.725– 1.10 μm	=	Near-infrared
C3	=	3.55 – 3.93 μm	=	Infrared
C4	=	10.5 –11.5 μm	=	Thermic infrared
C5	=	11.5 –12.5 μm	=	Thermic infrared

important in ecological monitoring. This is particularly the case when assessing the green herbaceous biomass in the Sahel, a subject to which we shall return in some detail. Satellite-borne sensors also permit the selection of information using various bands or combinations of bands of the electromagnetic spectrum for interpretation.

Furthermore, the cost of images per unit surface area is relatively small compared to other techniques. However, at the present state of civilian satellite technology, ecologically useful stereoscopic vision of the images is not available, satellite images are also incapable of providing information on finer phenomena such as those needed for detailed evaluation of vegetation cover, the detection of early erosion stages (which can still be kept under control), detailed aspects of hydrology, wildlife and livestock numbers, and many aspects of human activities.

7.3.1.2 Aerial Photography

Conventional aerial photography is amenable to stereoscopic interpretation and can therefore be used to collect very detailed, large-scale information such as tree height, crown diameter, proportion of canopy cover and, sometimes, the specific botanical composition of the tree layer, moisture in the upper soil layers, early erosion stages, density of human settlement, crops and so on.

The scale may vary from 1/2000 to 1/200 000. The cost has considerably decreased, with the combined use of high resolution emulsions and high altitude (10 000 m) photography from jet planes allowing for small-scale photographs (1/100 000) which can be enlarged four times or more. But the acquisition of an aerial cover is rather costly and can rarely be justified for areas smaller than 1000 km^2, because of its proportionally high cost. Their cost means that aerial photocoverages are not frequently renewed and surveys may be carried out once in 10 to 30 years depending on the countries and areas. This infrequent data collection limits their use in ecological monitoring, since the time gap between successive coverages, is too great. But aerial coverage constitutes currently irreplacable reference data often providing baseline information with which conditions in the field may be compared.

7.3.1.3 Low Altitude Systematic Reconnaissance Flights (Figs. 72–74).

Developed in East Africa in the 1960's, essentially for large game counts and censuses (Norton-Griffiths 1975/78; Gwynne and Croze 1975), the SRF makes it possible to record very fine, detailed and temporary phenomena such as wildlife and livestock numbers, spatial distribution of rain, hydrology, runoff, flooding, seasonal and annual changes in vegetation and so on.

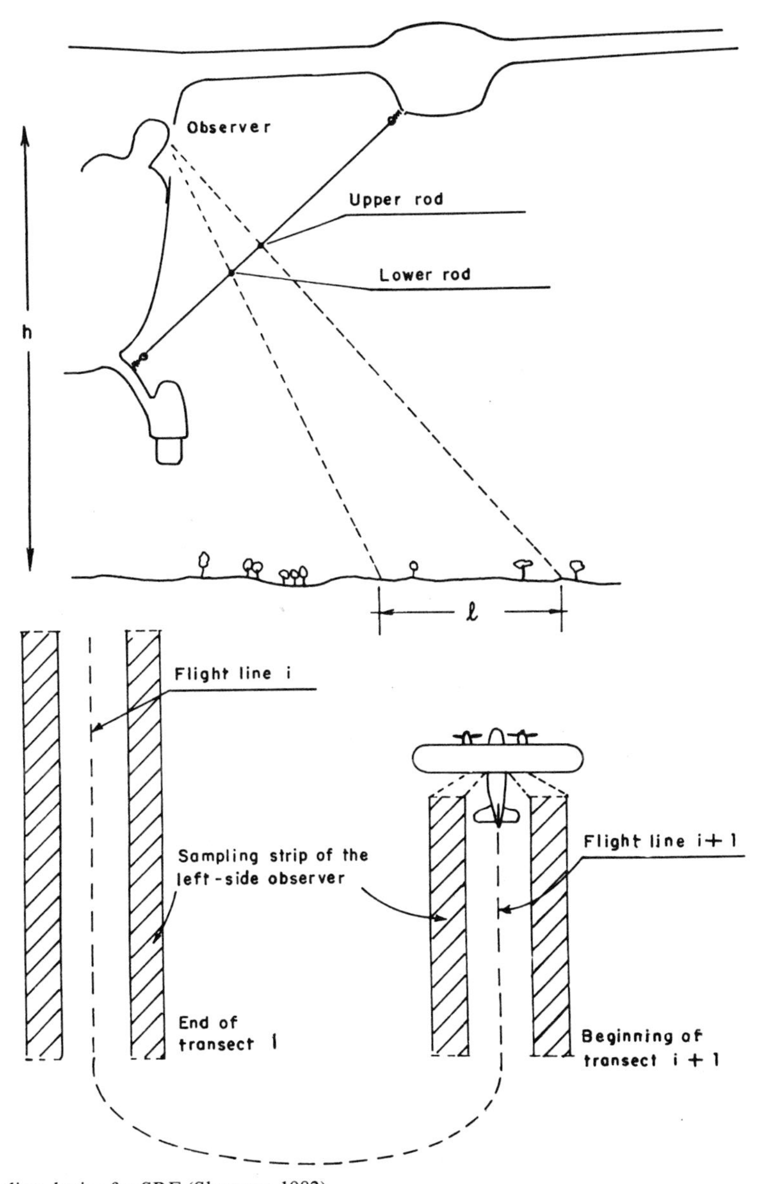

Fig. 72. Sampling device for SRF (Sharman 1982)

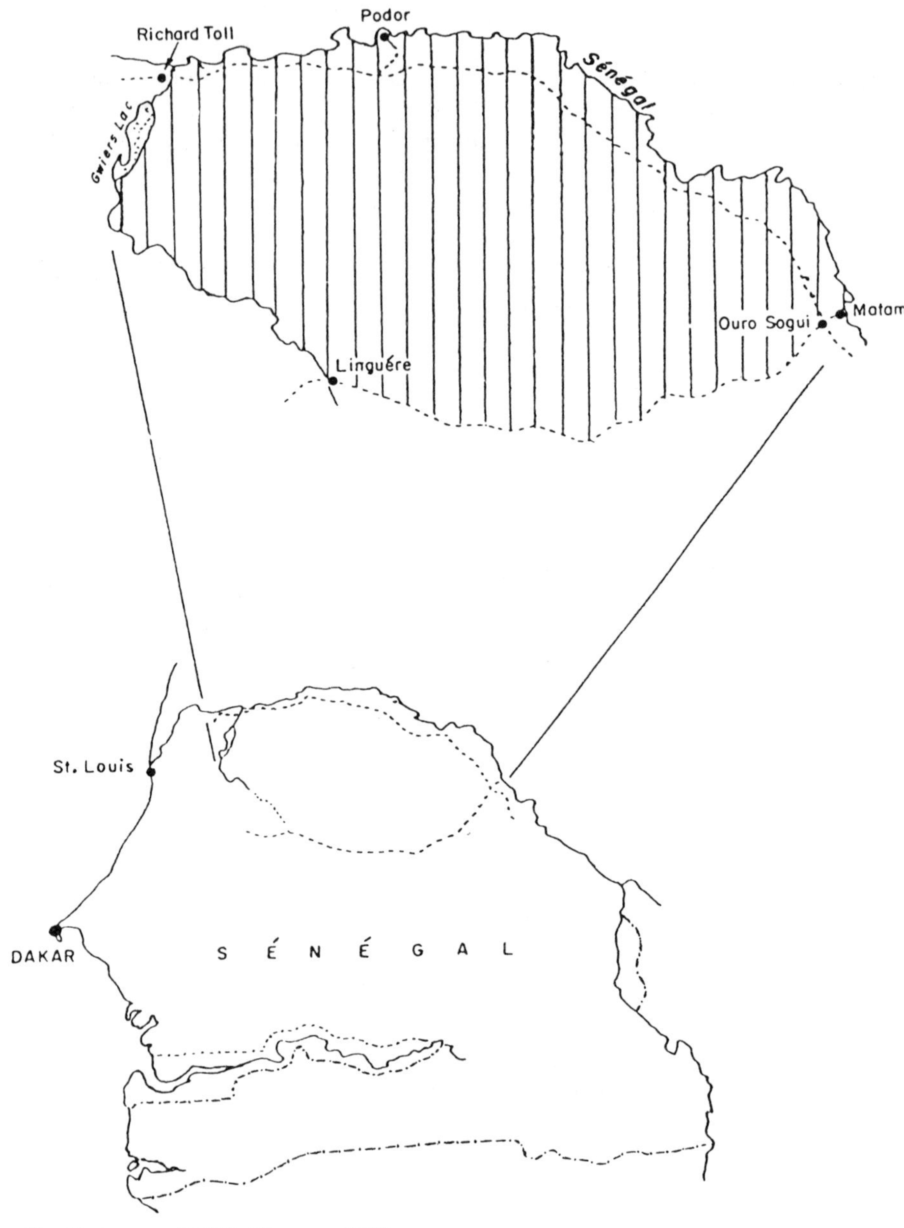

Fig. 73. Layout of SRF's (Sharman 1982)

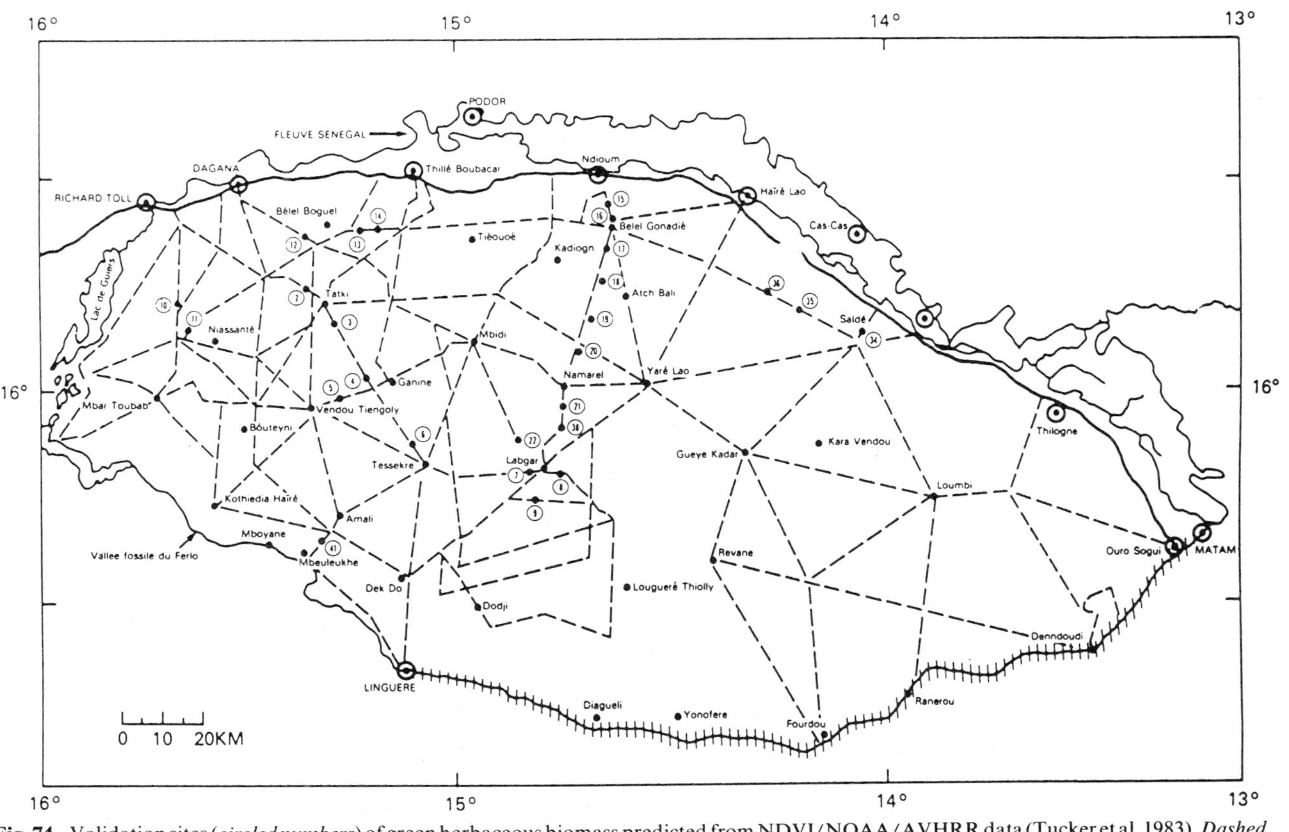

Fig. 74. Validation sites (*circled numbers*) of green herbaceous biomass predicted from NDVI/NOAA/AVHRR data (Tucker et al. 1983). *Dashed lines:* firebreaks, totalling 2700 km

Implementation is less costly than air photography, and as a consequence flights may be repeated at intervals as frequently as appropriate. But SRF's need precise and sophisticated sampling procedures and ground controls for validation. Furthermore, it may be a physically dangerous tool in inexperienced hands.

7.3.1.4 Ground Truth Validation (Fig. 75)

The three remote-sensing methods mentioned above have one common factor: they require validation from the ground in order to calibrate the measurements made from the air or from space and to interpret remotely-sensed data. But collecting data on the ground is time-consuming and costly and a combination of remote sensing and ground control allows for better accuracy at lower cost.

The IMRES project developed its activities along three main lines:

1. Evaluation from satellites of the green herbaceous biomass on the range at the end of the rainy season (maximum standing crop).
2. Livestock census, evaluation of erosion, fire extension and land use via SRF.
3. Control of remotely-sensed data via ground sampling of the information gained from air and space.

Other studies were being carried out during the same period by other teams in the project area that helped to serve the IMRES project's need for ground validation. Of particular note were the diachronic studies of aerial photocoverages on land use, erosion and the evolution of the canopy cover of woody vegetation (De Wispelaere 1980a, b, 1981; Barral et al. 1983). These same organizations (DGRST/LAT/GRIZA, LNERV, ORSTOM, IEMVT, CTFT) were also conducting several ground studies on:

1. Herbaceous vegetation (Boudet 1980, 1981, 1983a; Dieye 1981, 1983; Valenza 1984).
2. Ligneous vegetation (Piot and Diaité 1983; Valenza 1984).
3. Livestock enumeration (Meyer 1980; Planchenault 1981, 1983).
4. Erosion (Valentin 1981, 1983).
5. Socioeconomic and cultural aspects (Santoir 1977, 1981; Barral 1982).
6. Human health and nutrition (Benefice 1980; Benefice et al. 1981).

7.3.2 Evaluation of Green Herbaceous Biomass by Orbital Remote Sensing

7.3.2.1 Introduction

The goal of such an evaluation is to produce at a moderate cost a global, detailed and geographically precise map of the amount of green herbage standing crop available at the end of the rainy season. Such an evaluation should make it possible to establish appropriate range and livestock management goals and to help in designing the strategies to reach these objectives for each geographical

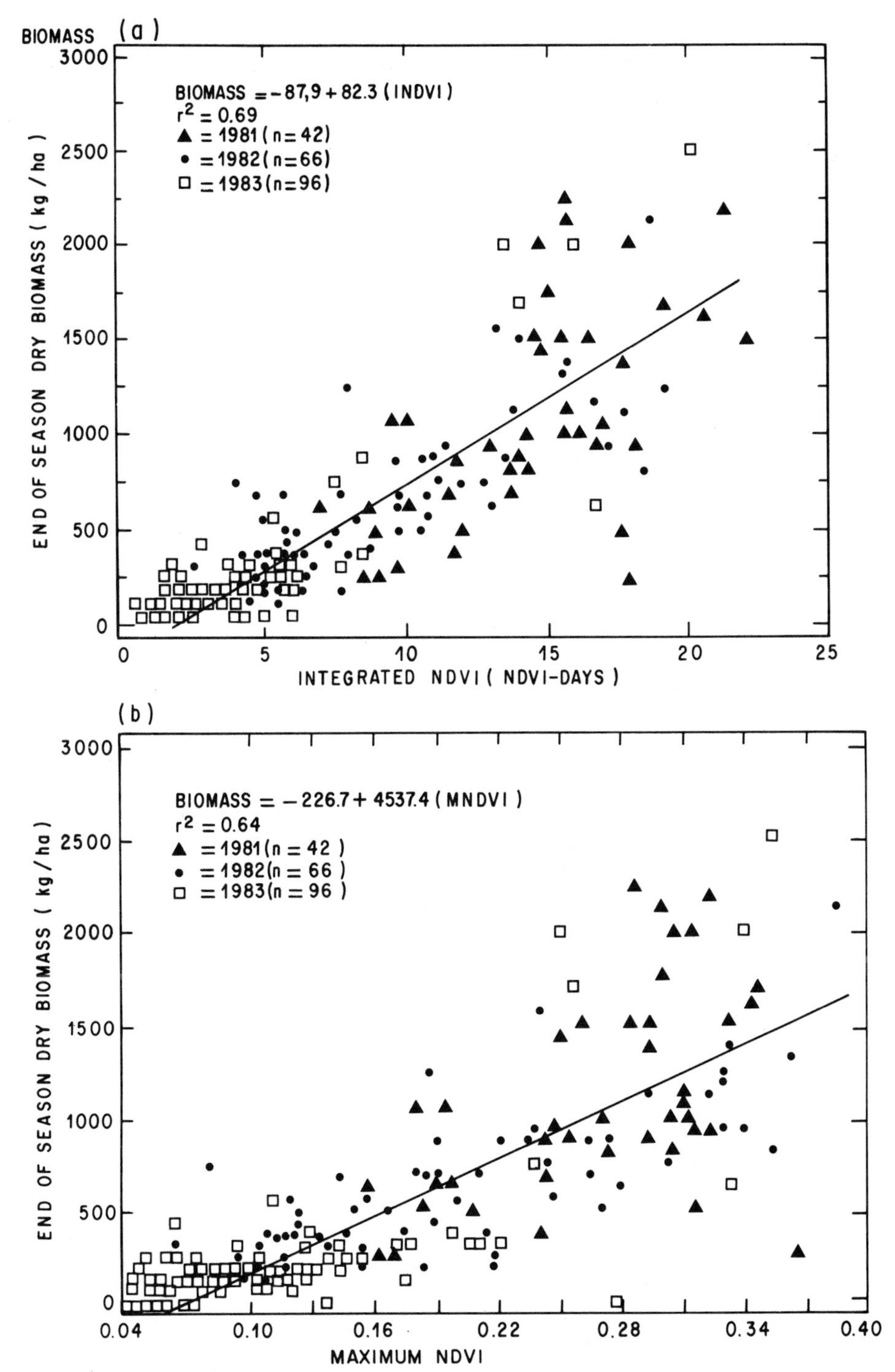

Fig. 75. Relationship between INDVI and MNDVI and ground-sampled biomass (Tucker et al. 1985c)

zone of the Ferlo for every dry season. Decisions which could be based on this information include transhumance dates, exclosures, dates of opening or closing pumping stations, destocking, purchase and marketing of stock, number of animals to be fattened, purchase, dispatching and delivery of concentrated feed, culling rates and stratification.

7.3.2.2 Methods (Figs. 75–79)

The method of measuring green biomass by remote sensing is based on radiation reflected from green plants, which in turn depends on the absorption spectrum of electromagnetic radiation by chlorophyll. Chlorophyll only absorbs radiation included in the visible part of the spectrum, i.e. wavelengths from 0.4 to 0.7 μm, with two areas of maximum absorption towards 0.43 and 0.65 μm, i.e. in the blue and in the red. The area of maximum photosynthetic efficiency for chlorophyll alpha is located in the red, i.e. at wavelengths between 0.6 and 0.7 μm (6000–7000 Å) (Rabinowitch 1959). It follows that reflectance of chlorophyllous plants is minimal for these wavelengths. As a consequence the ratio between the low reflectance of λ of 0.6–0.7 μm (red) to the high reflectance of λ 0.75–6.1 μm (near-infrared) is a good indicator of photosynthetic activity. Furthermore, it has been shown that this ratio is a reliable indicator of the Leaf Area Index (LAI) and of the photosynthetically active biomass (Tucker 1979; Tucker et al. 1981; Holben et al. 1980; Kimes et al. 1981; Markham et al. 1981; Wiegand 1979, etc.).

However, if the use of this ratio is reliable and straight-forward for monospecific and homogeneous plant covers with a simple structure either with erect or horizontal leaves, such as crops on certain types of forests, the situation is more difficult in mixed heterogenous plant populations, the structure and architecture of which are variable and usually complex (Tucker et al. 1983). Such a complexity requires relatively precise and numerous ground checks for each main vegetation type in order to calibrate the relationships between reflectance indices and green biomass actually present. The IMRES project provide a particularly interesting and favourable opportunity to apply the method of diachronic integration proposed by Tucker and co-workers (1981) to the concrete project of rangeland monitoring over a large territory.

It seems appropriate to recall at this stage that this innovation, resulting from the collaboration of the IMRES Project and NASA, was greeted as a break-through by the world's scientific community.

The first attempts made by the IMRES Project used the LANDSAT MSS images. But this was impractical because there were not enough cloud-free images usable for two rainy seasons since 1972. Furthermore, since there is no reception station in West Africa, no images were available from 1981 onward. Moreover, to cover the whole project area and its adjacent Mauritanian zones, it would have required two simultaneous LANDSAT scenes (Tucker et al. 1983). The interpretation of data on the photosynthetically active biomass requires frequent information at a maximum interval of 5–14 days so as to be able to estimate the temporal dynamics of the accumulation of green biomass. An example of this occurred in Central Mauritania from 28 August to 13 September 1982. This period witnessed a clear-cut, but short, burst of green biomass which does not appear on the images of 19 August or 21 September. This herbaceous growth pulse was recorded by AVHRR (individual images and integrated composites) but it would have escaped recording by LANDSAT, even if the cloud cover had permitted. The interval of 18 days between LANDSAT flybys therefore could not serve the project's goal (Tucker et al. 1983).

It was then decided to utilize the Advanced Very High Resolution Radiometer (AVHRR sensor embarked on NOAA 6 and 7 satellites). NOAA 6 and 7 are heliosynchronous, polar-orbiting operational satellites in the TIROS-N series of spacecraft. NOAA operates at an altitude of 850 km with local crossing times of 0230/1430 local solar time. The AVHRR instrument has $\pm$ 56° field of view and a 1.1-km Nadir spatial resolution (Kidwell 1979). The wide field of view (LANDSAT's by comparison is $\pm$ 5.6°) results in effective coverage 3 of 9 days; the orbital period is 9.2 days. The smallest vegetation unit of present interest (i.e. primary production) is constituted by the dune and interdune areas which are usually 1–2 km wide. Thus, the 1.1-km Nadir spatial resolution was well suited to the project's needs. In spite of wider spectral bands than LANDSAT, NOAA 7's AVHRR represented the only useable source of satellite data for conducting large-scale study of grassland green leaf biomass and total dry matter accumulation at high temporal frequency (Tucker et al. 1983; see Figs. 75–81 and Tables 42–46).

Simulation studies using ground-collected reflectances in the AVHRR 0.55–0.68 and 0.725–1.10 μm bands, coupled with the atmospheric models of Dave (1979), indicated an approximately 10–15% uncertainty using uncorrected data over the $\pm$ 20° scanning range of the NOAA-7 AVHRR for a situation typical of the Ferlo (Tucker et al. 1983).

Seven GAC (Global Area Coverage $\simeq$ 4-km resolution) images were available from 12 July to 25 October 1980 from the AVHRR/NOAA-6 data. In 1981 two GAC images from NOAA-6 were available from 13 July to 10 August and six LAC images (Local Area Coverage, resolution $\simeq$ 1.1 km pixels) from NOAA-7. Sixteen images were used in 1982: seven LAC and nine GAC. Sixteen again were used in 1983: 16 LAC and no GAC; while 22 were available for 1984: 14 LAC and 8 GAC (Tucker et al. 1985a, b, c).

7.3.2.3 Data Interpretation (Figs. 80–84)

Interpretation of electromagnetic signals was carried out using the method perfected by Tucker (1979, 1980), called by him the "Normalized Difference of Green Vegetation Index" (NDVI), which is now routinely referred to as the Tucker Index. This index reads NIR-R/NIR + R or else referring to the AVHRR sensor Ch_2-Ch1/Ch2 + Ch1 (see Tables 41, 42, 46 and Fig. 71).

Channel 5 (11.5–12.5 μm) may be used as an interactive cloud mask allowing for the elimination of parasite reflectance from clouds (Tucker et al. 1982). The digital data were analyzed at the Laboratory of Terrestrial Physics of the Goddard

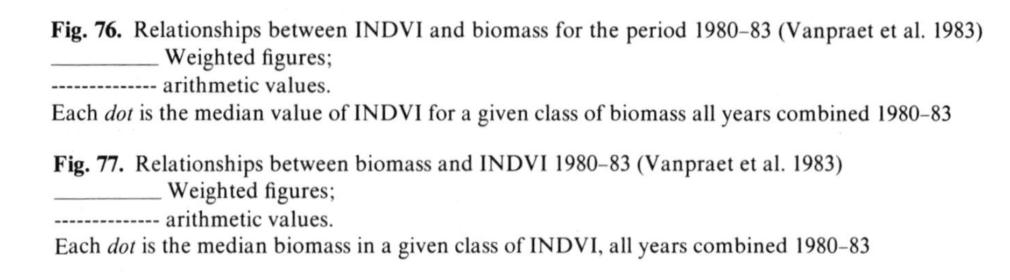

Fig. 76. Relationships between INDVI and biomass for the period 1980–83 (Vanpraet et al. 1983)
__________ Weighted figures;
-------------- arithmetic values.
Each *dot* is the median value of INDVI for a given class of biomass all years combined 1980–83

Fig. 77. Relationships between biomass and INDVI 1980–83 (Vanpraet et al. 1983)
__________ Weighted figures;
-------------- arithmetic values.
Each *dot* is the median biomass in a given class of INDVI, all years combined 1980–83

Fig. 78. Relationships between INDVI and biomass over 4 years, nonweighted figures (Vanpraet et al. 1983)
Each digit is the median value of INDVI for each class of biomass values per annum

Fig. 79. Relationships between INDVI and biomass over 4 years, weighted figures (Vanpraet et al. 1983)
Each digit is the median value of INDVI for each class of biomass values per annum

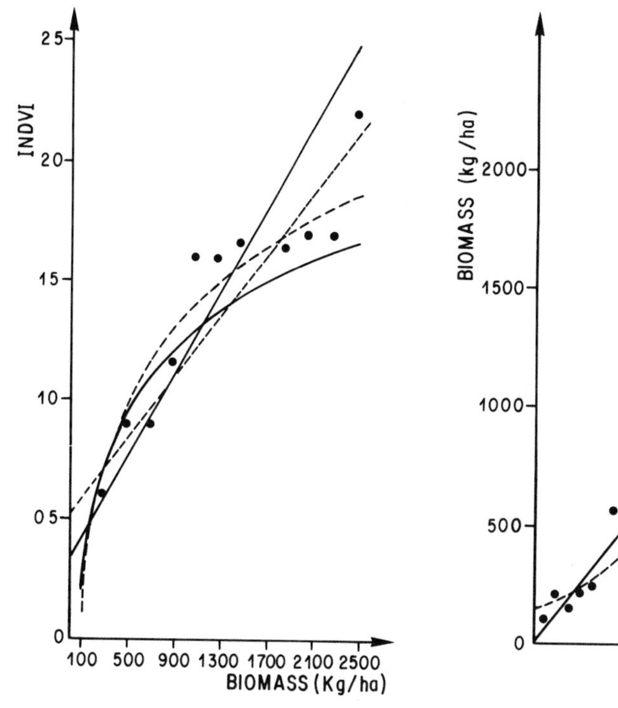

Fig. 76

Fig. 77

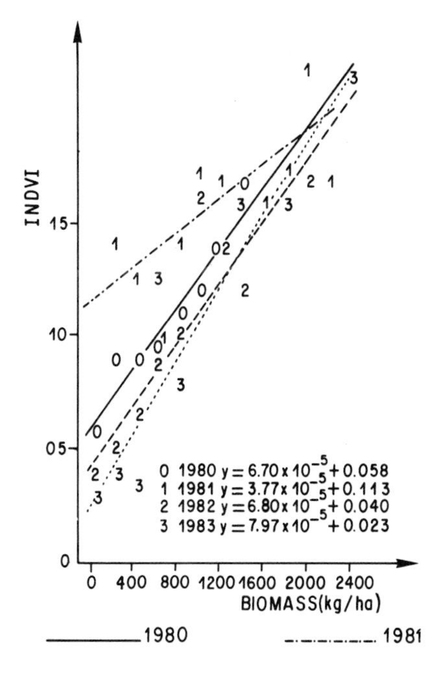

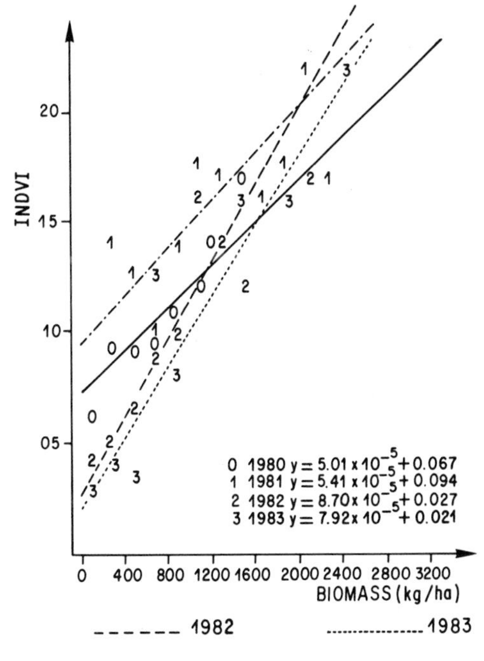

Fig 78

Fig. 79

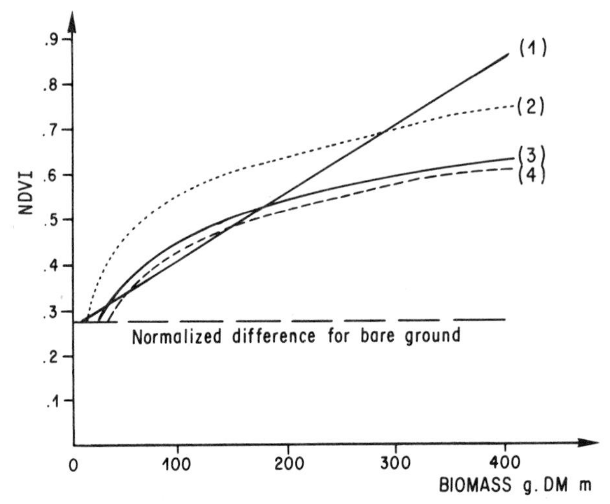

Fig. 80. Theoretical mean normalized difference indices with hand-held radiometers according to linear and logarithmic models for grasses and legumes (Vanpraet et al. 1983). *1* Linear regression all species combined; *2* logarithmic regression, legumes only; *3* logarithmic regression, all species combined; *4* logarithmic regression, grasses only

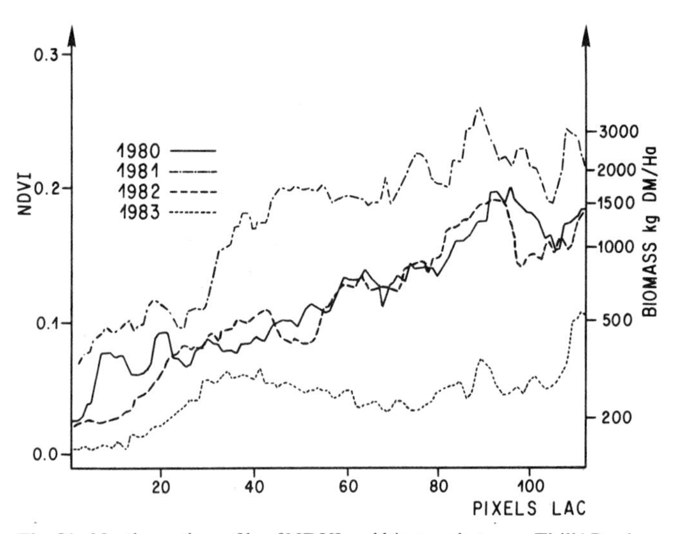

Fig. 81. North-south profile of NDVI and biomass between Thillé Boubacar and Linguére (112 LAC pixels) (Vanpraet et al. 1983)

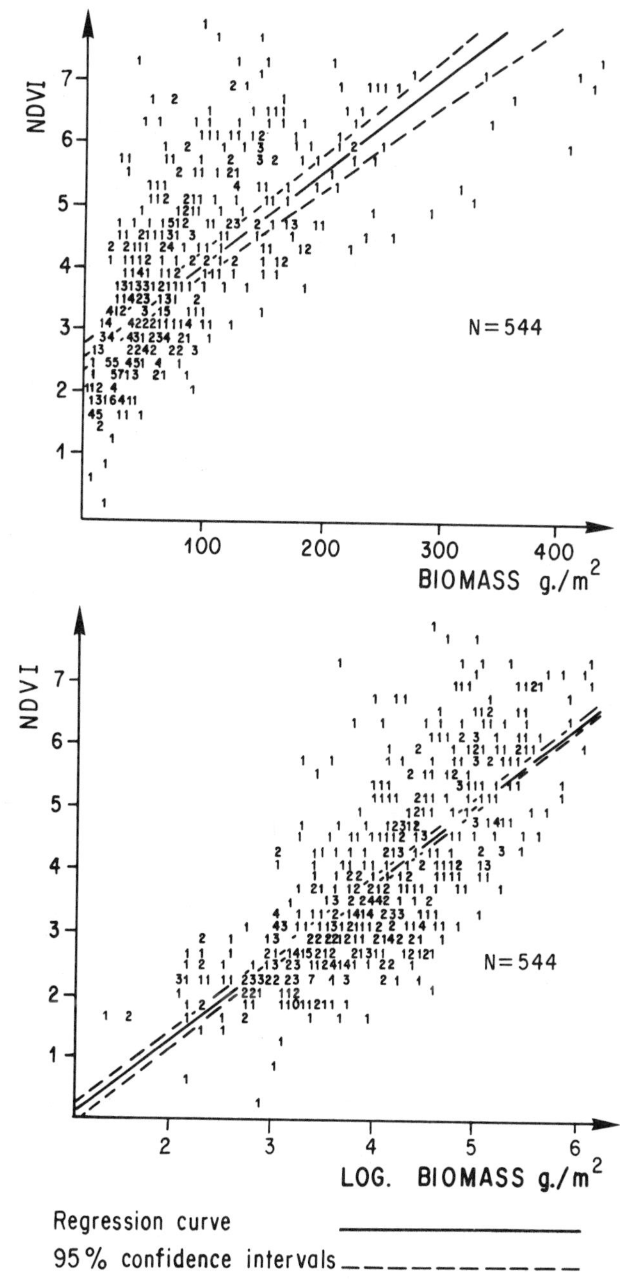

Fig. 82. Relationship between NDVI and green herbaceous biomass, all species pooled together (*Aristida mutabilis, Chloris prieurii, Schoenefeldia gracilis* and *Zornia glochidiata*) 1981–83 (Sharman and Vanpraet 1983)

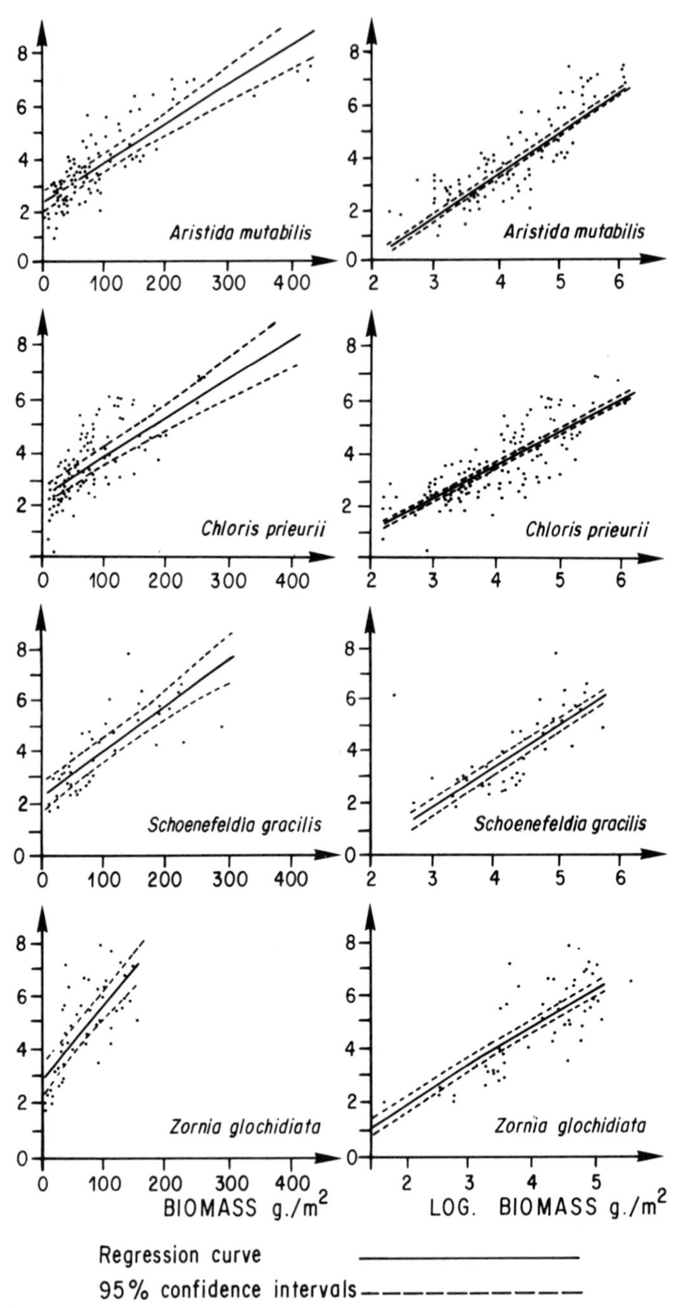

Fig. 83. Relationships between biomass and NDVI as measured with hand-held radiometers, all years pooled together (1981–83) (Sharman and Vanpraet 1983)

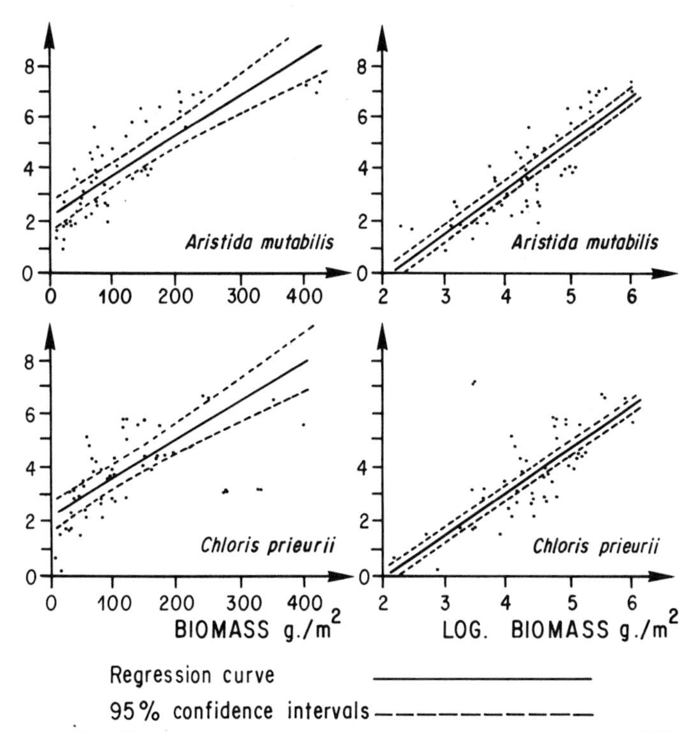

Fig. 84. Relationship between NDVI and biomass for two dominant species of annual grasses in 1981, using hand-held radiometers (Sharman and Vanpraet 1983)

Table 41. Coefficients of correlation between ground-measured and satellite-sensed INDVI and MNDVI (Tucker et al. 1985c)

Years	1980	1981	1982	1983	1981–83	1980–83
No. of observations	68	42	66	96	204	272
INDVI	0.46	0.58	0.78	0.82	0.83	0.79
MNDVI	0.33	0.52	0.77	0.73	0.80	0.76

Table 42. Normalized differences of vegetation index (NDVI) in the Ferlo (Tucker et al. 1985c)

Years	Mean INDVI	Standard deviation	Mean INDVI	Prediction of mean accumulated phytomass (kg DM ha^{-1})
1980	560.7	23.6	0.094	643
1981	590.9	24.5	0.153	1093
1982	553.3	23.0	0.080	536
1983	528.4	11.2	0.033	178
1984	520.7	11.9	0.017	55

Overall regression equation: Y = 74.3 + 7631.5 INDVI

Space Flight Center on a Hewlett Packard 1000 interactive image display system. These images constituted the polychromic visualization (Mercator Projection) of NDVI slides. Integrated images were constructed from a chronological series of images throughout a rainy season for each pixel ($1.1 \times 1.1 = 1.21$ km^2 for LAC and $4 \times 4 = 16$ km^2 for GAC). These images are the Integrated Normalized Difference of the Vegetation Index (INDVI). Images can also be constructed using the maximum NDVI for each pixel over a given period. These are images of the Maximum Normalized Difference = MNDVI. Correlations between ground-measured biomass, on the one hand, and INDVI or MNDVI, on the other hand, have been calculated. The correlation coefficients for the two indices are quite similar at 0.79 and 0.76 respectively, for the overall data between 1980 and 1983, as shown in Tables 43–46 (Tucker et al. 1983, 1985a, b, c). The regression equation between ground-measured biomass and INDVI is

$$Y = -74.3 + 7631.5 \text{ INDVI}.$$

LAC were available every 3 days and GAC every day, which considerably increases the probability of obtaining cloud-free pictures during the rainy season. The correlation between the two indices is high at r2 = 0.97; GAC = 0.95 LAC (Tables 43–45).

The differences between predicted biomass and measured values may arise from:

1. Less than optimal ground sampling;
2. Biomass consumption by livestock (grazing);
3. Presence of atmospheric haze;
4. Crown cover of woody vegetation above 10%;
5. Presence of less than 250 kg DM ha^{-1} of green herbaceous biomass;

Table 43. Normalized difference values from GAC and LAC for 13 Ferlo areas in 1981 (Vanpraet et al. 1983)

Areas	25 August		21 September		30 September		8 October	
	LAC	GAC	LAC	GAC	LAC	GAC	LAC	GAC
Podor	0.019	0.017	0.053	0.052	0.053	0.049	0.042	0.043
Tatki	0.100	0.103	0.115	0.114	0.092	0.096	0.074	0.077
Mbidi	0.090	0.086	0.116	0.119	0.094	0.095	0.072	0.074
Ganine Erogne	0.087	0.084	0.160	0.162	0.135	0.138	0.100	0.101
Namarel	0.076	0.078	0.131	0.134	0.118	0.116	0.083	0.086
Tessekre	0.108	0.106	0.153	0.153	0.143	0.143	0.098	0.098
Labgar	0.086	0.084	0.154	0.140	0.137	0.133	0.092	0.088
Mbeulekhe	0.095	0.099	0.167	0.165	0.143	0.144	0.099	0.101
Lougere Tioli	0.076	0.068	0.113	0.122	0.093	0.093	0.066	0.065
Linguère	0.089	0.092	0.161	0.159	0.145	0.143	0.091	0.091
Salde	0.064	0.045	0.105	0.105	0.090	0.057	0.065	0.049
Matam	0.102	0.097	0.122	0.103	0.106	0.098	0.070	0.065
Dagana	0.059	0.063	0.097	0.099	0.084	0.085	0.068	0.068

Table 44. Median raw values of the normalized difference of vegetation index (NDVI) for given green phytomasses (Vanpraet et al. 1983)

	1980		1981		1982		1983		All years combined	
Median phytomass (kg ha^{-1})	NDVI	n	NDVI	n	NDVI	n	NDVI	n	NDVI	n
100	0.06	5			0.04	4	0.03	49	0.03	58
300	0.09	22	0.14	2	0.05	19	0.04	26	0.06	69
500	0.09	21	0.125	4	0.065	10	0.035	2	0.09	37
700	0.095	12	0.10	6	0.09	11	0.125	2	0.09	31
900	0.11	4	0.14	8	0.10	9	0.08	1	0.115	22
1100	0.12	2	0.175	6	0.16	5			0.16	13
1300	0.14	1	0.17	5	0.14	3			0.16	9
1500	0.17	1	0.21	3	0.12	1	0.16	1	0.165	6
1700			0.16	1					0.16	1
1900			0.175	2			0.16	2	0.165	4
2100			0.22	3	0.17	1			0.17	4
2300			0.17	1					0.17	1
2500							0.22	1	0.17	1

Table 45. Sampling of hand-held radiometers on herbaceous vegetation (Gaston 1983)

Year	No. of biomass sampling sites	No. of sites used in Mark II calibration	No. of 1-m^2 quadrats sampled per site	Total number of calibration measurements
1981	42	27	13	198
1982	70	48	5	240
1983	112	18	10	180
Total	224	93	--	618

Table 46. Annual variation in precipitation and herbaceous phytomass predicted from satellite NOAA/AVHRR remotely-sensed data on six sites of 3 × 3 km centred on reliable rainfall record station and rain-use efficiency (RUE) (Tucker et al. 1985)

Stations	Years					
	1980	1981	1982	1983	1984	Average
Mbeuleukhe	349	465	412	79	181	297
Mbidi	210	317	141	139	84	178
Tatki	---	351	278	100	92	205
Tessekré	382	304	300	118	171	255
Vendo-Tiengoli	207	193	196	121	117	167
Mbar Toubab	245	298	193	119	77	186
Mean (mm)	279	321	236	113	120	214
Predicted Biomass	(700)	1375	780	285	112	650
RUE	2.51	4.28	3.31	2.52	0.93	3.04

6. Errors in the collection and processing of data;
7. Different people measuring ground samples, errors in site location. Finding exact sites on the ground is difficult in the Sahel because of the lack of clear landmarks.

The regressions shown in Fig. 77 suggest a lack of sensitivity of satellite data for green biomass below 250 kg DM ha^{-1} (Tucker et al. 1985c), but correlations between satellite data and ground validation are very good whenever annual rainfall reaches 250–300 mm or more (Tucker et al. 1985b).

7.3.2.4 Image Exploitation: Construction of Biomass Maps (Figs. 85–90)

Digital NDVI figures are used to construct images by the computer programme mentioned above. Images are available in various forms: as paper colour prints, colour transparencies (19 × 24 cm) and 35-mm slides. Computer analysis may provide images at various scales according to the area to be processed and the detail that is required.

Often the scale used is 1/3 000 000, which is sufficient to detect each colour pixel. Occasionally the scales of 1/1 000 000 or 1/2 000 000 are used in more complex areas. Within the IMRES Project, maps were constructed from 19 × 24 cm paper prints of INDVI's or from enlarged 35-mm slides. Maps used scales of 1/500 000, so that each GAC pixel covers 20 mm^2 (= 16 km^2 on the ground) and each LAC pixel represents 5 mm^2 (1.21 km^2 on the ground). Figures 86–90 depict the green herbaceous biomass for 1980–1984. These maps were made available to the users in October-November each year.

When delivered to the users (DSPA, SODESP, LNERV, etc.) each map was accompanied by an explanatory note and directions for use, including comments on the overall situation and with qualitative notes on pastoral value. This information was collected during the sampling for ground validation. The regional forage budget was established as follows:

1. Mapped isoproduction lines based on satellite data, preferably from integrated NDVI at the end of the growing season. This document allows the primary production of herbage to be evaluated spatially. Green biomass is divided into 12 classes of 200 kg DM ha^{-1} each corresponding to 12 brackets of Tucker's Index values and their matching colours (Gaston et al. 1983; see Tables 45–47) and Figs. 83–85.

2. From the digital data, one can derive the mean, the median or the mode of vegetation indices from each pixel within the area under consideration. Note that reflectance from woody cover is integrated in the satellite data, whereas it is not in the hand-held radiometer data. The contribution from the woody layer could in principle be subtracted from the final integrated seasonal index value by subtracting the early rainy season index values from the final ones. But this was not judged useful nor desirable because the woody layer con-

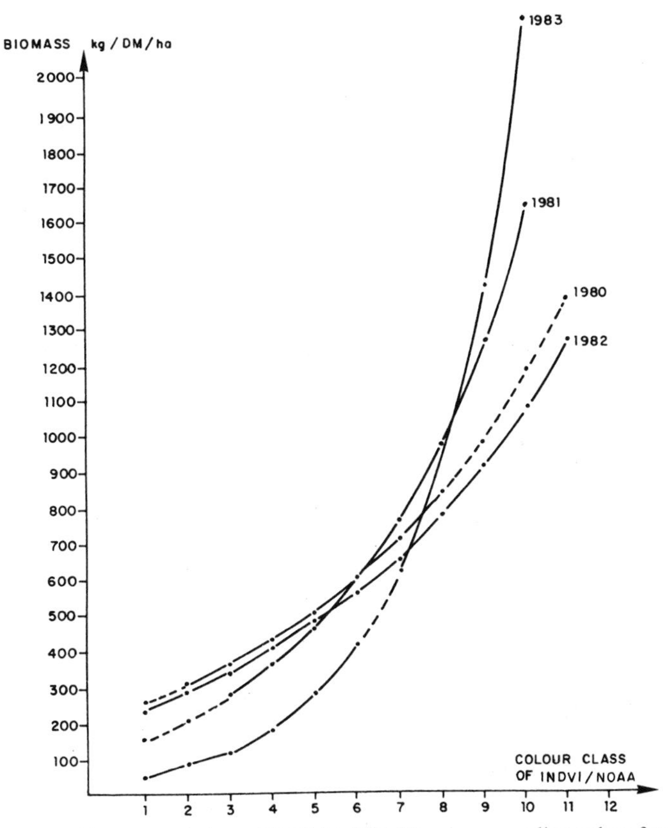

Fig. 85. Relationships between biomass and class of NOAAA/INDVI colours; median values for 1980–1983; *dashed parts* correspond to colour classes for which there were no ground validation data (Gaston et al. 1983b)

tributes significantly to the overall forage biomass and livestock feeding (Le Houérou 1980b; Sharman and Gning 1983).

It is regrettable that the percentage of crown cover of the woody layer and its foliar biomass was not recorded in the field-sampling procedures. Including green biomass of woody species would not have greatly increased the work load of the surveyors since foliar biomass can be related to allometric parameters which are easy to measure (stem diameter, height and crown diameter). The relationship between these parameters had been previously studied in the Ferlo (Bille 1977; Poupon 1980). Moreover, the "in-situ" study of the woody layer had been undertaken on 1-ha plots over 40 sites of the Ferlo from 1979 to 1981 (Piot and Diaité 1983; Valenza 1975). The data gathered by this LNERV-DGRST/CTFT were not used by the IMRES Project which instead undertook a study of 25 1-ha plots on the Mbidi dune and followed through some of the observations carried out by the IBP/ORSTOM Project at Fété-Olé from 1969 to 1978.

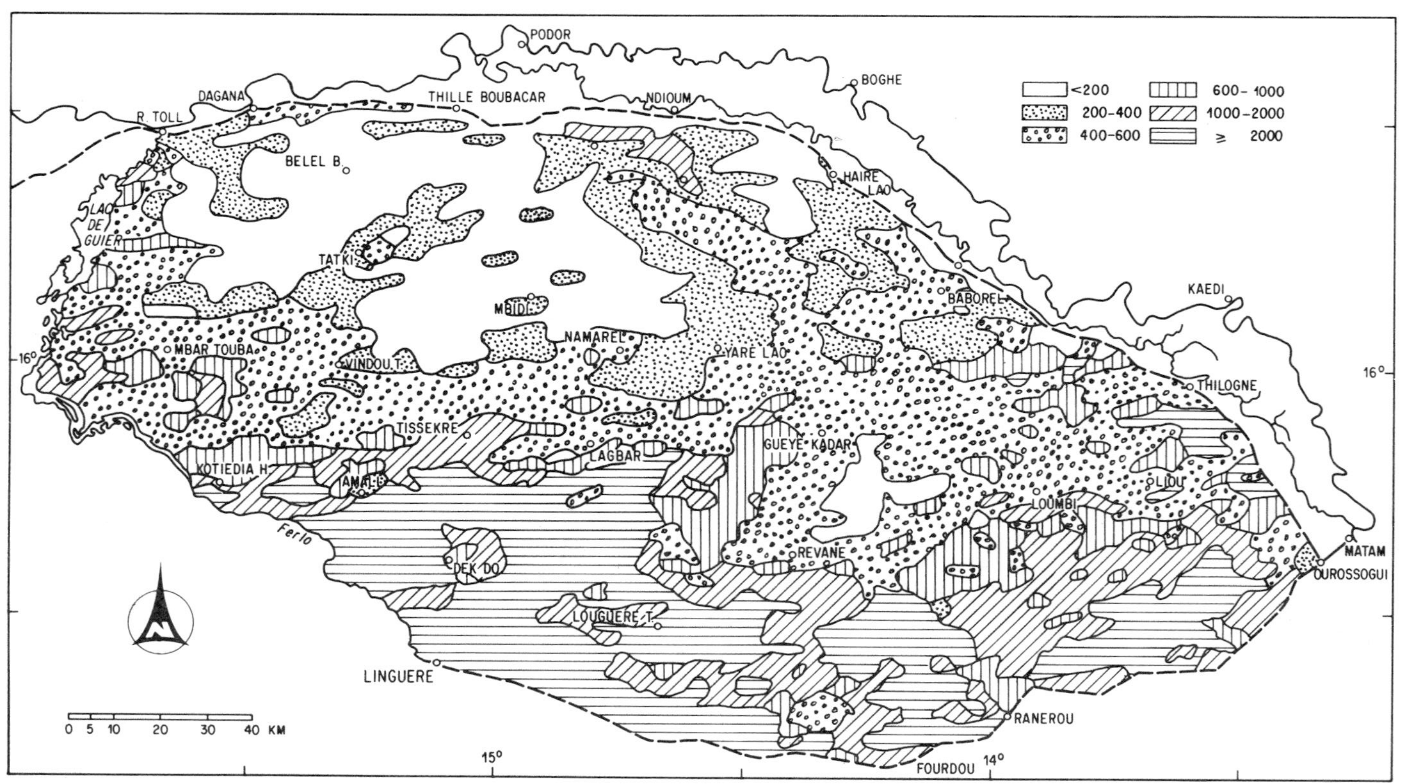

Fig. 86. Map of green herbage biomass in the Ferlo at the end of the 1980 growing season

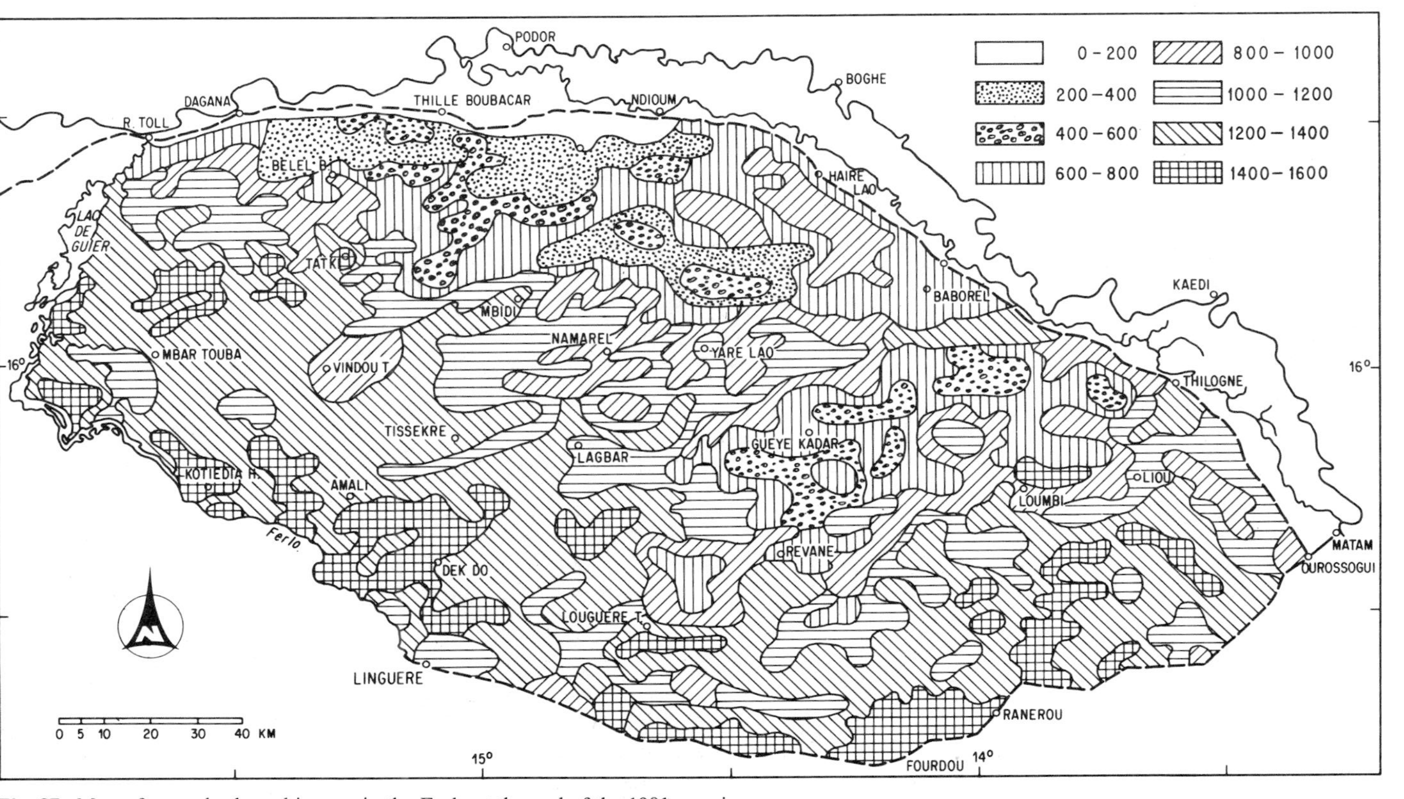

Fig. 87. Map of green herbage biomass in the Ferlo at the end of the 1981 growing season

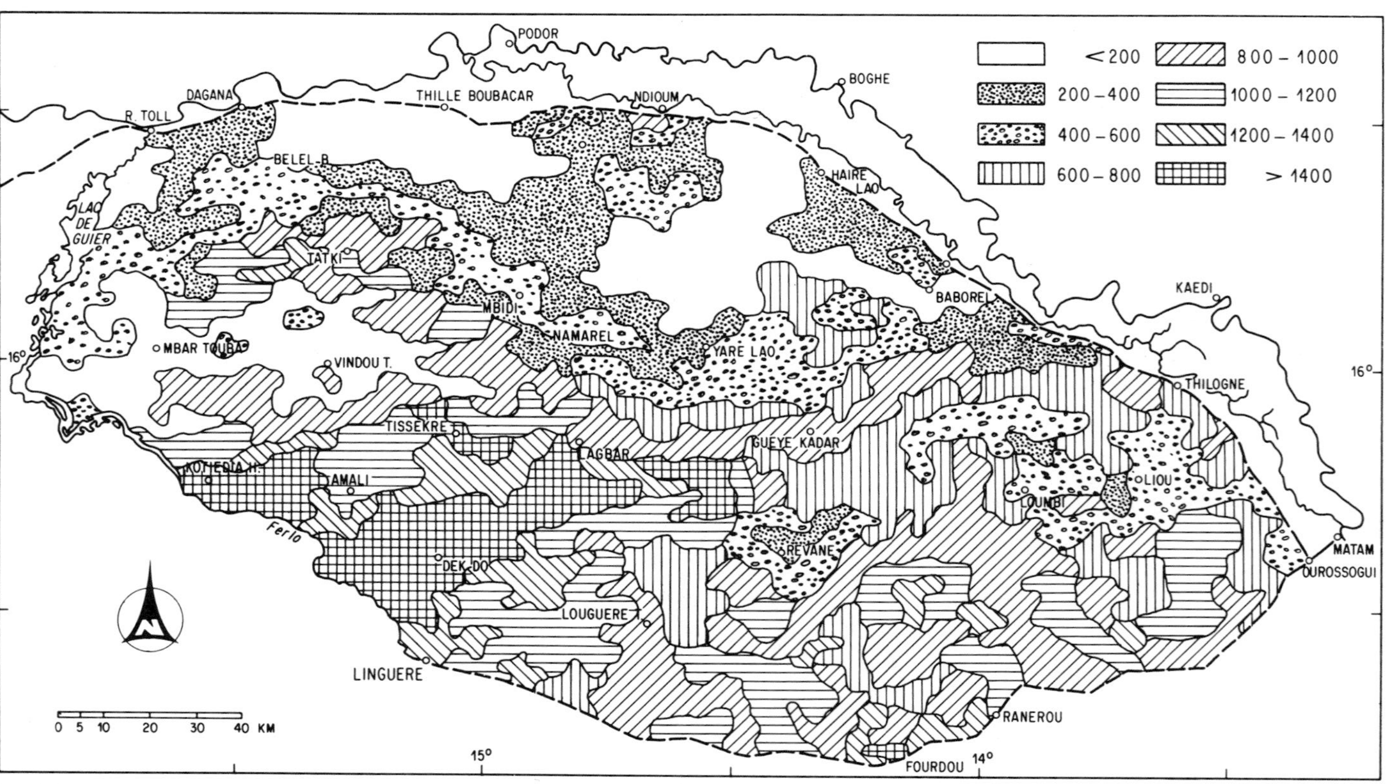

Fig. 88. Map of green herbage biomass in the Ferlo at the end of the 1982 growing season

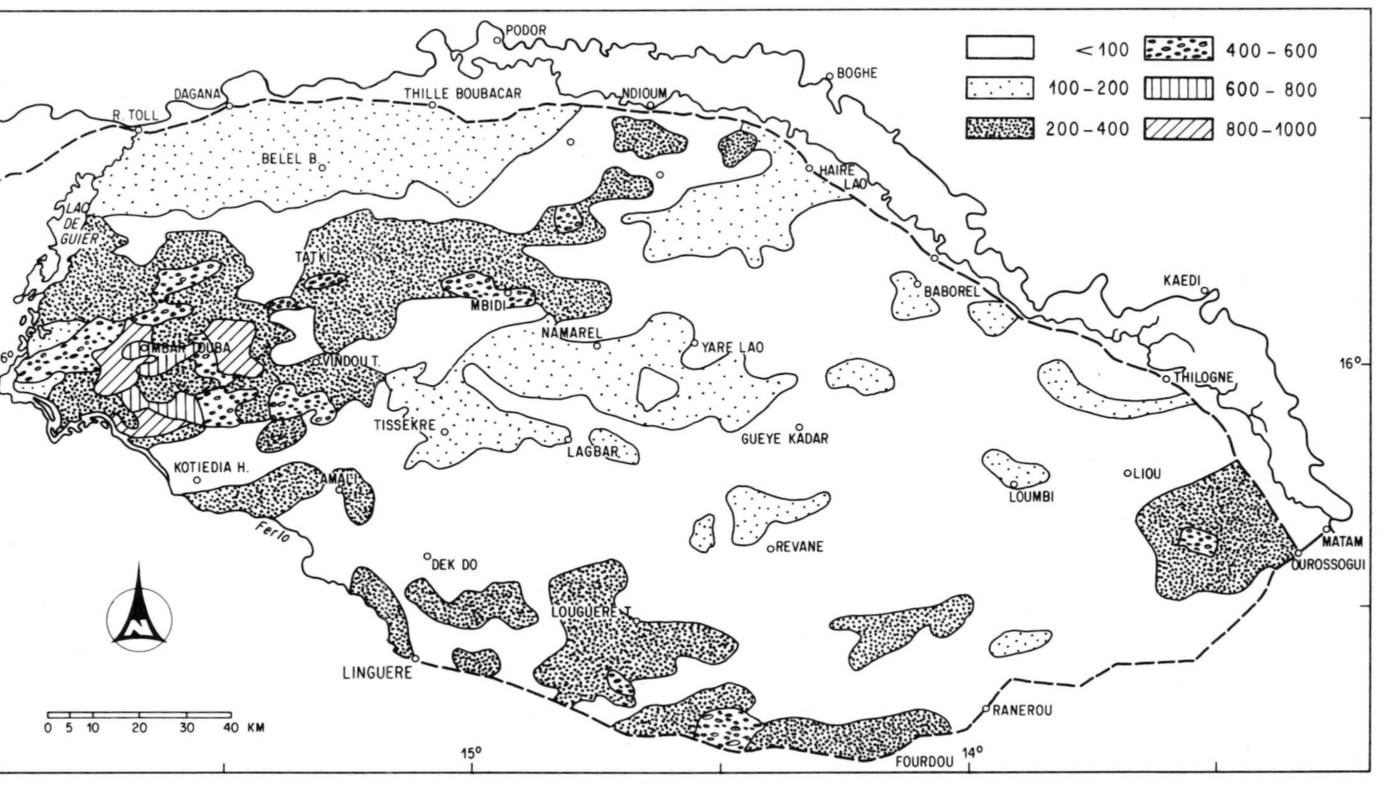

Fig. 89. Map of green herbage biomass in the Ferlo at the end of the 1983 growing season

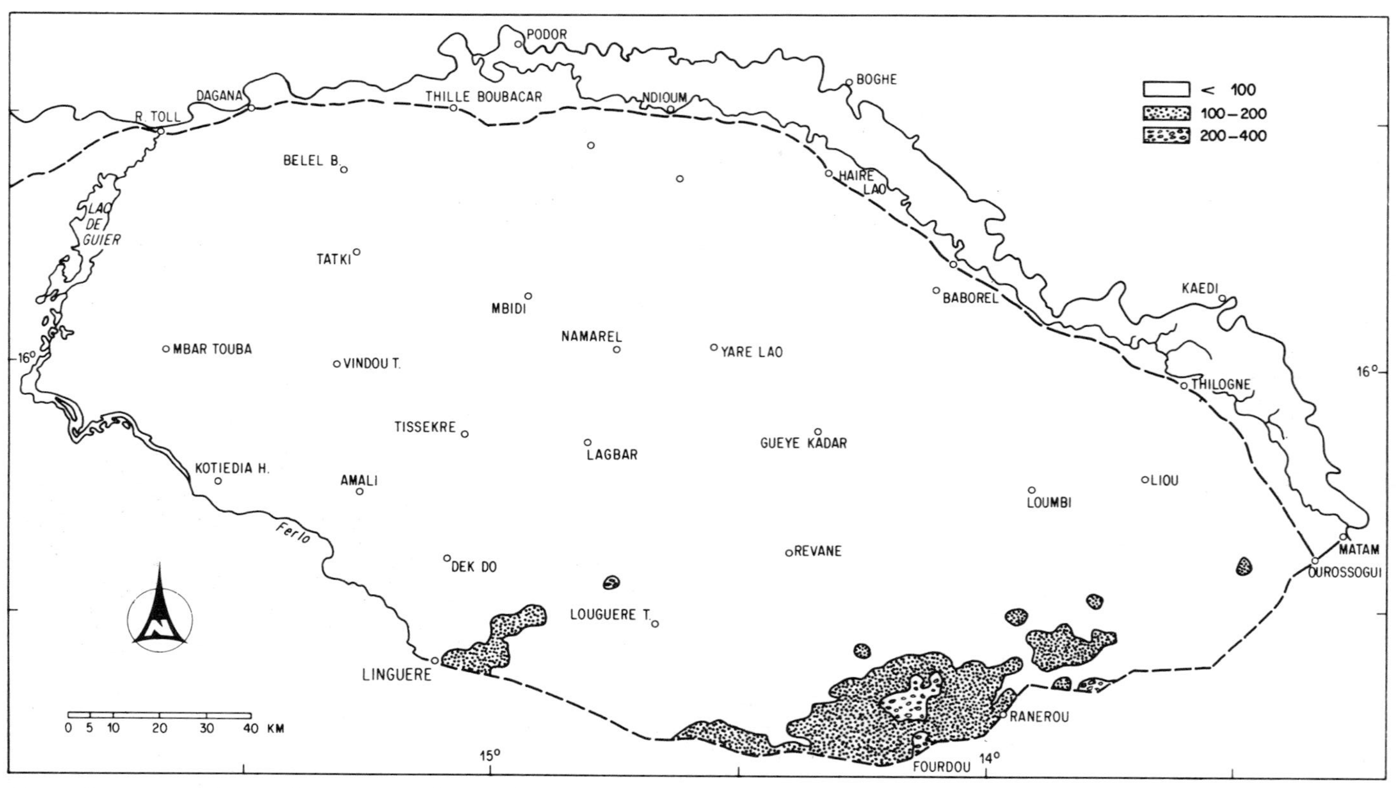

Fig. 90. Map of green herbage biomass in the Ferlo at the end of the 1984 growing season

Table 47. Matching normalized difference of vegetation index (NDVI) colours with biomass classes (kg DM ha^{-1}) as established at optimum of photosynthetic activity (Gaston et al. 1983a)

Phytomass class	0–200	200–400	400–600	600–800	800–1000	1000–1200	1200–1400	Over 1400
MNDVI class								
GAC 8/9/80	1	2–3	4–5	6–7	8	9–10	11–12	
LAC 21/9/81	1–2	3	4–5	6	7	8	9	10–1
Composite LAC up to 28/8/82	1–2	3–4	5–6	7	8–9	10	11	
Composite LAC 27/8 and 1 to 3/9/83	1–3	4–5	6	7	8			

7.3.3 Evaluation of Range Production from Ground Sampling

7.3.3.1 Introduction

Ground samples were used to calibrate the satellite data from the AVHRR. Ground sampling was made in two ways: (1) by the classical method of hand clipping and weighing a number of 1-m^2 plots at a number of sites and (2) by using a hand-held radiometer, the Mark II apparatus, designed by NASA, the characteristics of which are similar to the AVHRR embarked on NOAA 6 and 7 (Tucker et al. 1981). The field radiometer thus has the same three channels as AVHRR in the red, near-infrared and thermal infrared wavelengths. The device is powered by batteries available locally. Hand clippings were made in order to calibrate the radiometer and thus improve its utility in Sahelian rangelands. During the same period, other diachronic studies on primary production were being carried out in the Ferlo by two different research teams in close cooperation with LNERV (Boudet 1981, 1983a; Piot and Diaité 1983; Valenza 1984).

7.3.3.2 Sampling

Ground sampling was stratified. It was first undertaken in 1974 in order to monitor the impact of boreholes on the vegetation of the pastures in the Ferlo.

Permanent plots were selected on aerial photographs on the following criteria:

1. Pasture type based on the 1/200 000 map by Valenza and Diallo (1972).

2. Distance from boreholes along specific directions oriented as a function of the range types crossed (as indicated on the pasture land vegetation map and checked on the site). Permanent plots were located at 0.5, 2–3, 4–6 and 8–10 km

from each borehole under study. Three boreholes were monitored from 1974 to 1976 and six from 1976 to 1984: Tatki, Mbidi, Viendou-Tiengoli, Amali, Tessekré and Labgar. This resulted in a total of 16 transect lines and 60 sites (Valenza 1975, 1981, 1984). The IMRES Project monitored 30 sites along a transect from Ndioum to Namarel, from 1981 to 1983 (Tables 41, 45, 46; Fig. 74). In addition, as mentioned before, 25 1-ha plots were established near Mbidi to monitor woody vegetation. At the same time, the teams mentioned above monitored 112 sites for herbaceous production and 40 sites for browse.

3. Some of the results on herbaceous production are summarized in Tables 14–46. A total of 202 sites were sampled in the Ferlo for herbaceous data, while 68 other sites were concerned with the monitoring of woody vegetation. The monitoring took place over periods varying from 3 to 10 years. Such a density of data over time and space is unique in the Sahel.

4. Observations were made once each year in September-October, at the time of the herbaceous Maximum Standing Crop and in December-February for the woody layer.

5. Observation plots are marked with a ring of brightly coloured paint (usually red or yellow) on the stem of four trees demarcating a quadrangle of about 1 ha. The position of plots is indicated by other landmarks along tracks and firebreaks, etc.

6. The observations and measurements taken were:

a) Canopy cover of herbaceous biomass and percentage of bare ground using line interception (Canfield 1941) along the perimeter of each 1 ha plot and its two diagonals. Three classes of ground cover are recorded (low, $< 20\%$; medium 20–80%; and high $> 80\%$).

b) Mean herbaceous height.

c) Botanical composition. Two 20-m lines were selected near the plot center, one of which lay parallel to the contour and the other perpendicular. The degree of precision (confidence interval) must be equal to or lower than 5% according to the formula:

$$\pm 2 \frac{n\ (N - n)}{N^3} \times 100,$$

where N is the total number of individual plants recorded and n is the number of individuals of the dominant species. This measurement gives the frequency of each species on each site for each year. In the DGRST/LAT/GRIZA studies the interception line was replaced by the Point-Quadrat Method (Levy and Madden 1933) with a reading every 20 cm, i.e. 100 readings per line and thus 200 readings per site. Along each transect line a number of 1-m^2 quadrats were selected for hand-clipping and weighing of herbage. The grass and herbs were clipped 5 cm above ground in the LNERV studies and at ground level in the DGRST sites. The number of quadrats clipped per site depended on the homogeneity of the grass layer, usually being 10–20 and occasionally up to 30 (Levang and Grouzis 1980). The number of quadrats was determined by the cumulated average weight per quadrat so as to reach a confidence interval of 5% or less:

$$p = 2\sigma/m^n,$$

where p is the confidence interval; σ = standard deviation; m = mean;

n = number of quadrats sampled.

In high and heterogeneous herbaceous layers, the parameter n + 1 was used (Boudet 1983). Herbage was clipped and weighed by cumulative weighings up to 1.5 kg DM so as to reduce manipulation errors arising from very light samples.

At the IMRES sites, the number of 1-m^2 quadrats varied from 5 to 13 (Table 47), a figure that seems less than optimal.

d) Grazing value of the herbaceous vegetation.
The plant species recorded and collected were classed into eight categories see Tables 14 and 15; Boudet 1983a):

Grasses of good grazing value	G3
Grasses of medium grazing value	G2
Grasses of low grazing value	G1
Legumes of medium grazing value	L2
Legumes of low grazing value	L1
Miscellaneous species of medium grazing value	D2
Miscellaneous species of low grazing value	D1
Unpalatable species	0

This method allows a synthetic grazing value index to be determined according to the method set up by De Vries and De Boer (1959) and modified by Daget and Poissonet (1965). The latter method results in a grazing index value from 0 to 100. Following this method the mean GVI of the Ferlo grasslands varies from 69 to 83% depending on range type (see Table 14).

e) The woody layer. Shrubs and trees were inventoried for each site over an acreage of 1 ha, i.e. a 56.40-m radius circle. The centre of this circle is the centre of the site plot (Piot and Diaité 1983; Gaston and Boerwinkel 1982). The individual "trubs" are divided into ten height classes of 25 cm up to 2 m, 2–4 m and above 4 m. Observations are made in December-February. For each individual, its species, stem circumference at 10 cm above ground and vegetative stage and vigor were recorded. In multistemmed individuals only the larger stem was recorded, but the total number of stems was mentioned. At the Mbidi sites the 1-ha plot method was compared with the classical Point Centered Quadrat Method (Cottam and Curtis 1956). The two methods gave quite similar results (Gaston and Boerwinkel 1982).

Each individual "trub" in each plot is thus given an index card. All cards may be grouped in various ways according to site, species, year, serial type, vegetation type, vegetation stage or vigor or size, age, distance from borehole, etc. and finally according to ligneous communities and range types as a function of:

- Presence/absence
- Density/frequency
- Demography (age/size classes)
- Structure (vertical/horizontal)
- Vegetative stage
- Productivity. Woody/forage-perennial/deciduous
- Environmental condition: ecoclimatic zoning, soil type, drought, distance from borehole, etc.

Forty sites belonging to five range types and 14 soil types have thus been monitored by the DGRST/CTFT/LNERV team since 1979 (Piot and Diaité 1983; Valenza 1984).

7.3.3.3 Calibration of the Hand-Held Radiometer (Gaston et al. 1983a, b; Sharman and Vanpraet 1983) (Figs. 82–84)

Calibration of the field radiometer aims at fulfilling two main purposes:

1. To study the relationships between data from destructively sampled biomass (hand clipping) and the Tucker Index (NDVI). These correlations should make it possible to determine the green biomass from the radiometric indices produced by AVHRR of NOAA 6 and 7.
2. To find a quick, easy and reliable non-destructive evaluation of green herbaceous biomass which would constitute a particularly efficient monitoring tool since it would warrant a much larger number of data to be collected with given logistic means and personnel.

The radiometer used is the Mark II type, built by NASA, having the same sensing channels as AVHRR, as mentioned above. The principles behind these measurements are discussed in Section 4.2.2.

The outfit looks like a box 20 × 17 × 17 cm in size, held hanging from the neck like a camera. It is equipped with three dials corresponding to the three channels showing digital displays graded in W m^{-2} and powered by commonly available batteries. The radiometer is equipped with a handle by which it can be held facing the sun about 130 cm above ground when reading. Before and after each reading the device is calibrated above a plate coated with barium sulphate, thus providing the maximum reflectance under the prevailing conditions; the latter is amenable to changes according to atmospheric transparency and sky brightness. Measurements are only taken under clear sunshine. The total number of calibrating measurements carried out was 618, as shown in Tables 45–47.

The calibration studies led to the following conclusions:

1. The radiometer data are all the better as they come from monospecific or quasi-monospecific grass populations with a strong dominance of one single species. The regression curves are given in Figs. 82–85.
2. The codominant or subdominant species introduce a bias that may be very important compared to monospecific or quasi-monospecific stands.
3. The apparatus cannot be used in a reliable way in heterogeneous vegetative covers, particularly in years with deficient rainfall when herbaceous cover is scarce, sparse and irregular. Such a situation would often apply to the Saharo-Sahelian ecoclimatic zone with mean rainfalls below 200 mm.
4. Calibration may vary from one year to the next as a function of the physiognomy and botanical composition of the herbage, which is a reflection of rainfall and rainfall events.

One may conclude that the radiometer may be profitably, although cautiously, used in the evaluation of green herbaceous biomass in order to improve the output and efficiency of observers in the monitoring exercise of Sahelian rangelands. But the radiometer calibration should be continuously monitored and its utilization by non-specialized personnel should be avoided.

Similar conclusions have been drawn from a study in the Sahel of Burkina-Faso (Grouzis and Méthy 1983). The same remarks emerged from measurement campaigns in northern Kenya by an IPAL team (Herlocker and Dolan 1980).

The correlations evidenced warranted a satisfactory statistical interpretation of orbital NOAA 6 and 7 data for the evaluation of annual green herbaceous biomass in the Ferlo of Senegal (Tucker et al. 1983, 1985a, b; Vanpraet et al. 1983).

Further research is envisaged using SRF at an altitude of 300–500 m so as to integrate the woody layer in the radiometric measurements and thus come closer to the NOAA data using three levels of perception: ground, air and space. This should permit a refinement of the method (De Leeuw 1983).

7.3.4 Low Altitude Systematic Reconnaissance Flights (SRF)

7.3.4.1 Introduction

This method was perfected in the 1960's by zooecologists in East Africa for the inventory of large mammals. Several synthetic studies on this method have been published (Jolly 1969; Pennicuick et al. 1977; Norton-Griffiths 1975/78; Gwynne and Croze 1975, etc.). The IMRES project carried out four SRF over the whole project area at the end of the dry seasons: 1980–1983. Two more were planned that could not be achieved, the first due to a breakdown in the navigation system of the airplane usually utilized and the second in 1984 when most of the stock fled the project area due to extreme drought. Several IMRES project reports were devoted to SRF (Sharman 1982, 1983a, b). The main themes examined were:

1. Game numbers;
2. Livestock numbers;
3. Ligneous cover;
4. Bare soil;
5. Bushfires;
6. Aeolian erosion;
7. Water erosion;
8. Distribution of standing hay.

7.3.4.2 Methods and Means

The aircraft used is a light, twin-engine, high-wing six-seater (Partenavia P68) belonging to ILCA. The craft, based in Mali, was equipped with a sophisticated navigation system (Global Omega) coupled with an automatic pilot system which, in principle, permits accurate repeated flights along the same lines. The aircraft is also fitted with a radar altimeter with digital recording warranting elevation control of ± 10′ (3 m). This altimeter may also be coupled with the automatic pilot system. Flying altitude was 150 m ± 10 (510′ ± 30). The wings' struts were fitted with two wood or fiberglass rods that permitted the delineation of a 200-m-wide sampling belt on each side of the flight line (Fig. 72). Flight lines were 10 km distant in a N-S direction (Fig. 73). The sampling rate is thus 4%. These lines may be transcribed to a UTM Universal Mercator Transverse grid. The selection of the sampling procedures results from many statistical considerations that cannot be developed here (see Jolly 1969, 1979, 1981; Jolly and Watson 1979; Pennicuick et al. 1977; Norton-Griffiths 1975/78; Gwynne and Croze 1975; Smith 1981; Watson et al. 1977; Watson and Tippett 1981 a, b; Milligan et al. 1979; Grimsdell et al. 1981, etc.).

The N-S direction is dictated by the geomorphology and particularly the sand dune ridges and longitudinal depressions and also by the borehole layout which commands stock distribution patterns in the dry season. The mean flying speed was 212 km h^{-1} ± 20. Deviations from preset transect lines did not exceed 2 km over a maximum distance of 140 km except for 4 of 30 lines where they reached 2 km. These deviations do not seem significant where the stratification of the sampling of spatial distribution of livestock is concerned. Flights always took place in the morning between 07.00 and 11.30 AM. The total length of lines flown in each SRF was 3070 km with a flying time of 14.30 h (i.e. an average speed of 212 km h^{-1}). Flying is virtually impractical in the afternoons due to the high temperatures during that season (June > 40°C), poor visibility due to haze, aircraft instability resulting from thermal turbulence, observer weariness and also because animals are more difficult to see as they tend to rest in the shade during the hot hours.

Statistical interpretation is ticklish, mainly because of the contagious distribution of livestock, particularly in the dry season when they tend to concentrate around boreholes. This type of distribution produces a high variance and therefore poor precision of the results. This variability may be reduced, to some extent, by dividing the flight lines into segments during the data processing and by substratifying the samples in areas of high density of livestock. This subdivision allows for the definition of digitalized quadrats of 10 × 10 km. The stock densities on the borderlines of each quadrat (segments of flight lines) are assumed to represent the density within each of the 300 quadrats of the project area. But the number of animals ascribed to each quadrat is "smoothed" by balancing it against the density in its eight borderline quadrats. The central quadrat is ascribed a 4/10.8 ratio, the quadrats N,S,E and W are valued at 1/10.8, while the quadrats NE, SE, SW and NW are allocated a 0.7/10.8 value as shown in the following:

0.7	1	0.7
1	4	1
0.7	1	0.7

Besides the pilot in charge of navigation and control of altitude, SRF's include three observers: one in the front seat alongside the pilot and two in the middle seats, the rear seats being occupied by various equipment. The front recorder uses a tape recorder and a digital, bell chronometer, which rings regularly every minute. He records general ecological data, e.g. geomorphology, soils, woody cover, herbaceous cover, burnt areas, erosion, presence of humans, etc. The two rear recorders are equipped with a tape recorder to register animals and a 35-mm camera with a 50-mm lens in order to shoot groups of ten or more animals. The overall number of data collected in any one SRF is about 4000. The data from the first two flights were processed in Nairobi by a specialized firm (Ecosystems Ltd.). Those from the last flights were processed in the IMRES Project at Dakar. The computer programmes used would need some change in order to allow for a greater flexibility regarding the number and nature of entries, including particularly complementary information from other sources (satellite, ground). Small amounts of data may be processed on programmable pocket calculators.

7.3.4.3 Reliability of SRF Data

SRF techniques are liable to a number of shortcomings, limitations and biases, among which one may cite the following:

1. Difficulties in eliminating material errors due to human biases (unseen animals, erroneous estimates of the number of animals within groups); naturally, errors increase with the weariness of the observers.

2. Wide variability in animal densities and their contagious distribution, hence, high variance, large confidence intervals and therefore mediocre precision. The 95% confidence interval in IMRES SRF's led to a possible error of 20 to 30% in the evaluation of stock numbers. Therefore, variations in numbers of less than 20% around the mean may not be significant (Sharman 1983a). These drawbacks, however, are reduced during the rainy season when animals are much more regularly distributed. But the most important information remains animal numbers and distribution for the second half of the dry season (April/June). These difficulties can be overcome to some extent by increasing the rate of sampling, up to 10%, for instance, or even more in areas with high densities of animals and also by breaking down the transect lines into segments in the interpretation process as mentioned above.
3. Stock being very mobile and of a different mobility, depending on years and seasons, the evolutionary trends in numbers cannot be evaluated without a sufficient number of SRF surveys. Each SRF constitutes only an instantaneous image of a reality which is fluid, fluctuant, fugacious and difficult to circumscribe.

   Two populations, P1 and P2, obtained from n1 and n2 transects are different, if

   $$(P1 - P2) / var\ 1/n1 + var\ 2/n2^{t(a/b)-2)},$$

   where

   $$a = (var\ 1/n1 + var\ 2/n2)^2;$$
   $$b = \frac{(var\ 1/n1)^2 / (n1 - 1)\ (var\ 2/n2)^2}{n1 - 1}$$

   (Pollard 1977).

4. Ecological observations are difficult because of the fact that evaluations are made under an oblique angle and therefore instantaneous demarcation of any sample necessarily lacks precision.
5. Systematic biases in observer evaluations are unavoidable. Generally, it is accepted among SRF specialists that observers tend to underevaluate the number of animals, particularly in woodlands or tree savannas such as the *Mimosoideae* scrub and *Combretaceae* savanna of the Sahel (Norton-Griffiths 1975/78; Watson and Tippett 1981a). In the case of the Ferlo, Sharman (1983a) estimated this underevaluation to be 3.5%.

The main errors and biases in SRF's are the following:

1. Improper flight planning and/or execution.
2. Poor interpretation of the data, utilizing inadequate correction factors.
3. Insufficient control of altitude.
4. Crabwise flight resulting in a drift that tends to widen the transects.
5. Lack of precision in transects by absence of landmarks, a quite common situation in the Sahel.

6. Sharp turning at the end of each line may result in an ill-defined end of the line and therefore an erroneous transect length.
7. As a result of the noise made by the aircraft, fleeing animals enter the sampling strip and go out of it at the same time, which makes enumeration difficult.
8. High herd and flock densities lead to a permanent underevaluation of their numbers.
9. The sitting position of the observers tend to sink down as they grow tired. The width of the transect seen from the eyes of a weary observer is larger than that of a fresh one, because of the change in the angle of sight.
10. The number of unobserved animals grows with the density of the tree cover; the degree of underestimation therefore depends on the latter.
11. The orientation of the flight lines and the angle of these lines with the direction of the sun may considerably reduce the sharpness of sight of the observers in the direction of the sun but increase it in the opposite direction.

7.3.5 Practical Results

7.3.5.1 Satellite Data (Tables 44, 48; Figs. 71–81)

The correlation between INDVI and green herbaceous biomass at the end of the 1981–83 rainy seasons resulted in a determination coefficient r2 = 0.69, (n = 204; $p < 0.001$); the regression reads:

Y = 87.9 + 82.3 XI (Tucker et al. 1985a, b),
where Y = green biomass in kg DM ha^{-1};
XI = INDVI (see Fig. 69).

The correlation with MMDVI and biomass for the same period produced a coefficient of determination r2 = 0.64 (n = 204: $p < 0.001$); the regression reads:

Y = 226.7 + 4537 XM (Tucket et al. 1985),
where Y = green biomass kg DM ha^{-1}
XM = MMDVI (see Figs. 69 and 79).

Table 48. Livestock numbers recorded from SRF (Sharman 1983a)

Species	Numbers in 10^3						
	1980	1981	1982	1983	Corrected mean	TLU (No.)	TLU (%)
Small stock	726	1231	683	700	708	106.2	25.7
Cattle	367	602	325	385	358	286.4	69.4
Donkeys	25.6	27	12.7	23	19.2	10.2	2.5
Horses	9.7	10	6	8	7.4	7.4	1.8
Camels	1	2	2.5	6	2.7	2.7	0.6
Total	–	–	–	–	–	412.9	100.0
Density (ha TLU^{-1})						7.3	

In practical terms, the data from ground sampling are shown on tracing paper in classes of 200 kg. This tracing is then superimposed on the colour map of Tucker's Index; this gives mean and median values of biomass for each of the 12 classes of the index. Each class of colour is thus given a mean and median biomass equivalent. Five maps at the 1/500 000 scale were thus produced, showing the green herbaceous biomass and forage availability at the end of the rainy seasons of 1980–1984 (Figs. 86–90). In the present report, for reasons of printing convenience, these maps are reduced to approximately 1/1 000 000.

Six to eight classes of colour/biomass were recognized for each year. Comparisons between years are thus possible and the amount of forage available within the dependence area of each borehole can be mapped and measured with a surface integrator. This amount makes it possible to determine the number of stock that can be fed until the next rainy season; and therefore to know whether there is a gap to bridge or not, hence whether or not transhumance will have to take place or not, or whether supplementary feeding will be needed or not.

The biomass/colour classes used within the IMRES Project were as follows (Gaston et al. 1983 a, b):

INDVI class	Colour	Green biomass class (kg DM ha^{-1} yr^{-1})
1–2	Grey	$B < 200$
3–4	Sepia	$200 < B < 400$
5–6	Yellow	$400 < B < 600$
7	Pale green	$600 < B < 800$
8	Green	$800 < B < 10000$
9	Dark green	$10000 < B < 1200$
10–12	Red	$01200 < B$

These figures are provisional orders of magnitude which would need further refinement as additional data are recorded, stored and retrieved. Because, as mentioned above, they may be biased by the ligneous cover wherever canopy cover reaches 10% of the ground surface or more. Differences in signature may also take place according to the dominant species present as shown in hand-held radiometer studies (Gaston et al. 1983). Such a difference in signature also depends on the degree of greenness of the herbaceous layer at the time when the reflectance is sensed. This may vary considerably in a few days towards the end of the rainy season, a limiting factor which is not totally eliminated with the use of the integrated index (IMDVI). Great progress could be achieved with the setting up of a space-receiving station in West Africa and with the participation of a field research staff in the interpretation of the raw satellite data.

7.3.5.2 SRF Data (Sharman 1982, 1983a, b) (Figs. 91–106)

Four SRF's were carried out within the IMRES Project from October 1980 to June 1983. No flight took place in 1984 as most animals had fled the project area due to drought.

1. *Wildlife.* The June 1982 flight registered over a sample of 1200 km^2, 7 gazelles (*G. rufifrons*); 4 warthogs (*Phacochaetus aethiopicus*); 2 ostriches (*Struthio camelus*); 2 jackals (*Canus adustus*).

   Other large mammals, though known to be still present in the area, were not seen: spotted hyena (*Crocuta crocuta*), striped hyena (*H. hyaena*), Dorcas gazelle (*G. dorcas*), Dama gazelle (*G. dama*). (Poulet 1972, 1974). These numbers are only very crude orders of magnitude, however, they show how rare game animals have become in the Sahel, probably less than 0.4 kg km^{-2} live weight, i.e. 0.1% of the present ungulate biomass under very similar ecological conditions in the Kalahari and 0.020% of the theoretical potential (Bourlière 1978; Le Houèrou and Gillet 1985)

2. *Livestock* (Figs. 91–94). The livestock censused in the four SRF's is shown in Table 14. The average figures for the period 1980–83, converted in TLU's are shown in Table 48.

   The 1981 flight is obviously inconsistent with the three others. But the various species of stock have the same rate of increase with respect to the mean of the three others (except for camels); this coefficient is 1.7. The assumption of an influx of animals from Mauritania can only explain 5% of the over-evaluation. Moor cattle can easily be distinguished from the Senegalese "Gobra" breed of zebu since the former are reddish in colour while the Gobra have a uniformly white coat. This, of course, does not preclude that some Moor cattle belong to Senegalese pastoralists and vice versa.

   Various assumptions have been examined in an effort to explain these aberrant 1981 figures, including an afflux of animals from the south. The most likely hypothesis seems to be a sampling error (Sharman 1983a). The corrected figures retained are the number observed divided by 1.7 (Table 49).

   We tried to compare the SRF figures with the administration (DSPA) estimates for 1981. These estimates were based on vaccination and were used as the reference baseline in the official planning operations (National Scheme of Land Use, 1984). The comparison, however, is neither simple nor straightforward since the DSPA estimates are given for administrative units that are ignored in the IMRES Project boundaries and therefore do not compare with SRF figures. But, if the number of animals estimated in each administrative unit is multiplied by a coefficient equal to the ratio between the surface of this unit within the project area boundaries to the overall surface of this unit (a procedure that to be valid assumes a regular distribution of stock within each administrative unit), one then reaches the figures shown in Table 4, i.e. 407 000 TLU for the whole IMRES Project area.

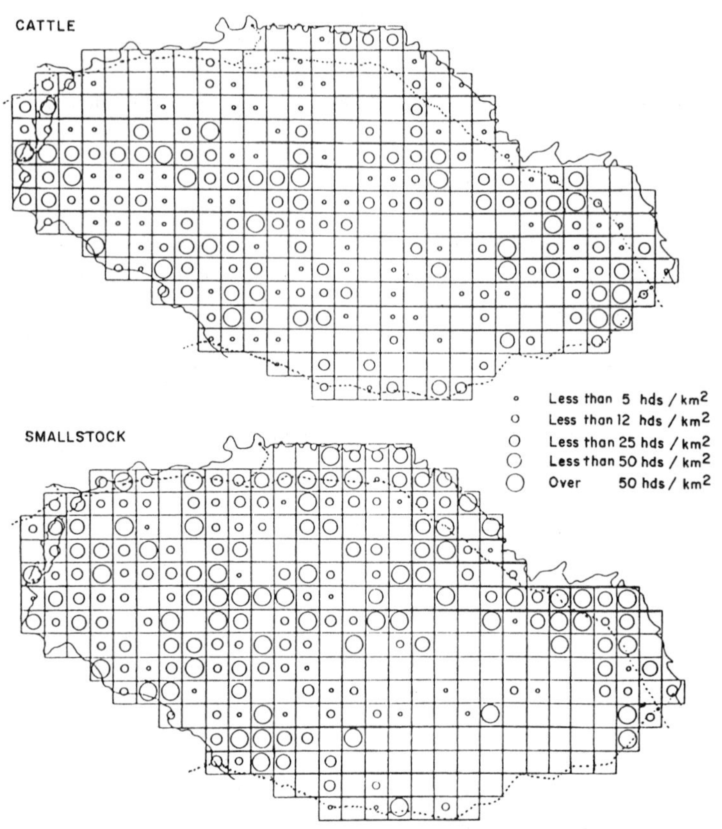

Fig. 91. Distribution of cattle and small stock according to SRF data, May 1983 (Sharman 1983a)

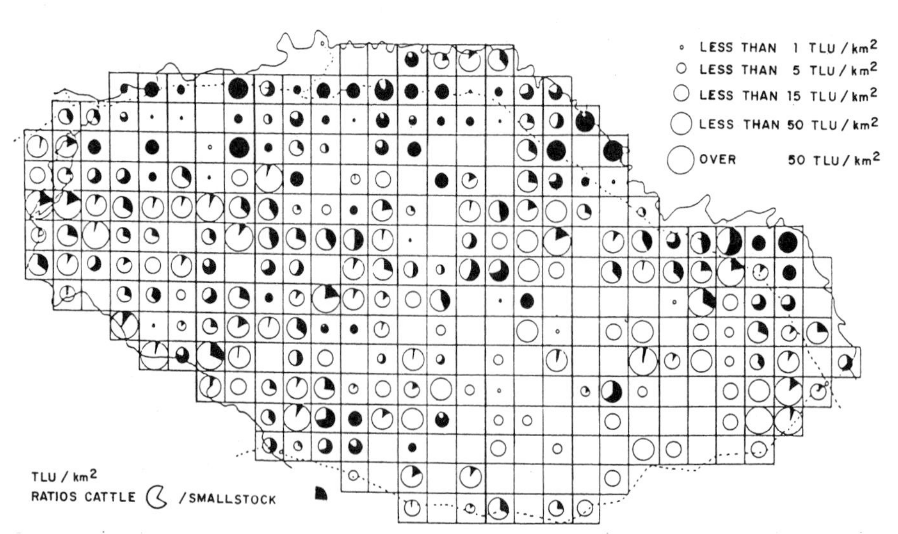

Fig. 92. Distribution of cattle and small stock according to SRF data, 1983, as expressed in TLU (Sharman 1983a)

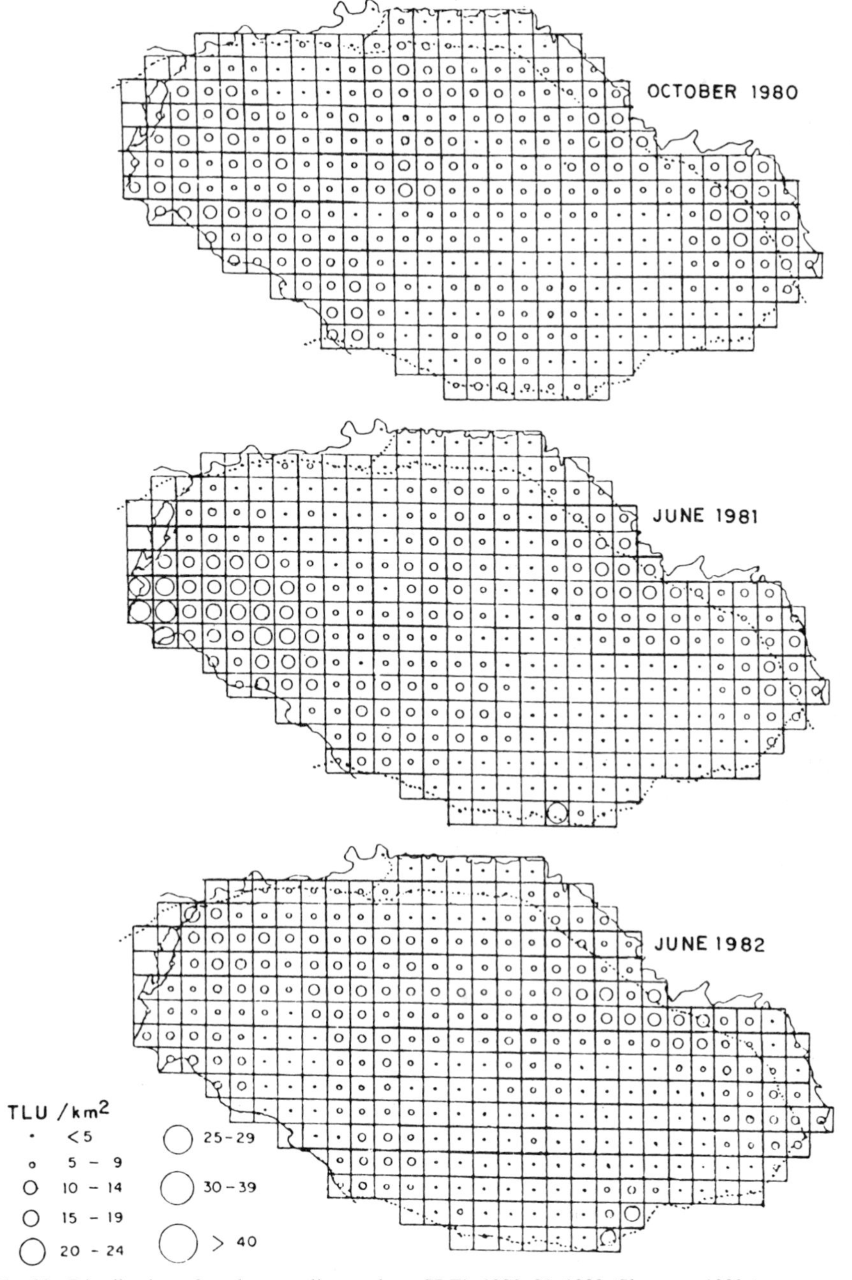

Fig. 93. Distribution of cattle according to three SRF's 1980–81–1982 (Sharman 1983a)

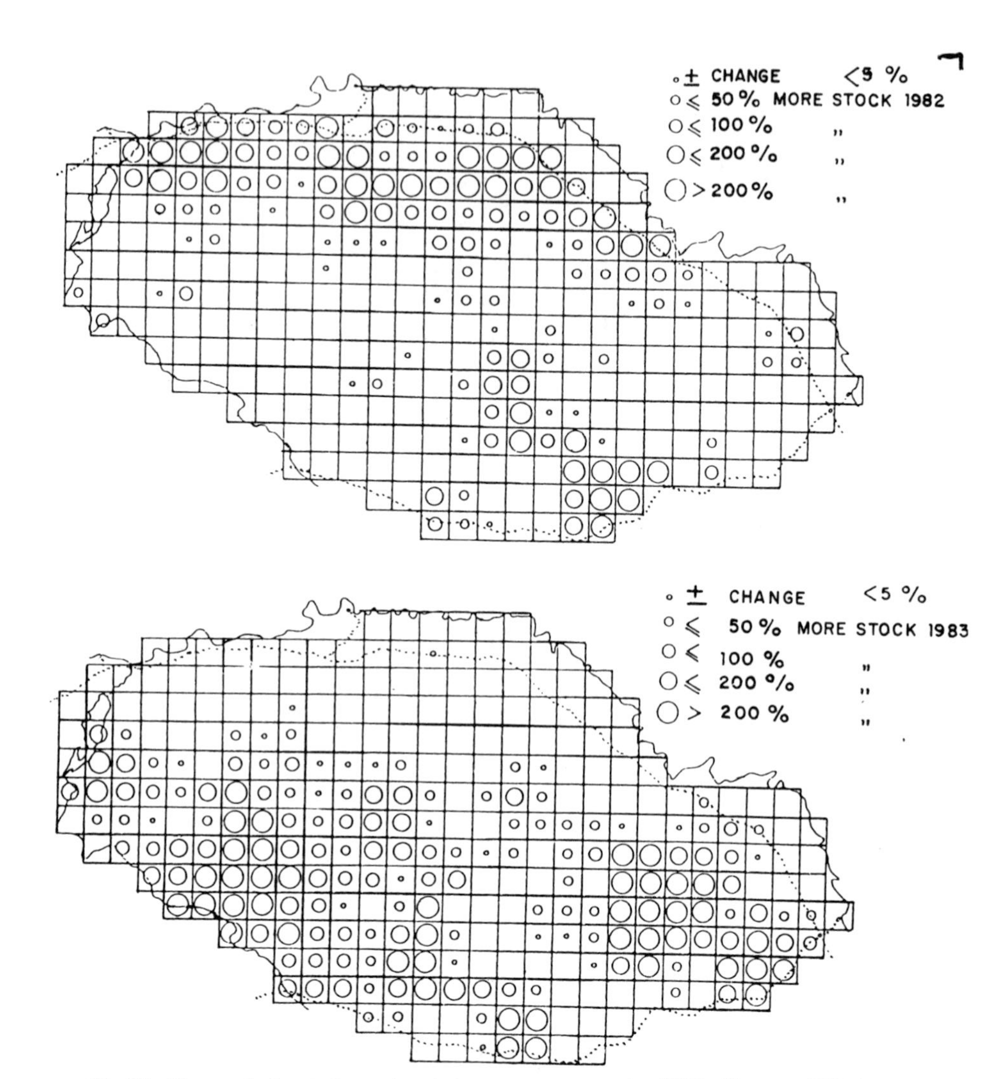

Fig. 94. Changes in livestock numbers between two consecutive SRF's (Sharman 1983a)

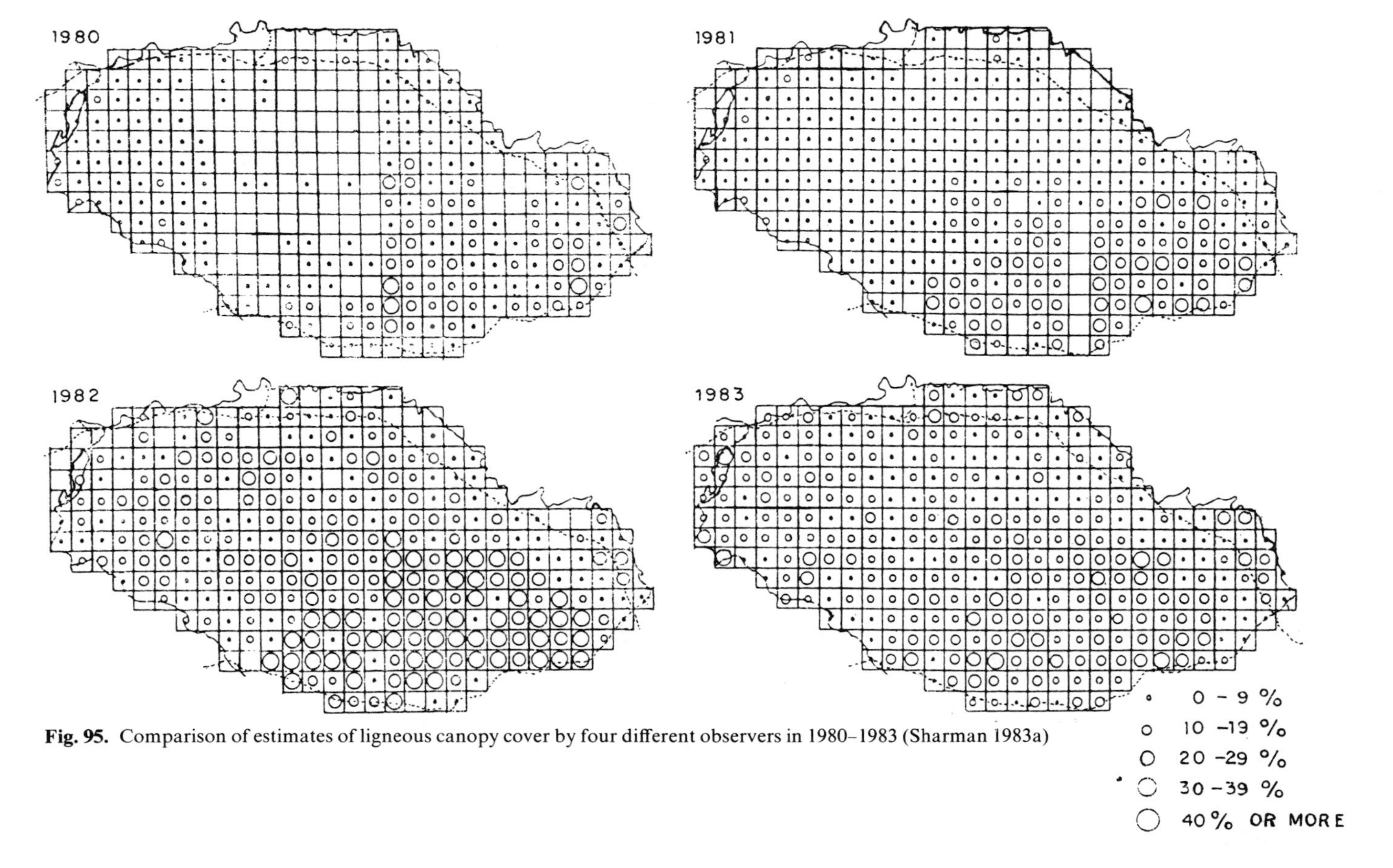

Fig. 95. Comparison of estimates of ligneous canopy cover by four different observers in 1980–1983 (Sharman 1983a)

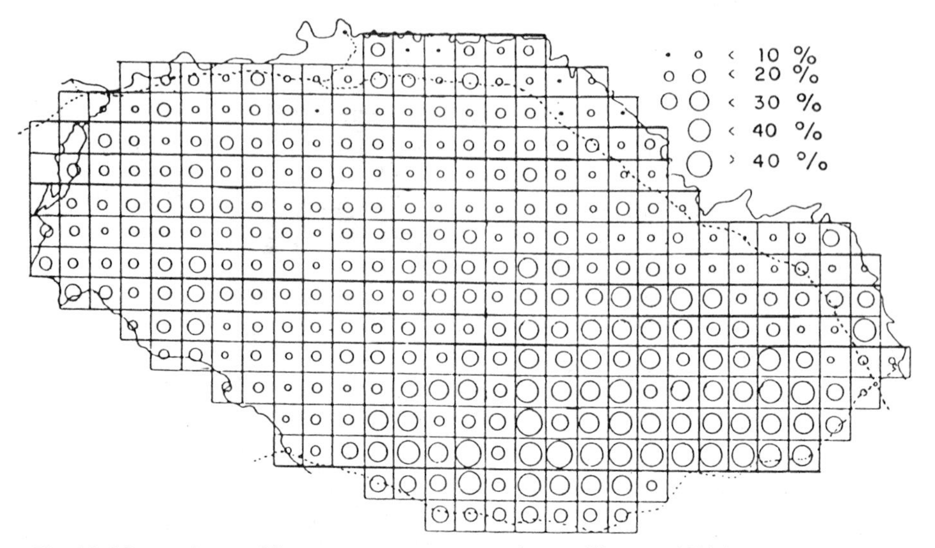

Fig. 96. Mean estimate of ligneous canopy cover over 4 years (Sharman 1983a)

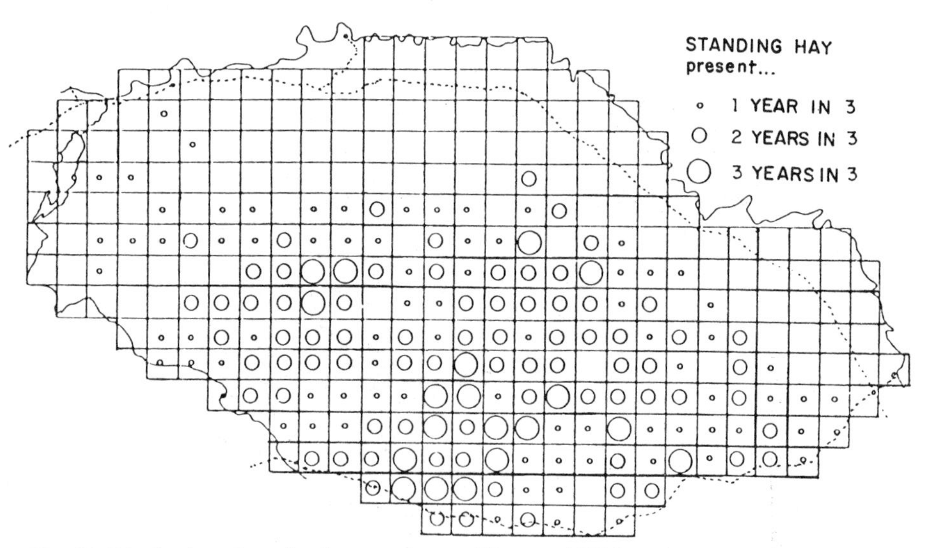

Fig. 97. Distribution of standing hay over 3 years (Sharman 1983a)

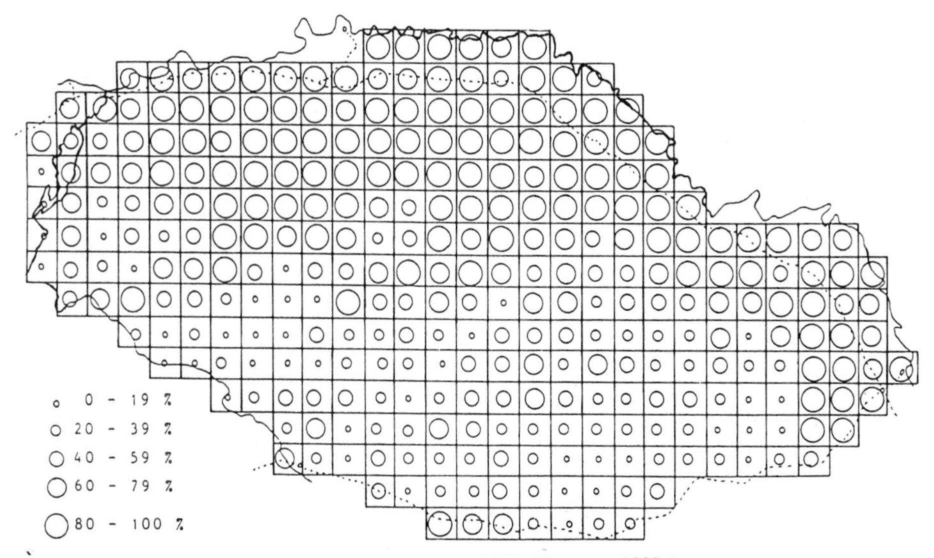

Fig. 98. Distribution of bare ground as from June 1983 (Sharman 1983a)

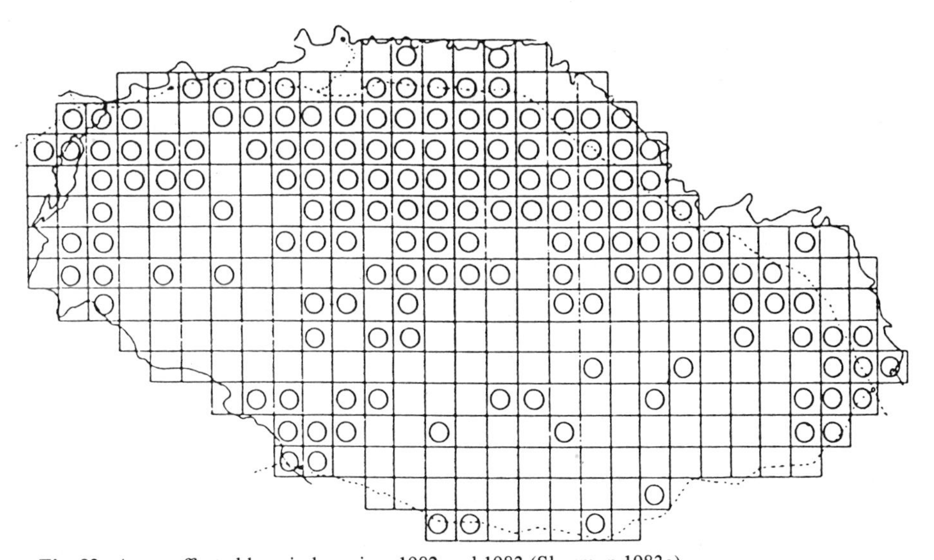

Fig. 99. Areas affected by wind erosion, 1982 and 1983 (Sharman 1983a)

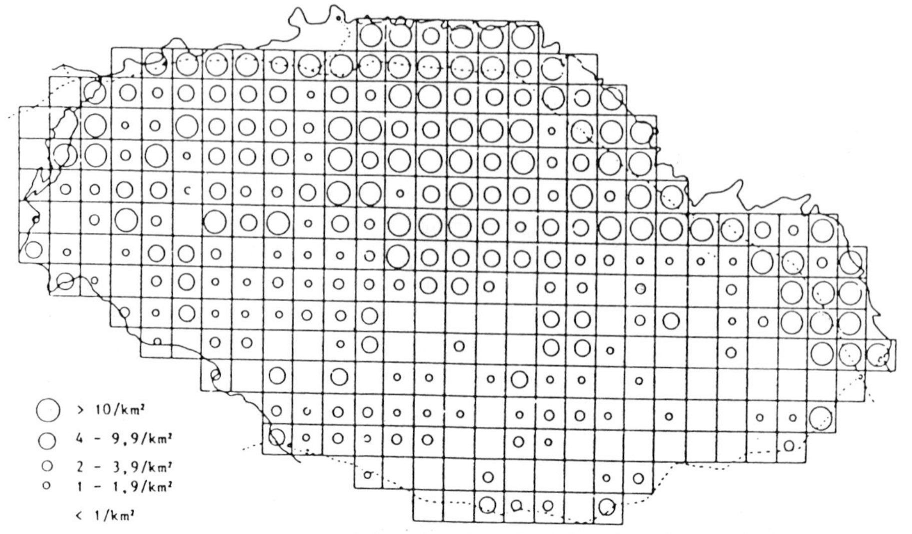

Fig. 100. Distribution of human population (data from the Office of Land Use and Management)

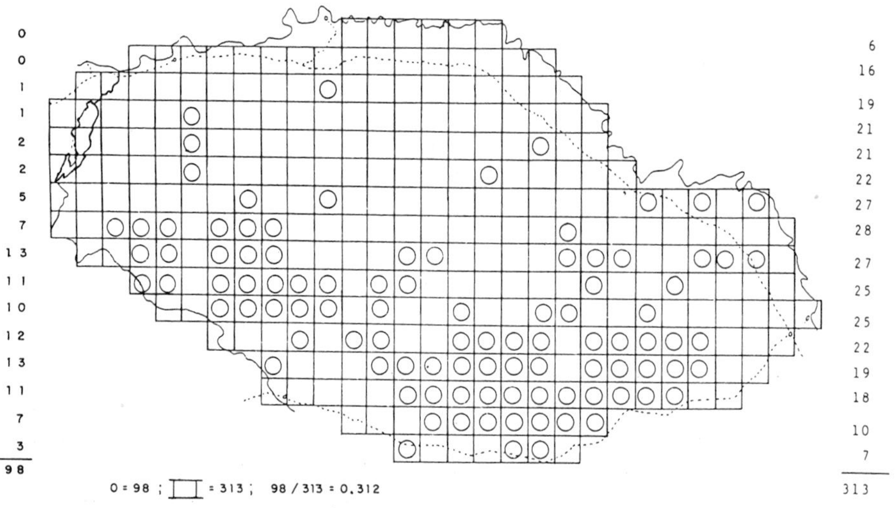

Fig. 101. Areas affected by bushfires, at least once in 4 years (Sharman 1983a). 0 = 98; ☐ = 313; 98/313 = 0.312

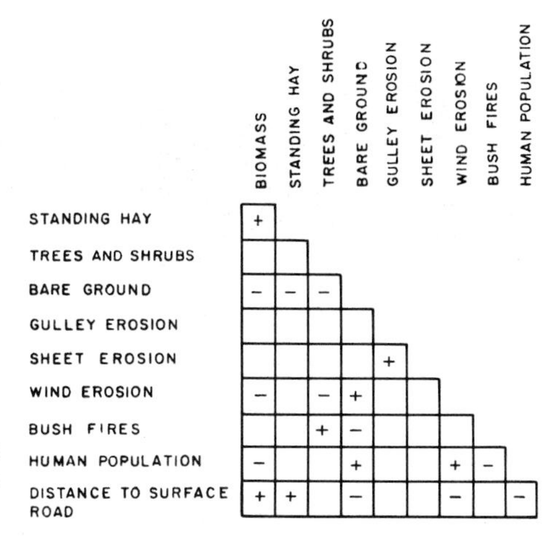

Fig. 102. Correlation matrix between some ecological variables from various sources (Sharman 1983b)

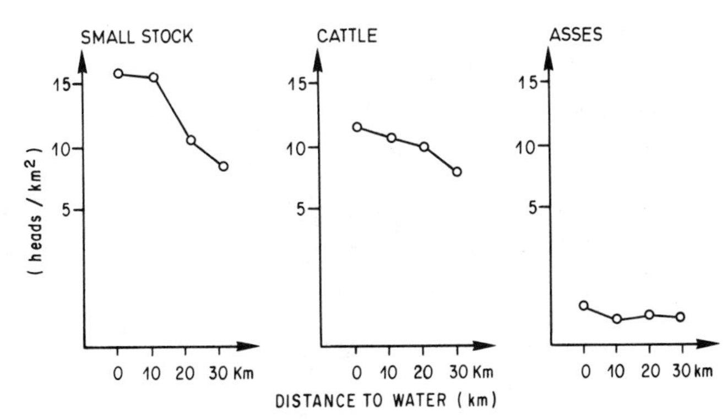

Fig. 103. Density of livestock as a function of distance from water in the dry season; SRF data (Sharman 1983b)

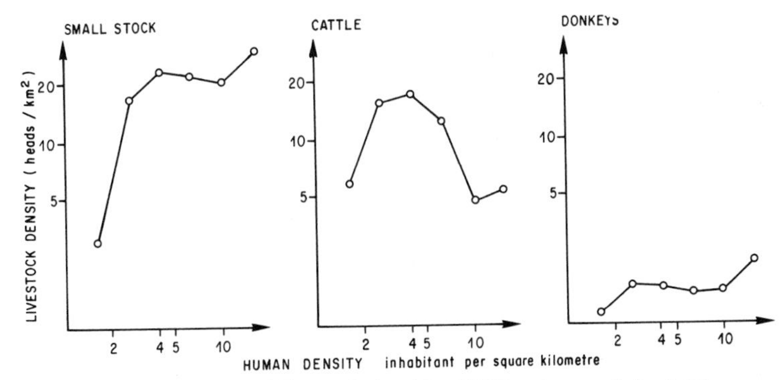

Fig. 104. Relationship between human and livestock densities (SRF and ground data) (Sharman 1983b)

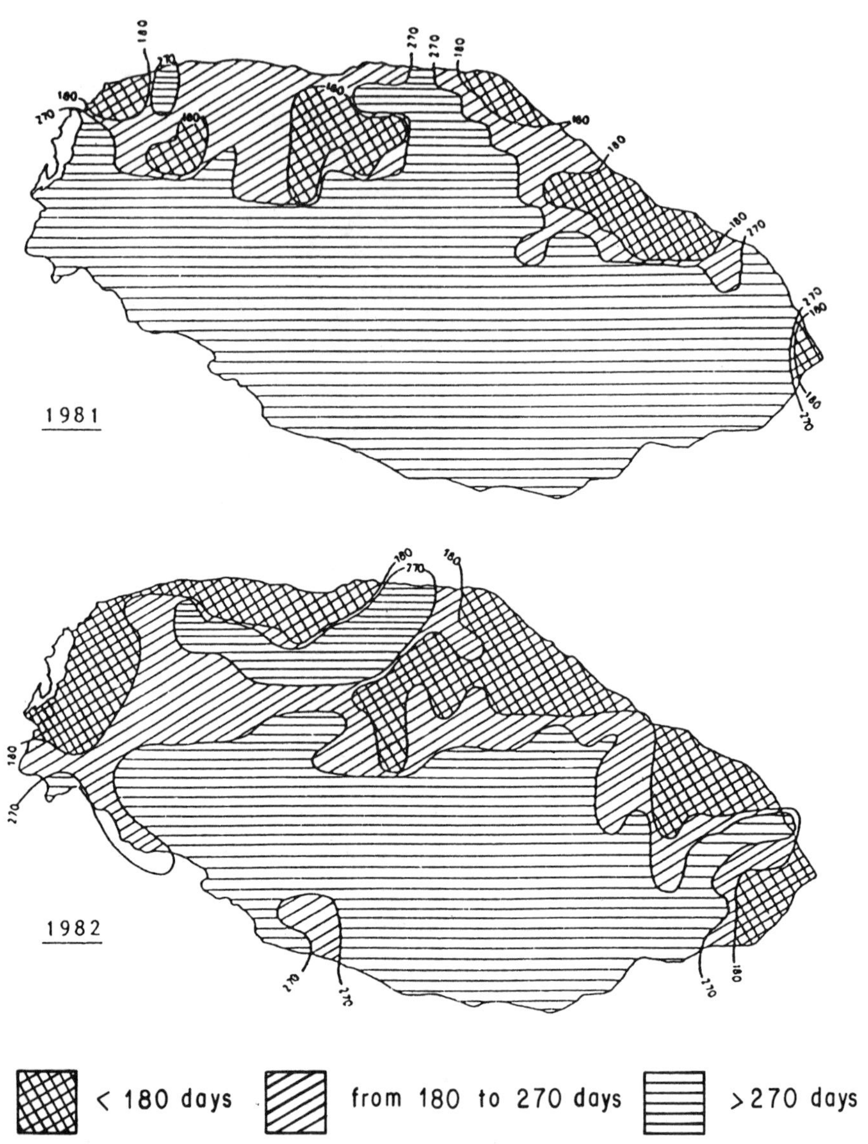

Fig. 105. Number of diet/day TLU available at the end of the rainy season taking into account the actual stocking rates as observed from SRF; forage production estimated from NOAA/AVHRR data complemented with ground truth validation. Global approach: combined data of satellite imagery, low altitude reconnaissance flights and ground control (Sharman 1983a)

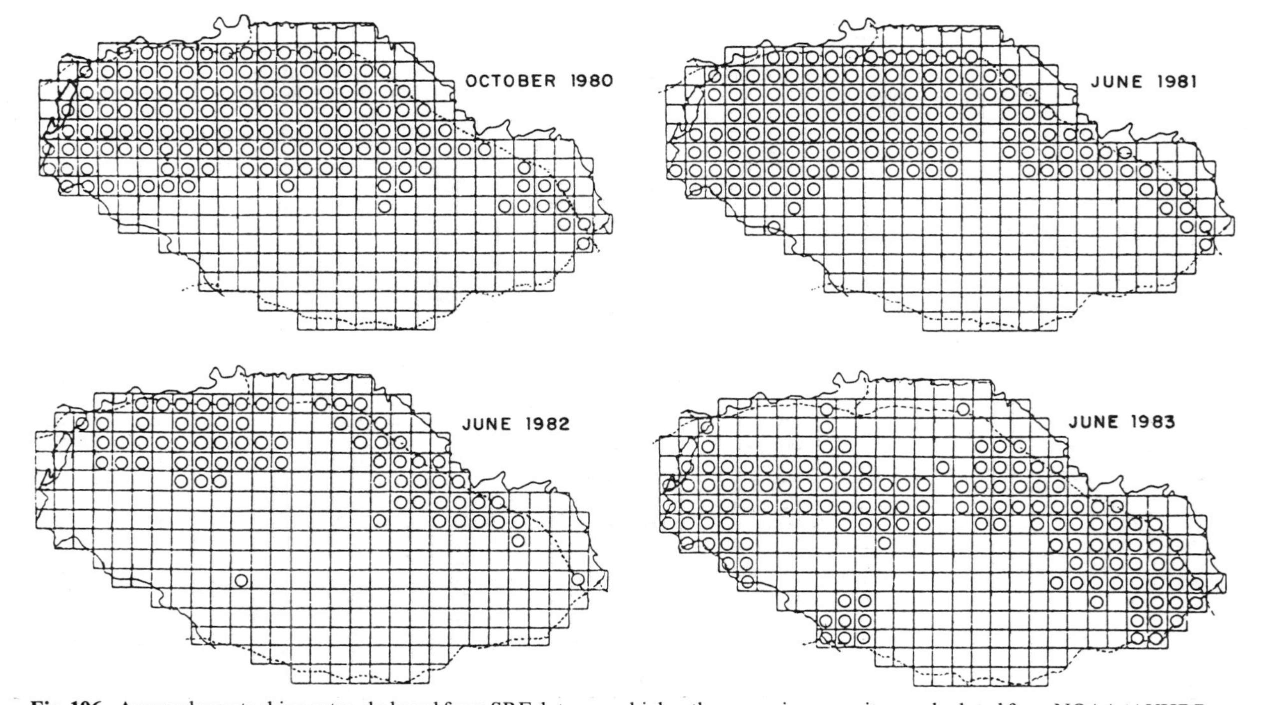

Fig. 106. Areas where stocking rates, deduced from SRF data, were higher than carrying capacity as calculated from NOAA/AVHRR data and ground sampling (Sharman 1983a)

Table 49. Variation around the corrected mean of livestock numbers (rounded figures)

Species	Numbers in 10^3				
	Mean (× 0.8)	Mean (× 0.9)	Mean	Mean (× 1.1)	Mean (× 1.2)
Small stock	570	639	710	781	850
Cattle	288	324	360	396	432
Donkeys	15	17	19	20	23
Horses	6	7	8	9	10
Camels	2	2.3	2.5	2.8	3

There is a 98.5% agreement with the latter figure and the overall SRF evaluation:

$$100 - (413 - 407)\, 413 \times 100 = 985.$$

The two figures correspond in an ideal fashion (which is not evidence of their correctness!).

But the same order of magnitude is confirmed by ground surveys in the DGRST/LAT/GRIZA-LNERV project that are based on two types of estimates: vaccinations and counts at the boreholes in the dry season (Meyer 1980; Planchenault 1981, 1983; Barral et al. 1983). These surveys result in an average density of 6.6 ± 1.3 ha TLU^{-1} over an area of 183 000 ha in the sandy Ferlo (boreholes of Tatki, Tessekré and Belel-Boguel).

For the Department of Podor, entirely included within the project area, the DSPA estimate for 1981 was 191 000 TLU over an area of 16 700 km^2, i.e. 8.74 ha TLU^{-1}, all species combined; and 12 ha TLU^{-1} for cattle alone (the cattle/livestock ratio being 0.728). One may therefore safely conclude that the SRF data, after correction for 1981, are quite plausible and most likely within a 20% interval from the actual figures.

3. *Variation of Numbers in Time and Space* (Figs. 93, 94). If we assume that the 95% confidence interval is of the order of magnitude of 20% around the multiannual mean (Sharman 1983a), we have the significant limits shown in Table 16, whereas the numbers censused in the various flights are given in Table 15. One can see that the differences between the various SRF's are less than 10% of the mean for cattle and small stock and less than 20% for equines. The differences found between SRF's are therefore not significant. The above, however, is not true for the camels whose significance threshold is probably much higher than 20% because of their small numbers. Furthermore, they represent only 0.6% of the total stock numbers expressed in TLU and have therefore a negligible pastoral impact on the Ferlo rangelands.

   Variations observed in space bear out facts that were already known, if not accurately quantified, i.e. (Sharman 1983a):

a) Large concentration of small stock along the Senegal River valley in the dry season, whilst cattle are, generally speaking, better represented in the southern half of the IMRES Project area. This distribution, incidently, is quite typical of the Sahel as a whole and fully justified on ecoclimatic grounds (Le Houérou 1980 f).

b) Cattle distribution varies substantially from one year to the next (see Figs. 93–94). One witnesses a tendency to concentration in the east and in the west in 1980–1982 and a more even distribution in 1983 (Figs. 91–94).

c) For all livestock combined, the distribution in 1983 is opposite to 1982. In 1982 there is a sharp increase in the north (50 to 200%), whilst in 1983 there is a similar increase in the southern half of the project area (Fig. 94).

d) The density of animals in the dry season is closely linked to the distance from boreholes (Fig. 103). Density in small stock is around 15 head km^{-2} between 0 and 10 km from boreholes, then it decreases to 8–10 head km^{-2} at a distance of 20 km and down to 6–8 head km^{-2} towards 30 km. The decrease is more regular when cattle are concerned: from 12 head km^{-2} near the boreholes to 8 head km^{-2} at 20 km and 6 head km^{-2} at 30 km. The distance from boreholes does not seem to influence the density of donkeys and camels, as one might have expected.

4. *The Woody Cover.* Figures 95, 96 show that the estimation of the ligneous cover is strongly influenced by the personal bias of the observers since the estimates in 1980–1983, carried out by four different observers, show large differences that could not possibly be explained by actual evolution. Figure 96 averages the four different estimates; it shows a rather high ligneous density in the ferrugineous Ferlo to the south-east, with canopy covers of 30–50%. Canopy cover is also rather high to the west in a strip some 60 km wide, to the east of Lake of Guiers with covers of the order of magnitude of 20%. In the other parts of the Ferlo ligneous cover varies from 5 to 15%, figures which are in agreement with ground surveys (Poupon 1980; Piot and Diaité 1983). SRF's do not permit, for the time being, one to draw any conclusions on the evolution of the woody layer in the Ferlo; but clear-cut conclusions emerge from other surveys that we shall discuss later.

5. *Importance of Bare Ground in the Ferlo.* The proportion of bare ground (less than 20% plant cover) was very high in the valley of the Senegal River and the northern half of the Ferlo in 1983: 60 to 100% bare ground compared to 20 to 40% in the sourthern half (see Fig. 98). The rate of bare ground was lower in the south-eastern ferrugineous Ferlo (10–20%), an estimate which corroborates the observations of the woody cover.

6. *Extension of Bushfires.* The areas having undergone some burning at least once in 4 years represent 32% of the project area, almost totally included in the sourthern half (see Fig. 101). However, this figure constitutes a minimum as fire scars are often difficult to see at the end of the dry season since the burnt land undergoes wind erosion and sand deposition which tend to obliterate the burnt areas (Sharman 1983a).

7. *Wind Erosion.* Wind and water erosion were not segregated in the two first SRF's. Apart from the Senegal River valley, the maps of wind erosion and soil denudation are in good agreement, as could be expected. Wind erosion affected 30% of the ferrugineous Ferlo, 40% of the water-logged soils and 70% of the sandy soils. This represents overall average of 50% of the IMRES Project area in 1983 [which was a drought year, as mentioned above (Fig. 99)].

8. *Water Erosion.* Generally, water erosion is less intense than wind erosion in the Ferlo. Sheet erosion occurs mainly in the ferrugineous Ferlo affecting to various degrees some 50% of the quadrats in the UTM grid compared to only 5% in the sandy Ferlo. Rill and gulley erosions are mainly active on the rim of the ferrugineous Ferlo and in some water-logged soils of the Senegal River valley. It also occurs to various degrees in some 20% of the most densely inhabited parts of the sandy Ferlo (Sharman 1983a).

9. *Distribution of Standing Hay Biomass at the End of the Dry Season.* The distribution of standing hay contrasts, as one may expect, with the distribution of bare ground, bushfires and erosion. SRF's evidenced the absence or the scarcity of standing dry grass at the end of the dry season in a 50-km-wide strip along the Senegal River valley, which altogether represents some 40% of the surface of the project area. The absence of standing dry grass at the end of the dry season is a clear indication of overgrazing and overstocking. In the central sandy Ferlo, in contrast, standing hay was present 1 to 3 years out of 3 during the month of June, i.e. over 60% of the project area. Overgrazing therefore occurred over 70% of the overall project area in 2 of 3 years from 1980–1983 (Fig. 97).

7.3.5.3 Conclusions on SRF's

The interpretation of the data gathered in the four SRF's carried out from 1980–1983 leads to the conclusion that the Ferlo has been overstocked since at least 1980, with a total elimination of standing hay at the end of the dry season over at least 45% of the IMRES Project area. Forty of the 45% are located in a 50-km-wide strip along the Senegal River valley and Lake of Guiers. Overgrazing is aggravated by bushfires, both resulting in considerable wind erosion.

The aerial study of livestock numbers leads to densities in agreement with ground data collected by various methods. The overall livestock density in the pilot area was of the order of 7.5 ha TLU^{-1} between 1980 and 1983. Assuming an annual proper use factor of 30%, this density corresponds to an overall average annual herbaceous biomass production of 1000 kg DM ha^{-1} yr^{-1}, excluding browse. But the integrated overall production figure deduced from the combination of NOAA data with ground truth figures was only 501 kg DM ha^{-1} yr^{-1} (55–1100) from 1980 to 1984 (Gaston et al. 1983a, b; Tucker et al. 1985a, b; see Table 43 and Figs. 73 and 74). For the period 1979–81, the mean intersite rainfall over ten reliable weather stations in the Ferlo was 379 mm, while the average herbaceous biomass recorded over 112 sites in the same area for the same period was 908 kg DM ha^{-1} yr^{-1} (Boudet 1983b; see Table 14 and Figs. 48–49). The rain-use efficiency (RUE) was 3.04 kg DM ha^{-1} yr^{-1} mm^{-1} in the former case and 2.9 in the latter (Tables 14 and 46).

The figure of 501 kg DM ha^{-1} yr^{-1} for the 1980–84 period results from two very dry years: 1983 and 1984 when intersite precipitation was less than 30% of the 1920–69 mean (116 mm compared to 400). Herbaceous biomass then dropped to 178 and 55 kg DM ha^{-1} yr^{-1} respectively, over the whole project area in 1983 and 1984. It follows that the present stocking rate is adequate in "normal" years, when rainfall is close to the long-term mean or above, but much above the carrying capacity of the rangelands in time of drought. It also follows that the gap between demand (1000 kg DM ha^{-1} yr^{-1}) and supply (500 kg DM ha^{-1} yr^{-1}) was filled by: (1) browse production (and overutilization) which in turn resulted in the receding of the woody layer, as examined below, (2) unusual transhumance to the south outside the IMRES Project area to the so-called Groundnut Belt (Bassin Arachidier). The interpretation of the SRF data shows that bushfires touched 12% of the Ferlo in 1982, while 32% was affected over the 4-year period. This is in complete agreement with previous observations (Naegelé 1971a, b), keeping in mind that the fuel load for this 4-year period was far below normal. This also suggests that firebreaks are insufficient in density and width or are not being properly maintained, a fact on which Naegelé already reported in 1971.

SFR's also evidenced the fact that the proportions of smallstock increase with the density of human population (Fig. 93 and 100) and with the concurrent depletion of vegetation. It would seem that the goat/sheep ratio would evolve in a similar manner, a plausible assumption which remains to be checked.

7.3.5.4 Evolution of Vegetation

The studies carried out by LNERV for over 10 years (1974) in the Ferlo lead to the following conclusions (Valenza 1981, 1984):

1. The density of the herbaceous layer depends more on the annual rainfall conditions than on range exploitation.
2. Above-ground herbaceous biomass is often greater and richer in nitrogen around boreholes than away from them, because of soil enrichment in organic matter and minerals (faeces, urine).
3. The botanical composition, in contrast, is strongly influenced by range utilization: dominant species, particularly annual grasses, depend to a large extent on the distance from boreholes. Areas close to boreholes have a dominance of *Cenchrus biflorus, Dactyloctenium aegyptium*, whilst further away other species dominate such as: *Aristida mutabilis, Chloris prieurii, Schoenefeldia gracilis* and *Eragrostis tremula*. Among the legumes, *Zornia glochidiata*, tends to proliferate around boreholes, which is a clear indication of their being increasingly used in the rainy season since this is a species promoted by rainy season grazing (Boudet 1975a). The latter fact is thus an

indication of increasing year-round settlement of pastoralists around bore-holes. Conversely, the contribution of forbs to the total herbaceous biomass tends to increase with the distance from boreholes: *Borreria* spp., *Merremia* spp., *Polycarpa lineariifolia, Fimbristylis hispidula, Cleome tenella*, etc.

The overall importance of woody species is in a steady decline: 15–25% decrease in numbers in the 40 site studies under the DGRST/LNERV programme between 1979 and 1981 (Piot and Diaite 1983). Similar facts were reported from remotely-sensed data via aerial photography (De Wispelaere 1980a, b; Barral et al. 1983).

At the same time, one witnesses a turnover of species by the replacement of more or less mesic species by more xerophilous ones. A fact which is clearly shown by comparing the rates of survival and the sclerophylly indices (cg of leaf DM cm^{-2}). *Boscia senegalensis* and *Balanites aegyptiaca* thus clearly appear as "increasers" (using a range management terminology). Conversely, more fragile species, having a lower sclerophylly index, are clearly decreasing: *Grewia bicolor, Sclerocaryo birrea, Combretum glutinosum.* In other words, in terms of bioclimatology and phytogeography, one observes a reduction in the proportion of species of Sudanian affinity and an expansion in the rate of taxa of Saharo-Sahelian origin. Not only do the absolute numbers of individuals in woody species decrease, but there is, in addition, an evolution towards more xerophilous woody populations.

IMRES studies at M'bidi and Fété-Olé, for instance, reached the following figures (Vanpraet and Van Ittersum 1983):

Species	Sclerophylly index ($cg\ DM\ cm^{-2}$)	Mortality rate (percent population/yr)
Boscia senegalensis	1.24	4.6
Balanites aegyptiaca	0.82	1.9
Guiera senegalensis	0.53	20.6
Acacia senegal	0.31	19.9
Sclerocarya birrea	0.20	High
Commiphora africana	0.18	10.4

Valenza (1984) published the following data from 40 sites in the project area monitored from 1979 to 1982 (before the last drought):

Species in expansion:

Combretum aculeatum
Commiphora africana

Stable species:

Ziziphus mauritiana
Dalbergia melanoxylon
Boscia senegalensis
Balanites aegyptiaca
Sterculia setigera
Acacia senegal
Acacia tortilis

Species on the decline:

Sclerocarya birrea	15.6% in 4 years
Combretum glutinosum	26.9
Calotropis procera	32.3
Guiera senegalensis	50.5
Grewia bicolar	50.0
Acacia seyal	63.6

It should be underlined that the latter conclusions are based over a 4-year period and regression rates are expressed in absolute numbers of individuals within species *and not* in their proportions in the overall woody populations. The latter may be quite different as a result of initial numbers and the differential rate of regeneration/mortality.

The decrease in population densities is more acute near boreholes: 25% decrease in 4 years at 2/3 km from the boreholes and 15% decrease at 5/6 km.

The correlation between survival rates and a number of parameters monitored by SRF have been calculated (Vanpraet and Van Ittersum 1983; see Fig. 58).

Survival rates are positively correlated with:

1. Distance from camping grounds;
2. Presence of standing hay.

They are negatively correlated with:

1. Rate of bare ground;
2. Density of track roads.

In other words, the rate of survival is inversely correlated with the intensity of human activities.

In conclusion one may say that the data emerging from the IMRES study fully support, if necessary, other data from other teams using different methods, i.e. a dramatic decrease in the woody population of the Ferlo, a consequence of the combined effects of long-lasting drought (15 years) aggravated by concurrent overexploitation. The consequences of this state of affairs has been analyzed elsewhere (Le Houérou 1979a, b, 1980b, f; Le Houérou and Gillet 1985; Le Houèrou 1988a, b).

7.4 Conclusions on Monitoring

The integrated approach combining orbital, SRF and ground-sampling data proved to be an efficient method for continuing evaluation of pastoral resources in the Ferlo. This method could be perfected thanks to the close cooperation of various organizations within the IMRES project: FAO (EMASAR), UNEP

(GEMS), NASA (LTP), LNERV and also due to the fact that the scientists involved had complementary experience and competence.

The perfection by NASA and IMRES of the method of determining green herbaceous biomass at the end of the rainy season has been a recognized breakthrough in the world scientific community. It warranted the evaluation, with an appropriate accuracy, of the amount of forage available at the end of each dry season. This made possible the comparison of primary production with the forage needs deduced from stock numbers censused via SRF's. That is matching supply and demand and therefore creating the feasibility of adequate and accurate planning of management of pastoral resources in the Ferlo, 9 months ahead every year. Maps of herbaceous biomass available at the scale of 1/500 000 were established and circulated to the users in October 1980–1984. A copy of these, reduced to the approximate scale of 1/1 000 000 are attached (Fig. 86–90).

Thus, by comparing supply and demand, it becomes possible to plan ahead of time the necessary actions needed to bridge the feed gap in case of drought, to initiate appropriate incentives for destocking, to make concentrate feed available, to organize rescue operations for the pastoralists, etc.

The information collected in the monitoring activities thus became an essential part of the overall strategy for reducing the effects of drought in the Sahel, as part of the "early warning system".

As to the long-term management of pastoral resources, the method tested by IMRES resulted in the conclusion that the Ferlo is already overstocked and that a reduction in stock numbers is advisable in the short term in order to rehabilitate the range resources, particularly the woody layer, the most seriously threatened.

These conclusions are supported by other studies using different techniques: a combination of aerial photography, LANDSAT images and ground surveys, particularly within the DGRST and LNERV programmes, the conclusions of which have already been published elsewhere.

The methods, particularly the SRF, also showed the disastrous impact of bushfires, particularly in good years when the herbaceous biomass and fuel load reaches 1000 kg DM ha^{-1} or more. As a consequence, it is advisable to dedicate more efforts in expanding firebreak networks and improving their efficiency by a watchful maintenance.

SRF showed the havoc of wind erosion as a consequence of overgrazing and wildfires.

It would seem that the method perfected in the Ferlo should be applicable to the Sahel zone as a whole (3 million km^2 from the Atlantic to the Red Sea). Several attempts towards this goal are underway, particularly in Mali and Niger, the preliminary results of which seem to support the Ferlo experience.

However, it should be mentioned that the utilization of AVHRR orbital data is also subject to limitations and constraints: vegetation indices become unreliable wherever green biomass is below 250–300 kg DM ha^{-1} or wherever the woody canopy reaches 10% of ground cover or more. It follows that the method is less accurate in years of drought and in the Saharo-Sahelian ecoclimatic zone where mean annual rainfall is below 200 mm. Vegetation indices are not highly reliable either in the Sudanian savannas where ligneous canopy cover is usually above 10%. It is therefore not likely that, at the present state of the art, orbital remote sensing of green herbaceous biomass could be applied to the whole Sudanian ecoclimatic zone, contrary to the Sahel. But it is also quite possible that somewhat different methods of space sensing, using recent higher resolution sensors (TM, SPOT) having an appropriate orbital period in order to overcome the cloud-cover problem in the rainy season, could be developed and bring about new breakthroughs in this particular field.

Low altitude systematic reconnaissance flights have made it possible to compare data from the aerial census of livestock with the classical ground estimates based on vaccination campaigns and/or count samples at boreholes. The degree of agreement between SRF and official estimates in 1981 was 98.5%. The 1981 aerial census, however, gave way to aberrant figures, the explanation of which could not be determined with precision (perhaps a sampling error), but the data could be corrected in a satisfactory manner, with a likely small edge of error. The agreement between SRF data and count samples at 13 boreholes is also satisfactory since it falls within 15%.

Comparison between SRF and official data on livestock numbers would be easier and more reliable if SRF could be organized in such a way that administrative limits could be integrated in the data processing and interpretation. SRF could then be favourably carried out by specialized private firms as was often the case in Kenya, Somalia, Ethiopia and Sudan, among others. But whenever regularly repeated SRF are planned, it may be more economical to set up a national organization as in Kenya (KREMU) and as envisaged in Senegal.

SRF's have also proved useful in other respects in the Ferlo: evaluation of bushfires, standing hay, bare ground, wind and water erosion and human settlements.

But SRF's are rather complex to organize, they require a critical analysis and must therefore be entrusted to senior professionals with a strong statistical background.

The experience gained in the Ferlo of Senegal in the inventory and monitoring of Sahelian rangeland ecosystems led to the recommendation of the government to set up a permanent operational organization for continuing the activities perfected and demonstrated in the IMRES Project. Such an organization has been set up at the end of the project under the label of Center for the Ecological Monitoring of Sahelian Pastoral Ecosystems (CEMSPE). The funding of CEMSPE has been provided by DANIDA and UNSO for an initial period of 4 years. The centre will reinforce national expertise in the field of ecological monitoring, whilst producing the required ecological information for the management of the Ferlo rangelands. One should incidentally mention that these harbour 50% of the livestock of Senegal, sustaining a population of some 100 000 pastoralists. One former expert from IMRES is at present involved in the centre.

It is furthermore suggested that other governments of Sahelian countries set up similar centres which would warrant establishing the foundations of their pastoral policies on sound and accurate information. In order to reduce the costs, several countries could perhaps join efforts in the enterprise.

Moreover, it is recommended that a joint Sahelian satellite receiving centre should be established in order to collect and process the data required for the evaluation of green herbaceous biomass of the Sahelian rangelands and also to better coordinate satellite data processing with ground sampling. The satellite receiving station could be part of the above suggested Ecological Monitoring Centre serving both the Sahelian and Sudanian ecoclimatic zones.

From a purely technical viewpoint, it is recommended to carry out measurements of vegetation indices from light aircrafts at an elevation of 300–500 m above ground in order to better integrate satellite, aerial and terrestrial data. This would warrant the integration of the leaf biomass of woody species which, in the present system, are poorly integrated since they are included in the satellite data but not in the ground sampling. It is expected that this procedure would permit a refinement and improvement of the presently used methodology.

When SRF's are concerned, it is suggested that flights be organized in such a way that administrative limits could be integrated in data processing and interpretation in order to secure a mutual control between SRF figures and routinely collected data at ground level, which are usually gathered on the basis of administrative units (vaccination, administrative surveys, and in some countries, taxes).

It is finally suggested to integrate in the global approach aerial photography which was unduely neglected in the IMRES Project methodology. Indeed, aerial coverage constitutes an extremely rich source of baseline information, most useful, if not mandatory, in any ecological monitoring of rangeland ecosystems.

8 General Conclusions: Towards an Ecological Management of the Sahelian Ecosystems

The Sahel ecological zone, a 3 million km^2 strip, stretches across the African Continent from the Atlantic Ocean to the Red Sea, between the 100–600 mm isohyets of mean annual rainfall. It thus appears as a transition belt separating the Sahara Desert from the Sudanian savannas.

In terms of temperatures, air dryness, evaporation and length of the dry season, the Sahel is one of the harshest, broad ecological regions in the world. Mean temperatures, for example, are higher than in the Sahara, while air humidity is considerably lower.

Natural vegetation is essentially a carpet of annual grasses dotted with sparse shrubs and small trees (see Fig. 40).

While the northern part is an area of extensive transhumant and nomadic pastoralism, agropastoralism and subsistence farming are the main, and almost unique, sources of livelihood in the southern half.

The Sahel, which seems to have remained unchanged for centuries, if we trust the early explorers and travellers, has undergone a fast and dramatic evolution since World War II. The change has been brought about by a rapid increase in human and livestock populations, as a result of improved health conditions.

The combination of an exponentially growing anthropozoic pressure on the land, with an almost continuous 20-year period of drought of various intensities, provoked a sharp imbalance between natural resources and their exploitation. Between 1950 and 1983, human population grew by a factor of 2.26, whilst the livestock population increased by a factor of 2.37, according to official statistics (FAO 1984).

This imbalance resulted from:

1. An acute overexploitation of the grazing ecosystems due to heavy and continuous overstocking. Grazing ecosystems thus progressively lost their potential resilience, became increasingly sensitive to depletion agents, leading to desertization.
2. A continuous expansion of cultivation encroaching on grazing lands at a pace similar to the demographical growth, i.e. some 2% per annum.
3. Rarefaction of wildlife and extinction of most large mammals, in particular, including those from areas where they were still common in the 1950's.
4. Reduction and even in many cases the total cancellation of fallowing in the cropping sequence of the agropastoral systems. The reduction or suppression of fallow inevitably results in declining fertility and therefore yields, since no

fertilization can be economically applied due to transport costs. Declining yields, in turn, means more land to be tilled for an appreciable amount of grain to be harvested in order to meet family needs. More tillage means new clearing of land which is increasingly inappropriate for farming (the best lands are cleared first). The spiral of land degradation is thus initiated.
5. The acceleration of water and wind erosion from overgrazing, overbrowsing, excessive woodcutting and overcultivation results in irremediable loss of formerly productive rangeland and farmland and the spread of desertlike conditions with extremely low productivity (10 to 20% of the productivity of the same land under pristine conditions).
6. Overexploitation of the woody layer both for browse and fuel wood, thus further leading to advancing destruction of the woody layer of 1–2% per annum as evidenced in several monitoring programmes in various Sahelian countries.
7. Expansion of desertlike conditions particularly in sandy areas (new dune systems) and on iron and lateritic hardpans ("tiger bush" and desert pavement).

The above diagnosis may seem overly gloomy; but it is a factual result of a large number of diachronic studies in virtually all Sahelian countries, using indisputable evidence from remote sensing [aerial photographs taken years apart, orbital imagery and systematic reconnaissance flight records (see Chap. 7)].

An in-depth, interagency, interdisciplinary study by a large number of specialists concluded that the Sahel had, already in 1980, many more people than it could possibly sustain in the long run under low to intermediate input conditions (Kassam and Higgins 1980; Gorse 1985; see Table 12).

Of particularly serious concern is the quick decline of ligneous vegetation which is the main insurance for ecosystem stability. Woody species ensure most of the turnover of the geobiogene elements, nitrogen fixation (legumes hardly make up 20% of the herbage biomass; see Tables 13–14), land fertility, wood, food and browse productions, supply of protein, phosphorus and carotene to herbivores during the long, dry season.

It has been shown that dietary requirements of herbivores imply a consumption of 20–25% browse (5% in the rainy season, 30% in the dry season); this balance between herbage and browse in the Sahel is threatened by the decline of shrub and tree species from overexploitation. It is clear that if the present trend of depletion of the ligneous vegetation is not discontinued, most of the browse would disappear within the next 20 to 25 years. In such a scenario herbivores could no longer exist through the dry season, except with the help of costly and presently uneconomic concentrate feed. The main economic activity in the Sahel, animal husbandry, would thus become impossible for the annual 9 months of the dry season. The Sahel livestock industry would thus be seriously jeopardized.

How can the present trend by halted or reverted? Is there any hope for improving the present situation of the Sahel? Or, to put it in another way, how could sound management be ensured on the basis of a sustained ecosystem productivity?

We have seen that this could be achieved, in principle, since small-scale experiments carried out in various parts of the Sahel show clearly that ecosystem productivity could be easily doubled or trebled with respect to the present situation, applying simple management techniques, both in grazing ecosystems as well as in agropastoral production systems (see Sect. 6.2). It is the transfer of scale from experiments to real-life situations that has failed to date, since no broad geographical scale development scheme has ever been successfully carried out so far.

At the conceptual level, in terms of planning, the philosophy and development objectives should to be clearly spelled out and the strategy and means to attain the selected goals should be carefully described in detail. This implies a dynamic and comprehensive livestock production policy including price policy, stratification of the industry and trade as well as encouraging appropriate range and livestock production techniques.

At the technical level, in the field, the problems are theoretically simple, if not easy; it is a question of adapting range stocking rates to the sustained long-term productivity of the grazing ecosystems. In practice this would mean that human sustenance and livestock carrying capacity would be controlled and that simple techniques such as deferred grazing, periodic exclosure and the adaptation of the watering regimes to the density and seasonal occurrence of watering points would be utilized. It would also include the rational pruning of browse and the selective gathering of fuel wood from non-forage species.

This kind of management naturally presupposes choice and daily decisions, i.e. the notion of responsibility. Yet the present situation in the Sahel is characterized by a generalized lack of responsibility at the resource management level (range and water).

Water and pasture are actually common, public resources, whereas animals are individually and privately owned; so it is in every user's interest to draw a maximum and immediate profit from communal resources without considering what may happen in the long run. Of course, such a situation results in the looting of the common resources, i.e. of the community to the individual's immediate benefit, this is "the tragedy of the commons" or the "looting strategy".

It is quite obvious that no rational system of any kind can be implemented on such premises, without the idea of responsible management, whether by individuals or groups. Meeting these responsibilities involves fundamental land reform in terms of land tenure and ownership and water usufruct.

Such responsible management used to be in existence in the Sahel long ago before the precolonial era and during early colonial times in the late 19th-early 20th centuries, but with much lower human and animal population densities, so that they resulted in a fairly steady balance between the resources and their utilization. This equilibrium was certainly not ideal with regards to productivity, social justice and human dignity. It has been frantically misused and overexploited everywhere. If drastic sociopolitical reforms are not implemented without delay in order to ensure the rational management of the Sahelian ecosystems, the Sahel will be faced with a deep crisis threatening its present main resources. The Sahara may thus expand several 100 km south of its present limit, as it has occurred on several occasions in the past 50 000 years; but this time the prime responsibility would rest on man, not on climatic change. Man has thus become a major factor of geological change in this part of the world.

Acknowledgements. The author wishes to express his deep appreciation to those colleagues and friends who assisted with information, publications, and, at times, unpublished data. Most of those are also the ones who "sweated the data out of the bush"! Particular gratitude has been earned by: E. Bernus, G. Boudet, A. Cornet, G. De Wispelaere, M.S. Dicko, C.R. Field, A. Fournier, A. Gaston, H. Gillet, G. Glaser, M. Grouzis, T. Ionesco, H.F. Lamprey, J.P. Lebrun, M. Mainguet, J.C. Menaut, Th. Monod, A. Naegelé, S.E. Nicholson, B. Peyre De Febrègues, A.R. Poulet, E.J. Roose, M.J. Sharman, B. Toutain, C.J. Tucker, G.E. Wickens and R.T. Wilson. Special thanks are due to C.J. Grenot for his essential contribution to the authorship of chapter 4 (Wildlife).

All the figures were drawn up in the CEPE's drawing workshop by Messrs A. Carrière, R. Ferris, D. Lacombe and J. Vilanova whose talents the author is pleased to give credit and appreciation to.

References

Adam JG (1966) Les pâturages naturels et post-culturaux du Sénégal. Bull IFAN 28:2:450–537
AETFAT (1959) Carte de la végétation de l'Afrique. Oxford University Press, London, 27 pp 1 carte coul 1/10 000 000
Albergel J, Carbonnel JP, Grouzis M (1984) Pluies, eaux de surface et production végétale en Haute Volta. ORSTOM, Paris, 63 pp
Andrews FW (1948) The vegetation of the Sudan. In: Tothill JP (ed) Agriculture in the Sudan, ch IV. Oxford Univ Press
Andrews FW (1950–56) The flowering plants of the Anglo-Egyptian Sudan. Vol I (1950) 237 pp; Vol II (1952) 485 pp; Vol III (1956) 579 pp
Anonymous (1977) The Sudan experience in the field of desert encroachment, control and rehabilitation. Country Rep UN Conf Desertification, UNEP, Nairobi, 27 pp
Asad T (1970) The Kabbabish Arabs: power, authority and consent in a nomadic tribe. Praeger, New York
Aubert G, Maignien R (1949) L'érosion éolienne dans le Nord du Sénégal. Bull Agr Congo Belge 40:1309–1316
Aubreville A (1949) Climats, forêts et désertification de l'Afrique tropicale. Soc Edit Géogr, Marit Col, Paris, 351 pp
Aubreville A (1973) Rapport de la mission forestière anglo-française Nigéria-Niger (déc 1936–fév 1937). Bois For Trop 148:3–26
Audibert M (1966) Etude hydrogéologique de la nappe profonde du Sénégal nappe Maestrichtienne. Mém BRGM, Paris
Audru J (1977) Les ligneux et sub-ligneux des parcours naturels Soudano-Guinéens en Côte d'Ivoire; leur importance et les principes d'aménagement et de restauration des pâturages. IEMVT, Maisons-Alfort, 267 pp
Audru J, Lemarque G (1966) Etude des pâturages naturels et des problèmes pastoraux dans le Delta du Sénégal. Et Agrostol 15, IEMVT, Maisons-Alfort, 359 pp 1 carte 1/100 000
Audry P, Rossetti C (1962) Observations sur les sols et la végétation en Mauritanie du Sud-Est et la bordure adjacente du Mali. Rappt UNSF/DL/ES/3, FAO, Rome, 267 pp 5 pl, 1 carte
Baasher MM (1961) Range and livestock problems facing the settlement of nomads. Proc Sudan Phil Soc, Khartoum, 11 p
Bagnouls F, Gaussen H (1953) Période sèche et végétation. C R Acad Sci 236:1076–1077, Paris
Bagnouls F, Gaussen H (1957) Les climats biologiques et leur classification. Ann Géogr, 335, LXVI:193–220
Bailly C, Barbier J, Clement J, Goudet JP, Hamel O (1982) The problems of satisfying the demand for wood in the dry regions of tropical Africa: knowledge and uncertainties. CTFT, Nogent-sur-Marne, 24 pp
Barbey C, Conte A (1976) Croûtes à Cyanophycées sur les dunes du Sahel Mauritanien. Bull IFAN, Ser A, 38:732–736
Barral H (1974) Mobilité et cloisonnement chez les éleveurs du Nord de la Haute Volta; les zones dites "d'endrodromie pastorale". Cah ORSTOM, Ser Sci Hum, Vol XI, 2:127–135
Barral H (1977) Les populations de l'Oudalan et leur espace pastoral. Trav Doc, No. 77, ORSTOM, Paris
Barral H (1982) Le Ferlo des forages: gestion ancienne et actuelle de l'espace pastoral. ORSTOM, Dakar, 85 pp, 10 figs

Barral H, Benefice E, Boudet G, Denis JP, De Wispelaere G, Diaité I, Diaw OI, Dieye K, Doutre MP, Meyer JF, Noel J, Parent G, Piot J, Planchenault D, Santoir C, Valentin C, Valenza J, Vassiliades G, (1983) Systèmes de production d'élevage au Sénégal dans la région du Ferlo. GRIZA/LAT-ISRA; GERDAT/ORSTOM, Paris, 172 pp
Barry JP et al. (1983) Etudes des potentialités pastorales et de leur évolution en milieu sahélien du Mali. GERDAT/ORSTOM, Paris, 116 pp
Barth H (1860) Travels and discoveries in North and Central Africa. Longmans and Green, London
Bartha R (1970) Fodder plants in the Sahel Zone of Africa. Ifo-Inst Wirtschaftsforsch München, Afrika Studien 48 Weltforum, München, 298 pp, 233 phot
Baulig H (1950) Essais de géomorphologie. Les Belles Lettres, Paris, 160 pp
Baumer M (1963) The ranges of Dar Meganin I. Sudan Notes Rec 44:120–13
Baumer M (1964) The ranges of Dar Meganin II. Sudan Notes Rec 45:143–147
Baumer M (1968) Ecologie et aménagement des pâturages au Kordofan. Thèse Dr-Ing, Fac Sci, Univ Montpellier, 560 pp
Baumer M (1987) Agroforesterie et désertification. ICRAF, Nairobi, 260 pp
Beaudet G, Michel P, Nahon D, Oliva P, Piser J, Ruellan A (1976) Formes, Formations superficielles et variations climatiques récentes du Sahara Occidental. Rev Geogr Phys Géol Dynam 2, XVIII, 2–3:156–174
Beaudet G, Coque R, Michel P, Rognon P (1977) Y a-t-il eu capture du Niger? Bull Ass Géogr Franç, 445–446:215–22
Bellocq A (1983) Climatologie de la pluviométrie au Sénégal. Suivi automatique de l'hivernage. In: Vanpraet CL (ed) Méthodes d'inventaire. Et de surveillance continue des écosystemes pastoraux sahéliens, ISRA, Dakar, pp 87–106
Benefice E (1980) Project DGRST dans la région du Ferlo. Rap préliminaire. ORANA, Dakar, 26 pp
Benefice E (1982) Lutte contre l'aridité en milieu tropical: systèmes de production d'élevage au Sénégal; alimentation et nutrition des éleveurs du Ferlo. ORANA, Dakar
Benefice E et al. (1981) Enquête sur l'état nutritionnel en zone tropicale sèche (Sahel 1976–1979): méthodologie et résultats. Etat nutritionnel de la population rurale du Sahel. Rap Groupe travail CRDI: 37–56, Dakar
Benoit M (1984) Le Seno-Mango ne doit pas mourir: pastoralisme, vie sauvage et protection au Sahel. Mémoire ORSTOM, 103, Paris, 143 pp
Berhault J (1967) Flore du Sénégal. Clairafrique, Dakar, 485 pp
Bernus E (1971/74) Possibilities and limits of pastoral watering plans in the Nigerian Sahel. Seminar on Nomadim, Cairo, FAO Rome, 1971; Cah ORSTOM, Ser Sci Hum 1974, vol XI, 2:119–126, 13 pp
Bernus E (1974a) L'évolution des relations entre éleveurs et agriculteurs en Afrique tropicale; l'exemple du Sahel nigérien. Cah ORSTOM, Ser Sci Hum XI, 2:137–143
Bernus E (1974b) Les Illabaken, une tribu touarègue sahélienne et son aire de nomadisation. Atlas des Structures Agraires au Sud du Sahara, Mouton, La Haye, ORSTOM, Paris, 116 pp, 4 cartes, 11 figs, 10 phot
Bernus E (1981) Touaregs nigériens: unité culturelle et diversité régionale d'un peuple de pasteurs. Mém 94, ORSTOM, Paris, 508 pp, 5 cartes, 30 figs
Bille JC (1977) Etude de la production primaire nette d'un écosystème sahélien. Trav Doc No. 65, ORSTOM, Paris, 82 pp, 1 carte
Bille JC (1978) Woody forage species in the Sahel: their biology and use. Proc First Int Rangelands Congr Denver, pp 392–395
Bille JC (1980) Measuring the primary palatable production of browse plants. In: Le Houérou HN (ed) Browse in Africa. ILCA, Addis-Abeba, pp 185–196
Blanck JP (1968) La boucle du Niger (Mali). Cartes géomorphologiques et notice. Projet d'aménagement: 11 cartes 1/100 000, 1 notice 26 pp, 1 fascicule 41 pp. Centre de Géomorphologie Appliquée, Univ de Strasbourg
Blancou J, Calvet H, Friot D, Valenza J (1977) Composition du pâturage naturel consommé par les bovins en milieu tropical. Note sur une technique d'étude nouvelle. Coll Recherches sur l'élevage bovin en zone tropicale humide; CNRZ, Bouaké, Côte d'Ivoire, 10 pp
Bonnet-Dupeyron F (1952) Déplacement saisonnier des éleveurs au Sénégal. ORSTOM, Dakar, 2 cartes 1/1 000 000

Bonte P (1973) L'Elevage et le commerce du bétail dans l'Ader Doutchi-Majya. Et Niger n° 23, Niamey-Paris, 219 pp

Bonte P, Bourgeot A, Digeard JP, Lefebure C (1979) Human occupation: pastoral economies and societies. In: Sasson A (ed) Tropical grazing-land ecosystems. Natur Resour Res, XVI, UNESCO, Paris, Ch 8, pp 260–302

Boudet G (1969) Etudes des pâturages naturels du Dallol-Maouri (Niger). Et Agrostol 26, IEMVT, Maisons-Alfort, 308 pp, 8 phot, 1 carte 1/200 000

Boudet G (1972a) Project de développement de l'élevage dans la région de Mopti. Et Agrostol 37, IEMVT, Maisons-Alfort, 309 pp, 1 carte 1/1 000 000

Boudet G (1972b) Désertification de l'Afrique tropicale sèche. *Adansonia*, Ser 2, 12, 4:505–524

Boudet G (1974a) Les pâturages et l'élevage dans le Sahel. Bibl 65, Reun Tech EMASAR, FAO, Rome, 44 pp

Boudet G (1974b) Les écosystèmes pâturés des régions tropicales; état des connaissances pour l'Afrique francophone. UNESCO, Paris, 67 pp

Boudet G (1975a) Manuel sur les pâturages tropicaux et les cultures fourragères. IEMVT, Maisons-Alfort 2 edn, 1984, 266 pp

Boudet G (1975b) The inventory and mapping of rangelands in West Africa Proceed. Symp Evaluation Mapping Tropical Afric Grasslands, ILCA, Addis-Ababa, pp 57–77

Boudet G (1976) Contribution à l'étude de faisabilité d'un projet transnational "Gestion du bétail et des terrains de parcours en régions Soudano-Sahéliennes"; UNEP, Nairobi, 42 pp

Boudet G (1977a) Désertification ou remontée biologique au Sahel. Cah ORSTOM, Ser Biol XII, 4:293–300

Boudet G (1977b) Contribution au contrôle continu des pâturages tropicaux en Afrique Occidentale. Rev Elev Medec Vét Pays Tropic 30(4):387–406

Boudet G (1979) Quelques observations sur les fluctuations du couvert végétal sahélien au Gourma malien et leurs conséquences pour une stratégie de gestion sylvo-pastorale. Bois For Trop, 184:31–44

Boudet G (1980) Systèmes de production d'élevage au Sénégal; étude du couvert herbacé. (1ère campagne). IEMVT, Maisons-Alfort, 48 pp, 19 tabl, 2 figs

Boudet G (1981) Systèmes de production d'élevage au Sénégal; étude du couvert herbacé (2 campagne). IEMVT, Maisons-Alfort, 21 pp, 5 figs, 11 tabl

Boudet G (1983a) Systèmes de production d'élevage au Sénégal, étude du couvert herbacé, compte-rendu de fin d'étude. IEMVT, Maisons-Alfort, 27 pp, 20 tabl, 10 figs

Boudet G (1983b) L'agropastoralisme en Mauritanie, perspectives de recherches. IEMVT, Maisons-Alfort; GERDAT, Paris, 47 pp

Boudet G (1984a) Projet de développement de l'élevage au Nord-Est du Mali, étude agro-sylvo-pastorale. GERDAT, Paris; IEMVT, Maisons-Alfort, 36 pp

Boudet G (1984b) Recherche d'un équilibre entre production animale et ressources fourragères au Sahel. Bull Soc Languedoc Géogr 18:3–4, 167–177

Boudet G (1985) Conservation et évolution des systèmes pastoraux. In: CIRAD, Sécheresse en zone Intertropicale. Paris, pp 477–485

Boudet G, De Wispelaere G (1976) Classification des pâturages tropicaux et niveaux de télédétection. IEMVT, Maisons-Alfort; FAO, Rome, 85 pp

Boudet G, Duverger E (1961) Etude des pâturages naturels sahéliens: Le Hohd (Mauritanie). IEMVT, Maisons-Alfort, 160 pp

Boudet G, Ellenberger JF (1971) Aménagement du berceau de la race N'Dama dans le cercle de Yanfolila. Et Agrostol 30, IEMVT, Maisons-Alfort, 175 pp

Boudet G, Leclercq P (1970) Etude agrostologique pour la création d'un ranch d'embouche dans la région de Niono (Rép du Mali). Et Agrostol 19, IEMVT, Maisons-Alfort, 269 pp

Boudet G, Rivière R (1968) Emploi pratique des analyses fourragères pour l'appréciation des pâturages tropicaux. Rev Elev Médec Vét Pays Tropic XXI, 2:227–266

Boudet G, Cortin A, Macher H (1971) Esquisse pastorale et esquisse de transhumance de la région du Gourma. Ges Ing Berat, Essen; IEMVT, Maisons-Alfort, 283 pp, un Atlas cartes 1/200 000 (39 600 km^2)

Boulet R (1970) La géomorphologie et les principaux types de sols en Haute-Volta septentrionale. Cah ORSTOM, Ser. Pédol 8 (3):245–271

Boulet R (1974) Toposéquences de sols tropicaux en Haute-Volta. Equilibres dynamiques et bioclimatiques. ORSTOM, Paris 272 pp

Boulet R (1978) Toposéquence de sols tropicaux en Haute-Volta. Déséquilibre pédo-climatique. Mém ORSTOM, Paris, 272 pp

Bourlière F (1962) Les populations d'ongulés sauvages africains: caractéristiques écologiques et implications économiques. Terre Vie 109:150–160

Bourlière F (1978) La Savane sahélienne de Fété-Olé, Sénégal. In: Lamotte M, Bourlière F (eds) Problèmes d'écologie: structure et fonctionnnement des écosystèmes terrestres: Masson, Paris, Ch 5, pp 187–229

Bourlière F, Morel G, Galat G (1976) Les Grands mammifères de la basse vallée du Sénégal et leurs saisons de reproduction. *Mammalia* 40:401–412

Brasseur G (1952) Le problème de l'eau au Sénégal. Et Séněg 4, IFAN, Dakar, 100 pp

Breman H, Clissé AM (1977) Dynamics of sahelian pastures in relation to drought and grazing. Oecologia, Berlin 28:301–315

Breman H, De Wit CT (1983) Rangeland productivity and exploitation in the Sahel. Science 221, 4618:1341–1347

Breman H, Diallo A, Traoré G, Djiteye MA (1978) The ecology of the annual migration of cattle in the Sahel. Proc First Int Rangelands Cong, Soc Rge Mgt, Denver, pp 592–595

Breman H, Cissé AM, Djiteye MA, Elberse WT (1980) Pasture dynamics and forage availability in the Sahel. Isr J Bot 28:227–251

Brenan JPM (1978) Some aspects of the phytogeography of tropical Africa . Ann Missouri Bot Gard 65:437–478

Brockelhurst HC (1931) Game animals of the Sudan. Their habitats and distribution. Gurney and Jackson, London

Bruneau de Miré, P Gillet H (1956) Contribution à l'étude de la flore du Massif de l'Air. J Agric Trop Bot Appl 3:221–247; 422–443; 701–760; 857–880

Bruneau de Miré P, Quézel P (1961) Remarques taxonomiques et biogéographiques sur la flore des montagnes de la lisière méridionale du Sahara et plus spécialement du Tibesti et du Djebel Marra. J Agr Trop Bot Appl 8:110–133

Brunet-Moret Y (1963) Etude générale des averses exceptionnelles en Afrique Occidentale: République de Haute-Volta. Comité Inter-Etats d'Etudes Hydrauliques, ORSTOM, Paris, 23 pp

Brunet-Moret Y (1967) Etude générale des averses exceptionnelles en Afrique Occidentale: République de Côte d'Ivoire. Comité Interafricain d'Etudes Hydrauliques, ORSTOM, Paris, 20 pp

Cabrera A (1932) Los mammiferos de Marruecos. Madrid

Calvet H et al. (1965) Aphosphorose et botulisme au Sénégal. Rev Elev Med Vet Pays Trop 18 (3):249–282

Canfield R (1941) Application of the line interception method in sampling range vegetation. J Forestry 39:388–394

Capot-Rey R (1961) Borkou et Ounianga. Etude de Géographie régionale. Mém. No. 5, Inst Rech Shar Univ d'Alger, 182 pp

Catinot R (1967) Sylviculture tropicale dans les zones sèches de l'Afrique. Bois For Trop 111:19–32; 112:3–29

CCE (1987) Suivi du bilan hydrique à l'aide de la télédétection par satellite. Application au Sénégal. Bruxelles, 151 pp

Chamard P (1973) Les paléoclimats du Sud-Ouest saharien au Quaternaire récent. C R Coll désertification Sud Sahara, Nelles Edit Afr, Dakar-Abidjan, pp 21–26

Chapelle J (1957) Nomades noirs du Sahara. Plon, Paris, 446 pp

Chapuis M (1961) Evolution and protection of the wildlife in Morocco. Afr Wild, 15 (2):107–112

Charre J (1974) Le climat du Niger. Thèse Géogr, Univ Grenoble, 188 pp

Charreau C, Vidal P (1965) Influence d'*Acacia albida* Del. sur le sol, la nutrition minérale et les rendements des mils *Pennisetum* au Sénégal. L'Agron Tropic, XX, 7:600–626

Chevalier A (1900) Les zones et les provinces botaniques de l'AOF C R Acad Sci, CXXX, 18:1205–1208

Chevalier A (1933) Le territoire géobotanique de l'Afrique tropicale Nordoccidentale et ses subdivisions. Bull Soc Bot Fr LXXX:4–26, 1 carte

Chevalier P, Claude J, Pouyaud B, Bernard A (1985) Pluies et crues au Sahel: Hydrologie de la Mare d'Oursi (Burkina-Faso). Trav Doc 190, ORSTOM, Paris, 251 pp
Chudeau R (1909) Le Sahara soudanais. In: Furon R (1950) La géologie de l'Afrique. Payot, Paris, 350 pp
Chudeau R (1918) La Dépression du Faguibine. Ann Géogr XXVII:43–60
Chudeau R (1919) La capture du niger par le Tafessassa. Ann Géogr 151:52–60
Cissé AM (1986) Dynamique de la strate herbacée des pâturages de la zone Sud-Sahélienne. Lab Crop Physiol, Agric Univ Wageningen, The Netherlands, 211 pp
Cissé MI (1976) Influence de l'exploitation sur la qualité d'un pâturage Soudano-Sahélien. Thèse de 3ème Cycle. Ec Norm Sup Bamako, 78 pp
Cissé MI (1980a) The browse production of some trees of the Sahel: relationships between maximum foliage biomass and various physical parameters. In: Le Houérou HN (ed) Browse in Africa. ILCA, Addis-Ababa, pp 211–214
Cissé MI (1980b) Effects of various stripping regimes on the foliage production of some browse bushes of the Sudano-Sahelian zone. In: Le Houérou HN (ed) Browse in Africa. ILCA, Addis-Ababa, pp 215–223
Cissé MI (1986) Les parcours sahéliens pluviaux du Mali Central. Caractéristiques et principes techniques pour une amélioration de leur gestion dans le cadre des systèmes de production animale existants. In: Touré IA, Dia PI, Maldague M (eds) La problématique et les stratégies sylvo-pastorales au Sahel. CIEM, Univ de Laval, UNESCO, Paris, pp 255–286
Clanet JC, Gillet H (1980) *Commiphora africana*: a browse tree of the Sahel. In: Le Houérou HN (ed) Browse in Africa. ILCA, Addis-Ababa, pp 443–448
Clos-Arceduc A (1955) Le fleuve fossile de Tombouctou. Tropiques 52, 371:36–40
Clos-Arceduc M (1956) Etudes sur photographies aériennes d'une formation végétale africaine: la "Brousse Tigrée" Bull IFAN, 18, 3:677–684
Cochemé J, Franquin P (1967) Etude agroclimatologique de l'Afrique sèche au Sud du Sahara en Afrique Occidentale. Projet conjoint OMN/UNESCO/FAO; FAO, Rome, 325 pp
Coe MJ, Cumming DH, Phillipson J (1976) Biomass and production of large African herbivores in relation to rainfall and primary production. Oecologia 22:341–354
Cornet A (1981) Le bilan hydrique et son rôle dans la production de la strate herbacée de quelques phytocoenoses sahéliennes au Sénégal. Thèse Dr-Ing, Univ Sci Tech Languedoc, Montpellier, 353 pp
Cornet A, Rambal S (1981) Simulation de l'utilisation de l'eau et de production végétale d'une phytocoenose sahélienne du Sénégal. Oecol Plant 3(17), 4:381–397
Cottam G, Curtis JT (1956) The use of distance measures in phytosociological sampling. Ecology 37:451–460
Coulibaly A (1979) Approche phyto-écologique et phytosociologique des pâturages sahéliens au Mali (région du Gourma). Thèse de 3ème cycle, Univ de Nice, 180 pp
Coulomb J (1971) Zone de modernisation pastorale au Niger. Economie du troupeau. SEDES/IEMVT, Maisons-Alfort, 178 pp
Coulomb J (1972) Projet de développement de l'élevage dans la région de Mopti. Etude du troupeau. IEMVT, Maisons-Alfort, 184 pp
Coulomb J, Serres H, Tacher G (1980) L'élevage en pays Sahélien. Tech Vivante, PUF, Paris, 193 pp
Courel MF (1977) Etude géomorphologique des dunes du Sahel. Les formes dunaires (Niger Nord-occidental et Haute-Volta septentrionale) Thèse 3ème cycle, Univ Paris VII, 284 pp
Courel MF (1985) Etude de l'évolution récente des milieux sahéliens à partir des mesures fournies par les satellites. Thèses Dr Lettres Sci Hum, Univ Paris I, 407 pp, 44 tab, 8 pl, 3 cartes
CTFT (1988) Faidherbia albida. Monographie. CTFT, Nogent sur Marne, 72 pp
Daget P, Poissonnet J (1965) Analyse phytologique des prairies; critères d'application. Ann Agron 22 (1):5–41
Daget P, Poisonnet J (1969) Analyse phytologique des prairies. Applications agronomiques. Doc 48, CEPE/CNRS, Montpellier, 67 pp
Dalebroux R (1972) Etude de l'utilisation des pâturages et des parcours au Niger. AGA: SF/NER 7, FAO, Rome, 29 pp
Dalziel JM (1955) The useful plants of West Tropical Africa. Crown Agents, London, 612 pp

Dancette C (1979) Agroclimatologie appliquée à l'économie de l'eau en zone Soudano-Sahélienne. L'Agron Tropic XXXIV, 4:331–355
Dancette C (1985) Contrariétés pédoclimatiques et adaptation de l'agriculture à la sécheresse en zone intertropicale. In: La sècheresse en zone intertropicale. Conseil International de la Langue française, Paris; CIRAD, Montpellier et ISRA, Dakar, pp 27–41
Dancette C, Poulain JF (1969) Influence of *Acacia albida* on pedoclimatic factors and crop yields. African Soils, 14, 1–2:182–184
Dancette C, Hall HE (1979) Agroclimatology applied to water management in the Sudanian and Sahelian zones of Africa. In: Hall HE, Cannell GH, Lawton HW (eds) Agriculture in semi-arid environments. Ecolog Study 34, Springer, Berlin, Heidelberg, New York, Tokyo, pp 98–113
Dave JV (1979) Extensive data sets of the diffuse radiation in realistic atmospheric models with aerosols and common absorbing goals. Solar Energy 21:361–369
Daveau S (1963) Etude des versants gréseux dans le Sahel Mauritanien. Mem Doc, CNRS, Paris, pp 3–36
Daveau S (1969) La découverte du climat d'Afrique tropicale au cours des navigations portugaises (XVème siècle et début du XVIème siècle). Bull IFAN B, 31:953–988
Daveau S (1970) L'évolution géomorphologique quaternaire au Sud-Ouest du Sahara (Mauritanie). Ann Géogr 431:20–38
Daveau S, Toupet C (1963) Les anciens territoires Gangara. Bull IFAN B, 25:193–215
Davendra C, Burns M (1970) Goat production in the tropics. Tech Commun n° 19, Common Bur Anim Breed Genet, Edinburg, Common Agr Bur Farnham Royal, England
Davy EG, Mattei F, Solomon SI (1976) An evaluation of climate and water resources for development of agriculture in the Sudano-Sahelian zone of West Africa. Special Environmental Report 9, WMO No 459, WMO Geneva, 289 pp
Debongnie B (1984) Socio-économie et demographie du Ferlo Nord et leurs suivis; un bilan des connaissances. Rapport de consultant. PISCEPS, Dakar; FAO, Rome
Dekeyser PL (1955) Les mammifères de l'Afrique noire française. IFAN, Dakar, 426 pp, 242 figs, 16 pl
Dekeyser PL, Villiers A (1954) Essai sur le peuplement zoologique terrestre de l'Ouest-Africain. Bull IFAN, XVI, 3:957–990
Delany MJ, Happold OCD (1979) Ecology of the African mammals. Longman, London
De Leuw PN (1983) Factors affecting remote spectral measurements at ground level and low altitude in herbaceous vegetation in Kenya. In: Vanpraet CL (ed) Méthodes d'inventaire et de surveillance continue des écosystèmes pastoraux sahéliens-application au développement, pp 273–286
Delibrias G, Petit-Maire N, Fabre J (1984) Age des dépots lacustres récents de la région de Taoudenni-Thraza (Sahara malien). CR Acad des Sces Paris, 219, II, 19:1343–1346
Delwaulle JC (1973a) Desertification au Sud du Sahara. Bois For Trop 149:51–68
Delwaulle JC (1973b) L'érosion au Niger. Bois For Trop 150:15–37
Delwaulle JC (1975) Le rôle du forestier dans l'aménagement du Sahel. Bois For Trop 160:3–22
Demange R (1975) Etude de la végétation de la zone d'inondation du Moyen Niger et des régions adjacentes. PNUD, SF/AML/ECO/3, FAO, Rome, 108 pp
Denis JP, Valenza J (1971) Exteriorisation des potentialités génétiques du zébu peuhl sénégalais (Gobra). Rev Elev Med Vet Pays Trop 24 (3):409–418
Depierre D, Gillet H (1971) Désertification de la zone sahélienne du Tchad. Bois For Trop 139:2–25
Devaux C (1973) Plantes toxiques ou réputées toxiques pour le bétail en Afrique de l'Ouest. IEMVT, Maisons-Alfort, 148 pp
De Vos A (1975) Africa, the devastated continent. Junk, The Hague, 236 pp
De Vries DM, De Boer TA (1959) Methods used in botanical grassland research in the Netherlands and their application. Herb Abstr 29 (1):1–7
De Vries Penning FWT, Djiteye AM (ed) (1982) La productivité des pâturages sahéliens. PUDOC, Wageningen, 525 pp
De Wispelaere G (1980a) Les photographies aériennes témoins de la dégradation du couvert ligneux dans un ecosystème Sahélien Sénégalais. Influence de la proximité d'un forage. Cah ORSTOM, XVIII, 3–4:155–166
De Wispelaere G (1980b) Systémes de production d'élevage au Sénégal:étude et cartographie de l'évolution de la végétation par la télédétection. Rapp lère année. IEMVT, Maisons-Alfort, 162 pp, 23 figs

De Wispelaere G (1981) Systèmes de production d'élevage au Sénégal:étude et cartographie de l'évolution de la végétation par télé-détection (2e campagne). IEMVT, Maisons-Alfort, 51 pp
De Wispelaere G, Toutain B (1976) Estimation de l'évolution du couvert végétal en 20 ans, consécutivement à la sécheresse dans le Sahel voltaique. Rev Photointerpret 76:3/2
De Wispelaere G, Toutain B (1981) Etude diachronique de quelques géosystèmes sahéliens en Haute Volta septentrionale. Rev Photointerpret 81:1/1-1/5
DHV Engineering (1980) Country-wide animal and range assessment. Report to the Government of Botswana. 5 vols. In: Williamson DT, Williamson JE (1981) An assessment of the impact of fences on large herbivores biomass in the Kalahar: Botswana Notes and Records, 13:107-110
Diallo A (1978) Transhumance: comportement, nutrition et productivité d'un troupeau de zébus de Diafarabé. Thèse 3ème cycle, Ec Norm Sup, Bamako, 88 pp
Diallo A, Wagenaar K (1981) Livestock productivity and nutrition in pastoral system associated with the flood plains. In: ILCA Systems research in the arid zones of Mali, Systems Study No 5, Addis-Ababa, pp 179-198
Diallo AK (1968) Pâturages naturels du Ferlo-Sud (Rép. du Sénégal). Et Agrostol. n° 23, IEMVT, Maisons-Alfort, 173 pp
Diallo AK (1983) Méthodes d'inventaire et de surveillance continue des écosystèmes pastoraux sahéliens-exemple du Sénégal. In: Vanpreat CL (ed) Méthodes d'inventaire et de surveillance continue des écosystèmes pastoraux sahéliens, ISRA, Dakar, pp 23-30
Diarra L (1976) Composition floristique et productivité des pâturages Soudano-Sahéliens sous pluvioisité annuelle de 1000 à 400 mm. Thèse 3ème Cycle, Ec Norm Sup, Bamako, 95 pp
Diarra L, Breman H (1975) Influence of rainfall on the productivity of Sahelian grasslands in Mali. Proc Symp Evaluation Mapping Tropic Afric Rangelands, ILCA, Addis-Ababa, pp 171-174
Dicko MS (1980) Measuring the secondary production of pasture: an applied example in the study of an extensive production system in Mali. In: Le Houérou HN (ed) Browse in Africa. ILCA, Addis-Ababa, pp 247-254
Dicko MS (1981) Commercialisation du lait par la femme Peuhle du Système agropastoral du mil et du riz. Doc et Trav, ILCA, Bamako
Dicko MS (1983) Rôle et niveau des études de nutrition animale dans le secteur agricole de l'Afrique de l'Ouest. Multigr CIPEA/ILCA, Niamey, 16 pp
Dicko MS (1984) Productivité et rôle du bétail en zones semi-arides au Niger CIPEA/ILCA, Niamey, 27 pp
Dicko MS (1985) Nutrition des bovins du Système Agropastoral du Mil en Zone Sahélienne: comportement et ingestion volontaire. CIPEA/ILCA, Niamey, 20 pp
Dicko MS, Sangaré M (1984) Feeding behaviour of domestic ruminants in Sahelian zones. Proc 2nd Int Rangelands Congr Adelaide, Australia, pp 388-390
Dicko MS, Sayers R (1988) Recherches sur le systeme agropastoral de production de la zone semi-aride du Niger. ILCA, Addis Ababa 148 pp
Dieye K (1981) Etude du tapis végétal d'un écosystème sahélien: zone du Ferlo sénégalais; estimation des potentialités, analyse des processus de dégradation. DEA d'Ecol Végétale, Univ Paris-Sud, Orsay, 64 pp
Dieye K (1983) Evaluation des ressources fourragères naturelles par la méthode du bilan hydrique; cas du Ferlo sénégalais. In: Vanpraet CL (ed) Méthodes d'inventaire et de surveillance continue des écosystemes pastoraux sahéliens, ISRA, Dakar, pp 43-58
Diop A, Gaston A, Sharman MJ, Van Ittersum G, Vanpraet C, Valenza J (1984) Inventaire des ressources pastorales en vue de leur gestion:zone d'encadrement III de la SODESP. ISRA, LNERV, Dakar
Dirschl HJ, Norton-Griffiths M, Westmore P (1978) Training of aerial observers and pilots for counting animals. KREMU, Tech Rep n° 15, Min Tourism Wildlife, Nairobi
Dommergues Y (1966) Contribution à l'étude de la dynamique microbienne des sols en zone semi-aride et en zone tropicale sèche. Ann Agron 113:265-324; 14:379-469
Doorenbos J, Kassam AH (1979) Yield response to water. Irrigation Drainage Pap No. 33, FAO, Rome, 193 pp
Dorst J, Dandelot P (1970) A field guide to the larger mammals of Africa. Delachaux Nestlé, Neuchatel, 286 pp
Doutresoulle G (1947) L'élevage en Afrique Occidentale française. Larose, Paris, 298 pp

Dregne HE, Tucker CJ (1988) Satellite determination of the Saharan-Sahelian border J of Arid Environment 15,3:245–252
DSPA (1982) Etudes sectorielles de l'élevage au Sénégal. Situation et perspectives. Rapports multigraphiés. Direction de la Santé et de la Production Animale, Ministère de l'Agriculture et de l'Elevage, Dakar
Dulieu D, Gaston A, Darley J (1977) La dégradation des pâturages de la région de N'Djamena (Rép du Tchad), en relation avec la présence de *Cyanophyceae* psammophiles. Etude préliminaire. Rev Elev Médec Vét Pays Trop 30(2):181–190
Dumas R (1977) Etude sur l'élevage des petits ruminants du Tchad. IEMVT, Maisons-Alfort, 355 pp
Dumas R (1980) Contribution à l'étude des petits ruminants du Tchad. Rev Elev Médec Vét Pays Trop 33:215–233
Dupire M (1957) Les forages dans l'économie Peuhl. In: Rap Grosmaire: éléments de politique pastorale au Sahel sénégalais, Serv For Min Agr, Dakar
Dupire M (1962) Peul nomades. Etude descriptive des Woodabe du Sahel Nigérien. Trav Mém Inst Ethnol, LXIV, Paris, 336 pp
Dupire M (1970) Organisation sociale des Peul: Etude d'ethnographie comparée. Rech Sci Hum, 32, Plon, Paris, 625 pp
Elouard P (1973/76) Oscillations climatiques de l'Holocène à nos jours en Mauritanie Atlantique et dans la vallée du Sénégal. C R Coll Nouakchott: Désertification Sud Sahara, Nelles Edit Afr, Dakar-Abidjan, pp 27–36
Eppstein H (1971) The origin of the domestic animals of Africa. 2 vols. Afr Publ Corp, New York, 573 pp; 719 pp
FAO (1950–1983) Production yearbooks. FAO, Rome
FAO/CILLS (1980–1986) Développement des cultures fourragères et améliorantes en zone Soudano-Sahélienne:Haute-Volta, Niger, Mali.12 Rap, 1081 pp
FAO/CILLS (1984) Petit manuel de vulgarisation des plantes fouuragères et améliorantes en zone Soudano-Sahélienne. 44, GCP/RAF/098/SWI, CILLS, Ouagadougou
FAO/SIDA (1975) Zone sahélienne. Stratégie à long terme et programme de protection de restauration et de développement. FAO/SWE/TF 117, FAO, Rome, 132 pp, 1 carte
FAO/UNEP (1974) Report of expert consultation panel on the formulation of an international programme on the "Ecological management of arid and semi-arid rangelands in Africa and the Near East. In: Le Houérou HN (ed) Pl Prod Prot Div, FAO, Rome, 68 pp
FAO/UNEP (1975) The ecological management of arid and semi-arid rangelands in Africa and the Near East (EMASAR). Rep Int Conf 3–8 feb 1975, FAO, Rome, 17 pp
FAO/UNEP (1976) Activités de surveillance en vue de l'évolution de certains problèmes d'utilisation d'environnements critiques liés aux pratiques agricoles et d'utilisation des terres (GEMS) consultation d'experts. 15–19 mars 1976, FAO, Rome
FAO/UNESCO (1973) Soil map of Africa, soil map of the world 1/5 000 000; Sheets VI-1, VI-2. FAO, Rome, UNESCO, Paris
Fauck R (1973) Les sols rouges sur sables et sur grès de l'Afrique Occidentale. Contribution à l'étude des sols des régions tropicales. Mém No. 61, ORSTOM, Paris, 258 pp
Felker P (1978) State of the art. *Acacia albida* as a complementary permanent intercrop with annual crops. Univ of California, Riverside, Dept Agr, 133 pp
Floret C, Pontanier R (1982) L'aridité en Tunisie présaharienne. Trav et Doc 150, ORSTOM, Paris, 544 pp
Floret C, Le Houérou HN, Pontanier R (1989) Climatic hazards and development: a comparative study of the arid zones north and south of the Sahara. 35 pp In: Roy J, Di Castri F (eds): Time scales of biological responses to water constraints. Ecological Studies, Springer (in press)
Flower SS (1932) Notes on the recent mammals of Egypt with a list of the species recorded from that kingdom. Proc Zool Soc, London, 2:369–450
Forest F, Poulain JF (1978) Etude du Ruissellement à la parcelle et de ses conséquences sur le bilan hydrique des cultures. CIEH/IRAT, Ougadougou, Haute Volta, 23 pp
Forget P, Barbault R (1977) Ecologie de la reproduction et du développement larvaire d'un amphibien déserticole, *Bufo pentoni* Anderson, 1893, au Sénégal. Terre et Vie, 31:117–125
Fotius G, Valenza J (1966) Etude des pâturages naturels du Ferlo Oriental. IEMVT, Maisons-Alfort, 180 pp

Frère M, Popov GF (1984) Données agroclimatologiques, Afrique. 2 Vol, Pl Prod, Prot Ser 22, FAO, Rome
Furon R (1929) L'ancien delta du Niger (contribution à l'étude de l'hydrographie ancienne du Sahel soudanais et du Sud saharien). Rev de Géogr Phys Géol Dyn, II, IV:265–274
Furon R (1950) La géologie de l'Afrique. Payot, Paris, 350 pp
Gallais J (1967) Le Delta intérieur du Niger, étude de géographie régionale. Mém No 79, IFAN, Dakar, 621 pp
Gallais J (1972) Essai sur la situation actuelle des relations entre pasteurs et paysans dans le Sahel Ouest-Africain. Etudes de géographie régionale offertes à P Gourou. Mouton, Paris, pp 301–313
Gallais J (1975) Pasteurs et paysans du Gourma: la condition sahélienne. Mém CEGET/CNRS, Paris, 239 pp. 15 cartes
Gallais J (1977) Stratégies pastorales et agricoles des sahéliens durant la sécheresse 1969–1974. Doc 30, Trav Doc Géogr Trop CGET/CNRS, Paris, 281 pp
Gaston A (1965) Etude agrostologique du Kanem (Tchad) Et Agrostol No 11, IEMVT, Maisons-Alfort, 155 pp
Gaston A (1967) Etude agrostologique des pâturages de la zone de transhumance de l'Ouadi Haddad (Rép du Tchad). IEMVT, Maisons-Alfort, 64 pp, 1 carte 1/500 000
Gaston A (1973) Etude des potentialités grainières de certains groupements végétaux. IEMVT, Maisons-Alfort, 122 pp, 1 carte 1/50 000
Gaston A (1974) Projet Assalé-Serbewel. Etude agrostologique des pâturages. Et Agrostol No 41, IEMVT, Maisons-Alfort, 143 pp
Gaston A (1975a) Etude des pâturages du Kanem après la sécheresse de 1973. IEMVT; Farcha, Tchad, 24 pp
Gaston A (1975b) Etude de la piste à bétail N'Djamena-Lai. IEMVT, Maisons-Alfort, 56 pp, 1 carte 1/200 000
Gaston A (1981) La végétation du Tchad, évolutions récentes sous les influences climatiques et humaines. Thèse Dr Sci Univ Paris XII, IEMVT, Maisons-Alfort, 333 pp, 1 carte
Gaston A (1983) Le couvert herbacé au Ferlo Sénégalais. In: Vanpraet CL (ed) Méthodes d'inventaire et de surveillance continue des ecosystèmes pastoraux sahéliens, ISRA, Dakar, pp 201–208
Gaston A, Boerwinkel E (1982) Essai de méthode de suivi continu du couvert ligneux. PISCEPS, Dakar; FAO, Rome, 61 pp
Gaston A, Botte F (1971) Etude de la réserve pastorale de Tin Arkachen (Haute Volta). Et Agrostol No. 31, IEMVT, Maisons-Alfort, 146 pp
Gaston A, Dulieu D (1975) Pâturages du Kanem. Effets de la sécheresse de 1973. IEMVT, Maisons-Alfort, 175 pp
Gaston A, Fotius G (1971) Lexique de noms vernaculaires de plantes du Tchad. 2 vol IEMVT/ORSTOM, N'Djamena, 173 pp; 182 pp
Gaston A, Lemarque G (1976) Bilan de quatre années de travaux phytoécologiques en relation avec la lutte contre *Quelea quelea*. Et Agrostol en sous-traitance No 25, IEMVT, Maisons-Alfort, 203 pp
Gaston A, Van Ittersum G, Vanpraet CL (1983a) Mesure au sol de la production primaire par utilisation du spectroradiomètre. Communication présentée au colloque sur les méthodes d'inventaire et de surveillance continue des ecosystèmes pastoraux sahéliens – application au développement, Dakar 16–18 nov 1983; PISCEPS, Dakar, FAO, Rome, ISRA/LNERV, Dakar, 7 pp
Gaston A, Van Ittersum G, Vanpraet CL (1983b) Utilisation des images NOAA 7 pour l'estimation de la production primaire du Ferlo; saisons des pluies 1980–83. Communication présentée au Colloque sur l'Inventaire et al surveillance continue des écosystèmes pastoraux sahéliens, Dakar 16–18 nov 1983. PISCEPS, Daker; FAO, Rome; ISRA/LNERV, Dakar, 12 pp, 3 figs
Gauthier-Pilters H, Dagg AI (1981) The camel. Univ of Chicago Press, Chicago, Ill, 208 pp
Giffard PL (1964) Les possibilités de reboisement en *Acacia albida* au sénégal. Bois For Trop 95
Giffard PL (1971) Recherches complémentaires sur *Acacia albida*. Bois For Trop 135
Giffard PL (1972) Le rôle de l'*Acacia albida* dans la régénération des sols en zone tropicale aride. VIIème Cong For Mond, Buenos-Aires
Giffard PL (1974) L'arbre dans le paysage sénégalais. Sylviculture en zone tropicale sèche. CTFT, Nogent/Marne, 431 pp

Gillet H (1957a) Quelques aspects biogéographiques du massif de l'Air. CR Somm Soc Biogéogr 294:20–25
Gillet H (1957b) Compte rendu sommaire d'une mission sur le massif de l'Ennedi (Nord Tchad) et au Djebel Marra (Soudan) J Agr Trop Bot Appl IV, 9–10:458–464
Gillet H (1960) Etude des pâturages du Ranch de l'Ouadi Rimé. J Agr Trop Bot Appl 8:465–536; 557–692, 21 pl phot, 7 cartes, 10 figs, 10 tabl
Gillet H (1962a) Agriculture, végétation et sols du Centre-Tchad. Feuilles de Mongo, Melfi, Bokoro, Guera. J Agr Trop Bot Appl, IX, 11–12:451–501, 4 pl
Gillet H (1962b) Agriculture, végétation et sols du Centre et Sud-Tchad. Feuilles de Miltou, Dagela, Koumra, Moussafayo. Ed J Agr Trop Bot Appl 1 vol, 108 pp, 18 pl
Gillet H (1964) Pâturages et faune sauvage dans le Nord du Tchad. J Agr Trop Bot Appl, XI, 5–7: 155–176
Gillet H (1965) L'Oryx algazelle et l'Addax au Tchad. Terre Vie, 3:257–272
Gillet H (1967) Essai d'évaluation de la biomasse végétale en zone sahélienne. J Agr Trop Bot Appl, XIV, 4–5:123–158
Gillet H (1968) Le peuplement végétal du massif de l'Ennedi (Tchad). Thèse Doct Sci, Impr Nat Paris, 206 pp, XXXIII pl, 1 carte
Gillet H (1969a) Rapport sur les conditions agroclimatiques et pastorales dans la région du Sahel. MAB 3, Niamey, UNESCO, 30 pp
Gillet H (1969b) L'Oryx algazelle et l'Addax. Distribution géographique. Chances de survie. C R Biogéogr, 405:117–189
Gillet H (1980) Observations on the causes of devastation of ligneous plants in the Sahel and their resistance to destruction. In: Le Houérou HN (ed) Browse in Africa. ILCA, Addis-Ababa, pp 127–130
Gillet H (1981) Girafe et Acacia: une heureuse association. Courr Nat 71:15–21
Gillet H (1983) Tchad: sécheresse et pastoralisme. Bull d'Info du Museum Nat d'Hist Nat Paris, 34:25–37
Gillet H (1984) La chèvre et la Gazelle: exploitation comparée des pâturages par la faune et le bétail en Afrique tropicale sèche. Courr Nat 90:17–26
Gillet H (1985) La sécheresse au Sahel. Encycl Universalis, Suppl 1985
Gillet H (1986) Les principaux arbres fourragers du Sahel sénégalais. In: Touré IA, Dia PI, Maldague M (eds) La problématique et les stratégies sylvo-pastorales au Sahel. CIEM, Univ de Laval, Québec, UNESCO, Paris, pp 37–56
Gillet H, Peyre de Fabrègues B (1982) Quelques arbres utiles en voie de disparition dans le Centre-Est du Niger. Rev Ecol Terre Vie 36:465–470
Gillon Y, Gillon D (1973) Recherches écologiques sur une savane sahélienne du Ferlo septentrional, Sénégal: données quantitatives sur les arthropodes. Terre Vie 27:297–323
Gillon Y, Gillon D (1974a) Recherches écologiques sur une savane sahélienne du Ferlo septentrional, Sénégal: données quantitatives sur les Ténébrionides. Terre Vie 28:296–306
Gillon Y, Gillon D (1974b) Comparaison du peuplement d'invertébrés de deux milieux herbacés Ouest-Africains: Sahel et Savane préforestière. Terre Vie 28:429–474
Gillon D, Adam F, Hubert B, Kahlem G (1983) Production et consommation de graines en milieu sahelo-soudanien au Sénégal: bilan général. Rev Ecol Terre Vie 38:3–35
Gorse J (1985) Desertification in the Sahelian and Sudanian zones of West Africa. The World Bank Washington DC, 60 pp
Granier P (1974) Rapport agrostologique sur la factibilité de deux ranchs au Niger. SEDES/IEMVT, Maisons-Alfort, 40 pp, 8 graph
Granier P (1977) Rapport d'activités agropastorales en République du Niger. IEMVT, Maisons-Alfort, 140 pp
Grenier P (1957) Rapport de mission dans le Ferlo Serv Hydraul d'Afrique Occidentale Française, Dakar
Grenot C (1974) Ecologie appliquée à la conservation et à l'élevage des ongulés sauvages au Sahara algérien. CNRS, Paris
Grimsdell JJR, Bille JC, Milligan K (1981) Alternative methods of aerial livestock census. In: ILCA Low level aerial survey techniques: Monograph n°4, Addis-Ababa, pp 103–112
Grosmaire (1957) Eléments de politique pastorale au Sahel sénégalais. Serv Eaux For, Dakar

Grouzis M (1979) Structure, composition floristique et dynamique de la matière sèche de formations végétales sahéliennes (Mare d'Oursi, Haute-Volta). DGRST/ORSTOM, Ouagadougou, 59 pp, 15 tabl, 17 figs
Grouzis M (1982) Restauration des pâturages sahéliens. Mise en défens et reboisement. ORSTOM, Ouagadougou, 37 pp
Grouzis M (1983) Problèmes de désertification en Haute-Volta. Notes et Documents voltaiques 15, 1-2:1-13
Grouzis M (1984) Pâturages sahéliens du Nord du Burkina-Faso. ORSTOM, Ouagadougou, 35 pp
Grouzis M (1988) Structure, productivité et dynamique des systèmes écologiques sahéliens (Mare d'Oursi, Burkina-Faso). Coll Etudes et Thèses; ORSTOM, Paris, 336 pp
Grouzis M, Méthy M (1983) Détermination radiométrique de la phytomasse herbacée en milieu sahélien: perspectives et limites. Acta Oecol Oecol Plant 4 (18) 3:241-257
Grouzis M, Sicot M (1980) A method for phenological studies of browse populations in the Sahel: the influence of some ecological factors. In: Le Houérou HN (ed) Browse in Africa. ILCA, Addis-Ababa, pp 233-240
Grouzis M, Legrand E, Pale F (1986) Aspects écophysiologiques de la germination des semences sahéliennes. Cpte-Rend Colloque sur les végétaux en milieu aride, Jerba, Tunisie, UNESCO, Paris 19 pp
Grove AT (1958) The ancien erg of the Hausaland and similar formation on the southern side of the Sahara. Geogr J, 124:528-533
Grove AT (1971) Africa south of the Sahara. Oxford Univ Press, London, 280 pp
Grove AT, Street FA, Goudie AS (1975) Former lake levels and climatic change in the rift valley of southern Ethiopia. Geogr J, 141:177-202
Guinea E (1945) La vegetacion leñosa y los pastos del Sahara Español. Inst For Invest Y Expert, Madrid, 152 pp
Guinea-Lopez E (1949) El Sahara Español. Cons Sup Invest Cient Est Afric, Madrid, 808 pp
Gwynne MD, Croze H (1975) Methodes de surveillance écologique dans l'Est-Africain. Actes Colloq Bamako Invest Cartogr Patur Tropic Afric CIPEA/ILCA, Addis-Ababa, pp 95-136
Haaland R (1979) Le rôle de l'homme dans l'évolution du milieu au Méma pendant l'ancient royaume du Ghana. Introduction par HN Le Houérou. CIPEA/ILCA, Bamako, Addis-Ababa, 21 pp, Image satellite
Hall HTB (1985) Diseases and parasites of livestock in the tropics. Longman, London, 328 pp
Haltenorth T, Diller H, Cusin M (1985) Mammifères d'Afrique et de Madagascar. Delachaux and Nestlé, Neuchatel/Paris, 397 pp 287 figs
Hamel O (1980) Acclimatation and utilization of phyllodineous Acacias from Australia in Sénégal. In: Le Houérou HN (ed) Browse in Africa. ILCA, Addis-Ababa, pp 361-374
Happold DC (1973) The distribution of large mammals in West Africa. Mammalia 37:90-93
Happold DC (1984) Small mammals. In: Cloudsley-Thompson JL (ed) Sahara Desert. Key Environments; Pergamon Press, Oxford, UK, Ch. 17, pp 251-275
Hardin G (1968) The tragedy of the commons. Science, 162:1243-1248
Harrison MN (1955) Report on a grazing survey of the Sudan. Min Anim Res, Khartoum, 280 pp
Harrison MN, Jackson JK (1958) Ecological classification of the vegetation of the Sudan. *Forest Bull, New Ser* n° 4, Khartoum, 46 pp, 1 map
Harroy JP (1944) Afrique, terre qui meurt. La dégradation des sols africains sous l'influence de la colonisation. Hayez, Bruxelles, 557 pp
Hassan HM (1974) An illustrated guide to the plants of Erkowit. Khartoum University Press, 106 pp
Haywood M (1981) Evolution de l'utilisation des terres et de la végétation dans la zone Soudano-sahélienne du project CIPEA au Mali. Doc Trav 3, CIPEA/ILCA, Addis-Ababa, 187 pp
Herlocker DJ, Dolan RA (1980) Comparison of different techniques for the determination of large dwarf shrubs biomass. Proc Sci Sem IPAL, UNESCO, Nairobi, pp 30-40
Heusch B (1975) La conservation des eaux et des sols dans la haute vallée de Keita (Rép du Niger). SOGREAH, Grenoble, 31 pp
Hiernaux P (1980) Inventory of the browse potential of bushes, trees and shrubs in an area of the Sahel in Mali: methods and initial results. In: Le Houérou HN (ed) Browse in Africa. ILCA, Addis-Ababa, pp 197-205

Hiernaux P (1984) Encore une saison des pluies très déficitaire sur le Ranch de Niono (Sud Sahel, Mali). Evolution de la végétation d'une cinquantaine de sites suivis depuis 1976. Doc de Progr No AZ 98, CIPEA, Bamako, 46 pp
Hiernaux P (1985) Distribution des pluies et production herbacée au Sahel: une méthode empirique pour caractériser la distribution des précipitations journalières et ses effets sur la production herbacée. Premiers résultats acquis le Sahel Malien. Doc de Progr No AZ 98, CIPEA, Bamako, Mali, 46 pp
Hiernaux P, Cissé MI, Diarra L (1978) Rapport de la section d'écologie Diff Restr, ILCA/CIPEA, Bamako, 92 pp
Holben BN, Tucker CJ, Fan CJ (1980) Assessing soybean leaf area and leaf biomass with spectral data. Photogramm Engineer Rem Sens, 46:651–656
Hoste C, Peyre de Fabrègues B, Richard D (1984) Le dromadaire et son élevage IEMVT, Maisons-Alfort, 162 pp
Hubert B (1977) Ecologie des populations de rongeurs de Bandia (Sénégal) en zone sahélo-soudanienne. Terre Vie, 31:33–100
Hubert B (1982a) Dynamique des populations de deux espèces de rongeurs au Sénégal: *Mastomys erythroleucus* et *Taterillus gracilis (Rodentia, Muridae and Gerbillidae)* I: étude démographique. Mammalia, 46:137–166
Hubert B (1982b) Ecologie des populations de deux rongeurs sahelo-soudaniens à Bandia (Sénégal). Thèse Doct Sc Univ de Paris-Sud, 448 pp
Hubert B, Gillon D, Adam P (1981) Cycle annuel du régime alimentaire des trois principales espéces de rongeurs *(Rodentia, Gerbillidae and Muridae)* de Bandia (Sénégal). Mammalia, 45:4–20
Hubert H (1920) Le desséchement progressif en Afrique Occidentale. Bull Com Et Hist Sci d'AOF, 3:401–467
Hufnagl E (1972) Libyan mammals. Oleander Press, Cambridge, 85 pp
Hugo HJ (1974) Le Sahara avant le désert. Ed Hespérides, Toulouse, France, 343 pp
Hunting Technical Services (1964) Land and water use survey in the Kordofan province of the Republic of Sudan. London; FAO, Rome, 349 pp, 9 tabl, 50 figs
Hunting Technical Services (1968) Reconnaissance vegetation survey of the Jebel Marra area. HTS London; FAO, Rome, 187 pp, 33 tabl, 13 pl
Hunting Technical Services (1974) Southern Darfur land-use planning survey Annex 3, Animal Resources and Range Ecology. Min Agric Food Natural Resources, Khartoum; HTS, London, 180 pp
Hutchinson J, Dalziel JM Flora of West tropical Africa. I, part 1 (1954) I, part 2 (1958) II (1963) III, part 1 (1968) III, part 2 (1972) Crown Agents Overseas Govern Administ, London
IEMVT (1958) Le Ranching Techniques rurales en Afrique, n° 15, SEDES/IEMVT, Maisons-Alfort, 278 pp
IEMVT (Institut d'Elevage et de Médecine Vétérinaire des Pays Tropicaux) (1971) Manuel vétérinaire des agents techniques de l'élevage tropical. IEMVT, Maisons-Alfort, 520 pp
IEMVT (1977) Manuel d'hygiène du bétail et de prophylaxie des maladies contagieuses en zone tropicale. IEMVT, Maisons-Alfort, 155 pp
IEMVT (1980) Les petits ruminants d'Afrique Centrale et de l'Afrique de l'Ouest. IEMVT, Maisons-Alfort, 295 pp
ILCA (International Livestock Centre for Africa) (1978a) Study of the traditions livestock production systems in central Mali (Sahel and Niger Inland Delta) (Le Houérou HN, Wilson RT (eds)). ILCA, Bamako, Addis-Ababa, 430 pp
ILCA (1978b) Evaluation of the productivities of Maure and Peul cattle breeds at the Sahelian Station, Niono, Mali. Monograph n° 1, ILCA, Addis-Ababa, 109 pp
ILCA (1981) Systems research in the arid zones of Mali. Initial results. Systems study n° 5, ILCA, Addis-Ababa, 250 pp
Jackson SP (1961) Climatological atlas of Africa. 55 pl
Jolly GM (1969) Sampling methods for aerial censuses of wildlife populations. East Afric Agric For J, Spec Iss, 3,34:46–49
Jolly GM (1979) Sampling of large objects. In: Cormack RM, Patil GP, Dobson DS (eds) Sampling biological populations. n° 5, Statistic Ecol Ser; Int Coop Publ House, Fairland
Jolly GM (1981) A review of the sampling methods used in aerial survey. In: Low level aerial survey techniques, monograph n° 4, ILCA, Addis-Ababa, pp 146–157

Jolly GM, Watson RM (1979) Aerial sample survey methods in the quantitative assessment of ecological resources. In: Cormak RM, Patil GP, Robson DS (eds) Sampling biological populations. n° 5, Statistic Ecol Ser, Int Coop Publ House, Fairland

Jung G (1970) Variation saisonnière des caractéristiques microbiologiques d'un sol ferrugineux tropical peu lessivé (Dior), soumis ou non à l'influence d'*Acacia albida*. Del. *Oecol. Plant,* V:113–136

Justice CO (1986) Monitoring the grasslands of semi-arid Africa using NOAA/AVHRR data. Int J Remote Sens 7, 11:1383–1622

Kahlem G (1981) La végétation de la forêt de Bandia. Evolution des populations végétales et de la production de graines pendant les années 1978–79–80. Bull IFAN, 43, 3–4:232–252

Kandel R, Courel MF (1984) Le Sahel est-il responsable de sa sècheresse? La Recherche, 15, 158:1152–1154

Kassam AH, Higgins GM (1980) Land resources for populations of the future. UNFA/FAO, FAO, Rome, 369 pp

Kassas M (1953) Landform and plant cover in the Egyptian desert. Bull Soc Geogr d'Egypte, 26:193–206

Kassas M (1955) Rainfall and vegetation belts in North-East Africa. Proc Symp plant ecol, Montpellier, UNESCO, Paris, pp 49–57

Kassas M (1956a) The mist oasis of Erkwit, Sudan. J Ecol 44:180–194

Kassas M (1956b) Land forms and plant cover in the Omdurman desert, Sudan. Bull Soc Geogr d'Egypte, 29:43–58

Kassas M (1968) Dynamics of desert vegetation. Proc IBP-CT Tech Meet Nature Conservation in the Mediterranean Basin. Hammamet, Tunisia, IBP, London

Kassas M (1970) Desertification potential for recovery in certain circum-Saharan territories. In: Dregne HE (ed) Arid lands in perspective. Americ Ass Adv Sci, Washington DC, pp 123–139

Keay RWJ (1959) Vegetation map of Africa, south of the Tropic of Cancer. Explanatory notes. Oxford Univ Press, London, 27 pp

Keita MN (1982) Les disponibilités de bois de feu en région sahélienne de l'Afrique Occidentale: situation et perspectives. Misc 82/15, Dept For, FAO, Rome, 79 pp

Kidwell KB (1979) NOAA polar orbiter data (Tiros-N and NOAA 6) user's guide. NOAA, National Climate Center. Satellite Data Services Division, Washington, D.C.

Kiekens JP (1984) Proposition de gestion agro-sylvo-pastorale pour les zones sahélienne et soudano-sahélienne. Diss Sect Agronomie, Lab Bot Sytém Ecol, Univ Libre de Bruxelles, 76 pp

Kimes DS, Markham BL, Tucker CJ, McMurtry JE (1981) Temporal relationships between spectral responses and agronomic values of a corn canopy. Remote Sensing of the Environment, 11:401–411

Knoess KH (1977) The camel as a meat and milk animal. World Animal Rev 22:3–8

Lamprey HF (1975) Report on the desert encroachment reconnaissance in northern Sudan. UNEP/UNESCO, Nairobi, 14 pp

Lamprey HF (1983) Pastoralism yesterday and today: the overgrazing problems. In: Bourlière F (ed) Tropical savanna. Ecosystems of the World, vol. 13, Elsevier, Amsterdam, ch. 31, pp 643–666

Landsberg HE, Lippman H, Paffen KH, Troll C (1965) World maps of climatology. Global radiation, sunshine and seasonal climate. Springer, Berlin, Heidelberg, New York, Tokyo, 28 pp, 5 maps 1/45 000

Lange RT (1969) "The Piosphere": sheep tracks and dung patterns. J Rge Mgt, 22:396–400

Lavauden L (1924) La chasse et la faune cynégétique en Tunisie. Serv For Dir Agr Tunis

Lawton RM (1980) Browse in the Miombo woodland. In: Le Houérou HN (ed) Browse in Africa. ILCA, Addis-Ababa, pp 25–34

Lebrun J (1947) La végétation de la plaine alluviale du Lac Edouard. Explor Parc Nat Albert, Inst Parcs Nat du Congo Belge, 800 pp

Lebrun J (1960) Sur la richesse de la flore de divers territoires africains. Bull Séances Acad Roy Sci Outre-Mer, ns, 6:-669–690

Lebrun JP (1973) Enumération des plantes vasculaires du Sénégal. IEMVT, Maisons-Alfort, 209 pp, 6 pl

Lebrun JP (1976) Richesses spécifiques de la flore vasculaire des divers pays ou régions d'Afrique. Candollea, 31:11–15

Lebrun JP (1977/78) Eléments pour un atlas des plantes vasculaires de l'Afrique sèche. IEMVT, Maisons-Alfort, vol I: 262 pp; vol 2: 255 pp
Lebrun JP (1981) Les bases floristiques des grandes divisions chorologiques de l'Afrique sèche. Et Bot No. 7, IEMVT, Maisons-Alfort, 483 pp
Lebrun JP (1983) Flore des massifs sahariens: espèce illusoires et endémiques vraies. Bothalia, 14, 3:511–515
Lebrun JP, Stork AL (1977) Index 1935–1976 des cartes de répartition des plantes vasculaires d'Afrique. Conserv Jard Bot, Genève, 138 pp
Legrand Ph (1979) Biomasse racinaire de la strate herbacée de formations sahéliennes (Etude préliminaire, Mare d'Oursi, Haute-Volta). ORSTOM, Ouagadougou, 28 pp
Le Houérou HN (1959) Recherches écologiques et floristiques sur la végétation de la Tunisie méridionale. Mém HS, Instit Rech Sahar Univ d'Alger, 2 vols + tables and maps, 520 pp
Le Houérou HN (1969) La végétation de la Tunisie steppique. Ann Inst Nat Rech Agro Tunisie, Plates and tables, 1 colour map 1.500 000 (128 000 km^2), 42, 5:1–640
Le Houérou HN (1970) North Africa: past, present and future. In: Dregne HE (ed) Arid lands in transition. Am Ass Adv Sci, Washington, DC, pp 227–278
Le Houérou NH (1972a) L'ecologia vegetale nella regione mediterranea. Parte I: I fattori climatici. Minervia Biologica, 1, 3:123–134
Le Houérou HN (1972b) Le développement agricole et pastoral de l'Irhazer d'Agadès (Niger). W S/D 7975, AGPC, FAO, Rome, 27 pp
Le Houérou HN (1973/76a) Contribution à un bibliographie écologique des régions arides de l'Afrique et de l'Asie du Sud-Ouest. CR Coll Nouakchott sur la désertification au Sud du Sahara, (1500 réf), Nelles Edit Afric, Dakar-Abidjan, pp 170–211
Le Houérou HN (1973/76b) Peut-on lutter contre la désertisation? CR Coll, Nouakchott sur la désertification au Sud du Sahara, Nelles Edit Afric, Dakar-Abidjan, pp 158–163
Le Houérou HN (1974) Rapport de mission au Sénégal. AGP:DP/RAF/68/114, FAO, Rome, 7 pp
Le Houérou HN (1976a) Nature and causes of desertization. In: Glantz M (ed) Desertification in and around arid-lands, Westview, Boulder, Colorado, pp 17–38
Le Houérou HN (1976b) Nature et désertisation. CR Consultation sur la Foresterie au Sahel. CILSS/UNESCO/FAO, FO: RAF /305/3, FAO, Rome, 21 pp
Le Houérou HN (1977) The grassland of Africa: classification, production, evolution and development outlook. Proc XIIIth Int Grassland Congr, Vol: I Akademie Verl, East Berlin, pp 99–116
Le Houérou HN (1978) The role of shrubs and trees in the management of natural grazing lands (with particular reference to protein production). Position Paper, Item n° 10; FFF/10-0; Eight World Forestry Congress, Jakarta; FAO, Rome, 24 pp. Also available in French and Spanish
Le Houérou HN (1979a) Ecologie et désertisation en Afrique. Trav Inst Géogr Reims, 39–40:5–26
Le Houérou HN (1979b) La désertisation des régions arides. La Recherche, 99:336–344
Le Houérou HN (1979c) Le rôle des arbres et arbustes dans les pâturages sahéliens. In: Le rôle des arbres au Sahel, CR Coll Dakar, IDRC/CRDI, Ottawa, pp 19–32
Le Houérou HN (ed) (1980a) Browse in Africa: the current state of knowledge. ILCA, Addis-Ababa 491 pp. Also available in French
Le Houérou HN (1980b) The role of browse in the Sahelian and Sudanian zones. In: Le Houérou HN (ed) Browse in Africa. pp 83–102
Le Houérou HN (1980c) Chemical composition and nutritive value of browse in West-Africa. In: Le Houérou HN (ed) Browse in Africa. pp 261–290
Le Houérou HN (1980d) Agroforestry techniques for the conservation and improvement of soil fertility in arid and semi-arid zones. In: Le Houérou HN (ed) Browse in Africa. pp 433–436
Le Houérou HN (1980e) Reboisement sylvopastoral et production fourragère aux Iles du Cap Vert. GCP/CVI/002/BEL, Dept For, FAO, Rome, 36 pp, 2 figs
Le Houérou HN (1980f) The rangelands of the Sahel. J Rge Mgt, 33, 1:41–46
Le Houérou HN (1981) How to calculate livestock conversion factors in the Libyan rangelands. Techn Doc No 9, Lib/018, FAO, Tripoli, 15 pp
Le Houérou HN (1982) Prediction of range production from weather records in Africa. In: Proc Tech Conf Climate Afr, WMO/OMM, Doc No 596, WMO, Geneva, pp 288–298
Le Houérou HN (1982/84) An outline of the bioclimatology of Libya. Bull Soc Bot Fr, 131, Actual Bot (2/3/4):157–178

Le Houérou HN (1984) Rain-use-efficiency: a unifying concept in arid-land ecology. J Arid Envir, 7:213–247
Le Houérou HN (1985) The impact of climate on pastoralim. In: Kates RW, Ausubel JH, Berberian M (eds) Climate impact assessment. SCOPE Study n° 27, Wiley, New York, pp 155–185
Le Houérou HN (1986) La variabilité de la pluviosité annuelle dans quelques régions arides du monde; ses conséquences écologiques. CR Coll Sahel/Nordeste Brésilien, Inst. Htes Et. Amér. Latine, Univ Paris III, Paris, 11 pp (in press)
Le Houérou HN (1987a) Agroforestry in the Peanut Basin of Senegal. Invest Center, FAO and FIDA, Rome, 75 pp
Le Houérou HN (1987b) Indigenous shrubs and trees in the silvo-pastoral systems of Africa. In: Steppler HA, Nair PKR (eds) Agroforestry, a decade of development. ICRAF, Nairobi, pp 117–140
Le Houérou HN (1987c) The Sahara from the bioclimatic viewpoint: definitions and limits. In: The Changing Sahara, Proc Int Sym, Wye College and Kew Royal Botanic Gardens, Richmond, 20 pp
Le Houérou HN (1988a) Meteorological aspects of plant growth in desert and desert-prone lands. WMO, Geneva, 59 pp, TD No 194
Le Houérou HN (1988b) Surveillance continue des écosystèmes paturés sahéliens au Sénégal. Rapport technique de synthèse. FAO, Rome and UNEP, Nairobi, 144 pp
Le Houérou HN (1988c) A comparative ecoclimatic study of rangelands in intertropical Africa and the hot arid zones of the Indian subcontinent. 3rd Int Rangeland Congr, Abst, vol II, Plenary Papers, Range Man Soc India, New Delhi, pp 675–685
Le Houérou HN (1988d) Forage diversity in Africa: an overview of the plant resources. Proc Int Sym Plant Gen Res Africa, Int Board Plant Gen Res, Rome, 25 pp
Le Houérou HN (1989) Relations entre la variabilité des précipitations et celles de la production primaire et secondaire en zone aride. 30 pp, multigr. In: Bille JC, Cornet A, Grouzis M, Le Floc'h E (eds) L'Aridité, contrainte au développement. ORSTOM, Paris (in press)
Le Houérou HN, Gillet H (1985) Conservation versus desertization in African arid lands. Proc 2nd Int Conf. Conservation Biology, Univ of Michigan, Ann-Arbor, pp 444–461
Le Houérou HN, Hoste CH (1977) Rangeland production and annual rainfall relations in the Mediterranean Basin and African Sahelian and Sudanian zones. J Rge Mgt, 30: 181–189
Le Houérou HN, Naegelé A (1972) The useful shrubs of the Mediterranean Basin and the Arid Tropical Belt south of the Sahara. AGPC: Misc/24, FAO, Rome 20 pp. Also In: McKell CM, Blaisdell JP, Goodin Jr (eds) Wildland shrubs their biology and utilization. Gen Rept, INT-I, USDA, For Serv, Intermount. For Range Exper Stn, Ogden, Utah, pp 26–36
Le Houérou HN, Norwine JR (1988) The ecoclimatology of south Texas. In: Arid lands today and tomorrow. Westview Press, Boulder, Colorado, pp 417–443
Le Houérou HN, Popov GF (1981) An ecoclimatic classification of intertropical Africa. Plant Prod Protec Pap No 31 FAO, Rome, 40 pp, 4 maps
Le Houérou HN, Bingham RE, Skerbek W (1988) Relationship between the variability of annual rainfall and the variability of primary production. J Arid Envir, 15:1–18
Lepage M (1972) Recherches écologiques sur une savane sahélienne du Ferlo septentrional: données préliminaires sur l'écologie des termites. Terre Vie, 26:325–372
Lepage M (1974a) Les termites d'une savane sahélienne (Ferlo septentrional, Sénégal), rôle dans l'écosystème. Thèse Doct Sci, Univ de Dijon
Lepage M (1974b) Recherches écologiques sur une savane sahélienne du Ferlo septentrional, Sénégal; Influence de la sècheresse sur le peuplement des termites. Terre Vie, 28:76–94
Leprun JC (1978a) Esquisse pédologique au 1/50 000ème des alentours de la Mare d'Oursi, avec notice et analyses des sols. Trav Doc ORSTOM, Paris, 53 pp, 4 figs, 1 carte
Leprun JC (1978b) Lutte contre l'aridité en milieu tropical, étude de l'évolution d'un système d'exploitation sahélien au Mali; Compte rendu de fin d'étude sur les sols et leur susceptibilité à l'érosion, les terres de cure salée, les formations de brousse tigrée dans le Gourma. DGRST/ORSTOM, Paris, 51 pp
Leprun JC (1979a) Lutte contre l'aridité en milieu tropical, étude de l'évolution d'un système d'exploitation sahélien au Mali. Volet pédologique, Rapport de campagne 1979, DGRST/ORSTOM, Paris, 28 pp

Leprun JC (1979b) Les cuirasses ferrugineuses des pays cristallins de l'Afrique Occidentale sèche. Genèse, Transformations, Dégradations, Sces Géol, ORSTOM, Paris, 224 pp
Leroux M (1983) Le climat de l'Afrique tropicale. Champion, Paris, 633 pp, 348 figs; Atlas 247 pl, Bibl; 1016 réf
Leroux P (1970) Etudes méthodologiques pour l'établissement d'une carte de la végétation. Progr Nat Unies Criquet Migrateur Afric UNDP (SF) AML/ECOL, FAO, Rome, 77 pp
Levang P, Grouzis M (1980) Méthodes d'étude de la biomasse herbacée de formations sahéliennes: application à la Mare d'Oursi, Haute-Volta. *Acta Oecol/Oecol Plant* 1 (15), 3:231–244
Levy EB, Madden EA (1933) The point method of pasture analysis. New Zeal J Agric, 46:267–279
Lhoste Ph (1977) Lutte contre l'Aridité dans l'Oudalan. Etude zootechnique, Inventaire du cheptel. ACC, IEMVT Maisons-Alfort, 49 pp
Lhoste Ph (1986) L'association agriculture-élevage. Evolution du système agro-pastoral au Siné-Salloum (Sénégal). Thèse Ing-Doct, INA-PG, Paris-Grignon, 314 pp
Lhote H (1958) A la découverte des fresques du Tassili. Arthaud, Paris, 268 pp
Maignien R (1965) Carte des sols du Sénégal 1/1 000 000 ORSTOM, Dakar Cambridge Univ press, London, 259 pp
Mainguet M (1972) Le modelé des grès, problèmes généraux. 2 vol. Inst Géogr Nation, Paris, 657 pp
Mainguet M (1976) Propositions pour une nouvelle classification des édifices sableux éoliens d'après les images landsat I, Gemini, et NOAA 3. Z Geomorph, NF, 20, 3:275–296
Mainguet M (1977) Analyse quantitative de l'extrémité sahélienne du courant éolien transporteur de sable au Sahara nigérien. CR Acad Sci 285, D:1029–1032
Mainguet M (1983) Dunes vives, dunes fixées, dunes vêtues: une classification selon le bilan d'alimentation, le régime éolien et la dynamique des édifices sableux. Z Geomorph, NF, Suppl Bd 45:265–285
Mainguet M (1984a) A classification of dunes based on Aeolian dynamics and the sand budget. In: El Baz F (ed) Deserts and Arid Lands. Ch 2. Martinus Nijhoff, The Hague, pp 31–58
Mainguet M (ed) (1984b) Le vent: mécanisme d'érosion, de dégradation, de désertification. Trav Inst Géogr Reims, 59–60:1–142
Mainguet M (1985) Le Sahel, un laboratoire naturel pour l'étude du vent, mécanisme principal de la désertification. In: Bandorff-Nielsen OE, Meller JT, Remer Rasmussen K, Willets BB (eds) Proc Int Workshop on the physics of blown sand. Dept Theoret Stat, Inst Mathematics, Univ Aarhus, Mem No 8, pp 545–561
Mainguet M, Callot Y (1978) L'Erg de Fachi-Bilma (Tchad-Niger). Contribution à la connaissance des ergs et des dunes des zones arides chaudes. Mem Doc, Vol n° 18, CNRS, Paris, 184 pp
Mainguet M, Canon L (1976) Vents et paleovents du Sahara. Tentative d'approche paleoclimatique. Rev Géogr Phys Géolog Dynam XVIII, 2–3: 241–250
Mainguet M, Chemin MC (1977) Les marques de l'érosion éolienne dans le Sahel du Niger, d'après les images satellite et les photographies aériennes. Coll Pédol Télédétec, AISS, Rome, pp 139–149
Mainguet M, Chemin MC (1983) Sand seas of the Sahara and Sahel: an explanation of their thickness and sand dune types by the sand-budget principle. In: Brookfield ME, Ahlbrandt TS (eds) Aeolian sediments and processes. Elsevier, Amsterdam, pp 353–363
Mainguet M, Cossus L (1980) Sand circulation in the Sahara: Geomorphological relations between the Sahara Desert and its margins. In: Van Zinderen Bakker EM, Coetzee JA (eds) Palaeoecology of Africa and the surrounding islands. Vol 12: Sarnthein M, Seibold E, Rogreon P (eds) Sahara and surrounding seas, AA Balkema, Rotterdam, pp 69–78
Mainguet M, Canon-Cossus L, Chemin MC (1979) Dégradation dans les régions centrales de la République du Niger: degré de responsabilité de la nature du milieu, de la dynamique externe et de la mise en valeur par l'homme. Trav Inst Géogr Reims, 39–40; 61–73
Mainguet M, Canon L, Chemin MC (1980) Le Sahara: geomorphologie et paléogéomorphologie éolienne. In: Williams MAJ, Faure H (eds) The Sahara and the Nile. Balkema, Rotterdam, pp 17–35
Maley J (1973) Mécanismes de changements climatiques aux basses latitudes. Palaeogeography, Palaeoclimatology, Palaeoecology, 14:193–227
Maley J (1977) Palaeoclimates of Central Sahara during early Holocene Nature 269:573–577
Maley J (1980) Les changements climatiques de la fin du Tertiaire en Afrique et leur conséquence sur l'apparition du Sahara et de sa végétation. In: Williams AJ, Faure H (eds): The Sahara and the Nile, A.A. Balkema, Rotterdam, pp 63–86

Maley J (1981) Etudes palynologiques dans le bassin du Tchad et Paleo-climatologie de l'Afrique Nord-Tropicale de-30 000 ans à l'époque actuelle. Mém Doc 129, ORSTOM, Paris, 586 pp
Maley J (1982) Dust, clouds, rain-types and climatic variation in tropical North Africa. Quat Res, 18:1-16
Maley J (1983a) Histoire de la végétation et du climat de l'Afrique Nord-Tropicale au Quaternaire récent. Bothalia, 14, 3-4:377-389
Maley J (1983b) Histoire de la végétation et du climat de l'Afrique Nord-Tropicale au Quaternaire récent. Bothalia, 14, 3:377-389
Mané Y (1986) Etude du couvert herbacé du pâturage naturel et essai de cultures fourragères. Rapp n° 1912, AGRHYMET, Niamey; OMM, Genève, 69 pp
Markham BL, Kimes DS, Tucker CJ, McMurtey JE (1981) Temporal spectral response of a corn canopy. Photogramm Engin Rem Sens 48 (11):1599-1605
Marty JP, Greigert J, Peyre de Fabrègues B (1966) Mise en valeur du complexe pastoral situé au Nord de l'axe Filingué-Tahoua, au Niger. Min Coop, Paris, 38 pp
Mason IL (1951) A classification of West African livestock. Techn Comm n° 7, Commonw Agric Bur, Farnham Royal, England
Mason IL, Maule JP (1960) The indigenous livestock of eastern and southern Africa. Techn Comm n° 14, Bur Anim Breed Genet, Edinburgh; Commonw Agric Bur, Farnham Royal, England
Mattei F (1977) An agroclimatological study of the Democratic Republic of Sudan. WMO, Geneva, 270 pp
Mauny R (1956) Préhistoire et zoologie: la grande faune éthiopienne du Nord-Ouest africain du Paleolithique à nos jours. Bull IFAN, 18 (1):246-249
Mauny R (1957) Répartition de la "grande faune éthiopienne" du Nord-Ouest Africain du Paleolithique à nos jours. CR III e Congrès Panafr de Préhist Livingstone, 1955, London, Chatto and Windus, 1957, 4 figs, pp 102-105
Mauny R (1961) Tableau géographique de l'Ouest-Africain au Moyen âge, d'après les sources ècrites, la tradition et l'archéologie. Mém 61, IFAN, Dakar, 587 pp
Mauny R (1978) Trans-Saharan contacts and the Iron Age in West Africa. In: Farge JD (ed) The Cambridge history of Africa. Cambridge University Press, London, vol 2:272-341
McIntosh RJ (1983a) Floodplain geomorphology and human occupation of the upper Inland Delta of the Niger. Geogr J, 149:182-201
McIntosh RJ, McIntosh SK (1981a) The Inland Niger Delta before the Empire of Mali: evidence from Jenne-Jeno. J Afric Hist, 22:1-22
McIntosh SK, McIntosh RJ (1981b) West-African prehistory from 10,000 BC to 1,000 AD Am Sci 69:602-613
McIntosh SK, McIntosh RJ (1983b) Current directions in West-African prehistory Ann Rev Anthrop, 12:215-258
McMichael HA (1912) The tribes of northern and central Kordofan
Menaut JC (1983) Structure et production de peuplements de ligneux fourragers. In: Toutain B et al. (eds) Espèces ligneuses et herbacées dans les écosystemes pâturés sahéliens de Haute-Volta. GERDAT, IEMVT, ENS, CNRST, IRBET Paris, Ougadougou, pp 17-55
Menaut JC, Cesar J (1979) Structure and primary productivity of Lamto savannas, Ivory Coast. Ecology, 60 (6):1197-1210
Meyer JF (1980) Etude des systèmes de production d'élevage au Sénégal; volet zoo-économie, lere année. DGRST, IEMVT, Paris/ISRA-LNERV, Dakar, 28 pp, 23 tables, 1 fig
Michel P (1973) Les bassins des fleuves Sénégal et Gambie. 3 vols. Mém No 63, ORSTOM, Paris, 752 pp
Michel P, Naegelé A, Toupet C (1969) Contribution à l'étude biologique du Sénégal septentrional. In: Le Milieu Naturel, Bull IFAN, sér A 31:755-839
Michon P (1973) Le Sahara avance t-il vers le Sud? Bois For Trop 150:69-80
Milleville P (1980) Lutte contre l'aridité dans l'Oudalan (Haute-Volta): étude d'un système de production agro-pastoral sahélien de Haute-Volta. lere partie: le système de culture. Progr ACC, ORSTOM, Paris, 64 pp
Milligan K (1983) Mode de répartition animale et humaine dans le Gourma du Mali en saison sèche. Doc Progr AZ 92, CIPEA, Bmako, 36 pp

Milligan K, Bourn D, Chachu R (1979) Aerial surveys of cattle and land-use in four areas of the Nigeria sub-humid zone. Consultant's Report to ILCA. ILCA, Addis-Ababa
Monod T (1937) Méharées. 300 pp, Paris
Monod T (1950) Autour du problème du dessechement africain. Bull IFAN, XII, 2:514–523
Monod T (1954) Modes contractés et diffus de la végétation saharienne. In: Cloudsley-Thompson JL (ed) Biology of deserts. Inst of Biology, Univ of London, pp 35–44
Monod T (1957) Les grandes divisions chorologiques de l'Afrique. Publ n° 24, Cons Scient l'Afrique Sud Sahara, Londres, 146 pp, 2 cartes
Monod T (1958) Majabat al Koubrâ. Contribution à l'étude de l'"Empty quarter" Ouest saharien. Mém n° 52, IFAN, Dakar, 407 pp, 135 figs, 3 cartes
Monod T (1973a) La dégradation du monde vivant. Flore et Faune. In: CR Coll Nouakchott sur la désertification au Sud du Sahara. Nelles Editions Africaines, Dakar-Abidjan, pp 91–95
Monod T (1973b) Les Déserts. Horizons de France, Publ, Paris, 247 pp, 200 phot, 16 figs
Monod T (ed) (1975) Pastoralism in tropical Africa. Oxford Univ Press, London, 502 pp
Monod T (1986) The Sahel zone north of the Equator. In: Evenari M, Noy-Meir I (eds) Hot deserts and arid shrublands, ch 6. Ecosystems of the world, vol 12 B, Elsevier, Amsterdam, pp 203–243
Monod T, Toupet C (1961) Land-use in the Sahara-Sahel region. In: Dudley-Stamp L (ed) A history of land-use in arid regions. Arid Zone Res, vol 17, UNESCO, Paris, pp 239–253
Montgolfier-Kouevi C de, Le Houérou HN (1980) Study on the economic viability of browse plantations in Africa. In: Le Houérou HN (ed) Browse in Africa, ILCA, Addis Ababa pp 449–464
Morel G (1968a) Contribution à la synécologie des oiseaux du Sahel sénégalais. Mém n° 29, ORSTOM, Paris, 180 pp, 8 pl
Morel G (1968b) L'impact écologique de *Quelea quelea* sur les savanes sahéliennes; raison du pullulement de ce plocéide. Terre Vie, 22:69–98
Morel G, Bourlière F (1962) Relations écologiques des avifaunes sédentaires et migratrices dans la savane sahélienne du Bas-Sénégal. Terre Vie, 16:371–393
Morel G, Morel MY (1972) Recherches écologiques sur une savane sahélienne du Ferlo septentrional, Sénégal: l'avifaune et son cycle annuel. Terre Vie, 26:410–439
Morel G, Morel MY (1974) Recherches écologiques sur une savane sahélienne du Ferlo septentrional, Sénégal: influence de la sécheresse de l'année 1972–73 sur l'avifaune. Terre Vie, 28:95–123
Morel G, Morel MY (1980) Structure of an arid tropical bird community. Proc IVth Panfrican Ornitol. Congr :125–133
Mori F (1965) Tadrart-Acacus: Arte rupestre e culture del Sahara preistorico. Einaudi, Torino, 257 pp
Moro D, Hubert B (1983) Production et consommation de graines en milieu sahélo-soudanien au Sénégal: les rongeurs. *Mammalia* 47:37–57
Mosnier M (1961) Les pâturages naturels sahéliens, région de Kaedi (Mauritanie). Et Agrostol n° 3, IEMVT, Maisons-Alfort, 169 pp
Mosnier M (1967) Les pâturages naturels de la région de Gallayel (Rép du Sénégal). Et Agrostol 18, IEMVT, Maisons-Alfort, 137 pp
Murat M (1937) Végétation de la zone prédésertique en Afrique Centrale (région du Tchad). Bull Soc Hist Nat Afr Nord, 28, 1:19–83
Murat M, Monod T, Rungs C, Sauvage C (1944) Esquisse phytogéographique du Sahara occidental. Mém 1, Off Nat Anti-Acridien, Alger 31 pp, 3 cartes
Myers N (1979) The sinking ark Pergamon Press, New York, 307 pp
Naegelé AFG (1959) Contribution à l'étude de la flore et des groupements végétaux de la Mauritanie. Les parcelles protégées IFAN-UNESCO, de la région d'Atar. Bull IFAN, 21:A, 4:1195–1204
Naegelé AFG (1967) Observations sur les pâturages naturels du ranch de Doli. FAO, Rome, 77 pp
Naegelé AFG (1969) Etude des pâturages naturels de la Forêt classée des Six Forages ou réserve sylvo-pastorale de Koya (République du Sénégal). Tome I: Généralités sur la région étudiée. FAO, Rome; Min Agric, Dakar, 117 pp, 5 cartes, 20 figs
Naegelé AFG (1971a) Etude et amélioration de la zone pastorale du Nord-Sénégal. Etude Pâturages et Cultures Fourragères n° 4, AGPC, FAO, Rome, 163 pp, 38 phot
Naegelé AFG (1971b) L'amélioration des conditions d'utilisation des pâturages naturels en Zone sahélienne. Rapp Gouv Sénégal. Rapp AT 2963, FAO, Rome, 15 pp

Naegelé AFG (1977a) Les Graminées des pâturages de Mauritanie. Pât Cult Fourr, Et 5, AGPC, FAO, Rome, 298 pp
Naegelé AFG (1977b) Plantes fourragères spontanées de l'Afrique tropicale sèche. Données techniques. EMASAR, Phase II, UNEP/FAO; FAO, Rome, 510 pp, bibl ca 250 réf
N'Derito PC, Adeniji KO (1976) Distribution of cattle in Africa. Map 1/10 000 000; Inter-afr Bur Anim Resour, Organisation of African Union, Nairobi
Nebout JP, Toutain B (1978) Etude des arbres fourragers dans la zone sahélienne (Oudalan voltaïque). CTFT, Nogent/Marne IEMVT, Maisons-Alfort, 119 pp
Newby JE (1975a) The addax and the scimitar-horned oryx in Niger. Report to WWF/IUCN/UNEP, Morges
Newby JE (1975b) The addax and the scimitar-horned oryx in Chad. Report to WWF/IUCN/UNEP, Morges
Newby JE (1981) Desert antelopes in retreat. World Wildlife News, 14–18 (Summer)
Newby JE (1982) Avant-projet de classment d'une aire protégée dans l'Air et le Ténéré (République du Niger), WWF/IUCN, Gland
Newby JE (1984) Large mammals of the Sahara. In: Cloudsley-Thompson JL (ed) Sahara Desert. Ch 18. Key environments; Pergamon Press, Oxford, UK, pp 277–290
Nicholson SE (1978) Climatic variation in the Sahel and other African regions during the past five centuries. J Arid Env 1:3–34
Nicholson SE (1979) The methodology of historical climate reconstruction and its application to Africa. J Afr Hist, 20 (1):31–49
Nicholson SE (1980) Saharan climate in historic times. In: Williams MAJ, Faure H (eds) The Sahara and the Nile. Balkema, Rotterdam, pp 173–200
Nicholson SE (1981a) The historical climatology of Africa. In: Wigley PG, Ingram MJ, Farmer G (eds) Climate in history. Cambridge Univ Press, London, pp 249–270
Nicholson SE (1981b) Climate and man in the Sahel during the historical period. In: Berry L (ed) Environmental change in the Sahel. BOSTID, Nat Acad Sci, Washington, DC, pp 15–23
Norton-Griffiths M (1975/78) Counting animals. Afr Will Lead Found, Handbook n° 1, Nairobi, 139 pp
Pélissier P (1967) Les paysans du Sénégal – Les civilisations agraires du Cayor et de la Casamance. Impr Fabrègues, St Yriex, 939 pp
Olsson K (1985) Remote sensing for fuelwood resources and land degradation studies in Kordofan, The Sudan. Dept of Geography, Royal Univ Lund, Sweden, 182 pp
Osborn DJ, Krombein KV (1969) Habitats, mammals and wasps of Gebel'Uweinat, Libyan Desert. Smith Contrib Zool, 11–1–18
Palausi G (1955) Au sujet du Niger fossile dans la région de Tombouctou. Rev Géogr Phys Géol Dyn 5:217–219
Pennicuick CJ, Sale JB, Stanley-Price M, Jolly GM (1977) Aerial systematic sampling applied to censuses of large mammals populations in Kenya. East Afr Wildl J, 15:139–146
Perreau P (1973) Maladies tropicales du bétail. PUF, Paris, 220 pp
Petit-Maire N (1986) Paleoclimatstes in the Sahara of Mali, a multidisciplinary study. Episodes 9, 1:7–15
Petit-Maire N, Gayet M (1984) Hydrographie du Niger à l'Holocène ancien. Comptes Rendus Hebdomadaires des Séances de l'Académie des Sciences de Paris, Sér II, V, 298, No 1:21–23
Petit-Maire N, Riser J (eds) (1983) Sahara ou Sahel? Quaternaire récent du Bassin de Taoudenni (Mali). Lab de Géol Quaternaire, Luminy, Marseille, 473 pp
Peyre de Fabrègues B (1963) Etude des pâturages sahéliens. Ranch du Nord-Sanam (Rép du Niger). Et Agrostol No 5, IEMVT, Maisons-Alfort, 135 pp
Peyre de Fabrègues B (1966) Etude des pâturages naturels sahéliens de la région du Nord-Gouré (Rép du Niger). Et Agrostol No 10, IEMVT, Maisons-Alfort, 163 pp
Peyre de Fabrègues B (1967) Etude agrostologique des pâturages de la zone nomade de Zinder. Et Agrostol No 17, IEMVT, Maisons-Alfort, 188 pp, 1 carte
Peyre de Fabrègues B (1970) Pâturages naturels sahéliens du Sud-Tamesna (Rép du Niger). Et Agrostol No 28, IEMVT, Maisons-Alfort, 200 pp, 1 carte
Peyre de Fabrègues B (1971) Evolution des pâturages naturels sahéliens du Sud-Tamesna (Rép du Niger). Et Agrostol; No 32, IEMVT, Maisons-Alfort, 135 pp

Peyre de Fabrègues B (1972) Lexique des noms vernaculaires de plantes du Niger. 2 vol IEMVT, Maisons-Alfort, 190 pp
Peyre de Fabrègues B (1973) Synthèse des études de la zone de modernisation pastorale du Niger. Amélioration de l'exploitation pastorale. SEDES/IEMVT, Maisons-Alfort, 50 pp
Peyre de Fabrègues B (1975) Etude de la piste à bétail Ati-N'Djamena. Et Agrostol n° 25, IEMVT, Maisons-Alfort, 111 pp
Peyre de Fabrègues B (1985) Quel avenir pour l'élevage au Sahel? Rev Elev Médec Vétér Pays Trop 38, 4:500–508
Peyre de Fabrègues B, Lebrun JP (1976) Catalogue des plantes vasculaires du Niger. IEMVT, Maisons-Alfort, 433 pp
Peyre de Fabrègues B, De Wispelaere G (1984) Sahel: fin d'un monde pastoral? Marchés Tropicaux, 12 Oct 1984, pp 2488–2491
Phillips JFV (1959) Agriculture and ecology in Africa; a study of actual and potential development south of the Sahara. Faber and Faber, London, 424 pp, 1 map
Phillips JFV (1965) Fire as master and servant: its influence in the bioclimatic regions of trans-Saharan Africa. Proc 4th Ann. Tall Timbers Fire Ecology Conf. Tall Timbers, Res Stn, Tallahassee, Florida, pp 7–109, Bibl 350, 1 map 1/20 000 000 of Africa
Pias J (1970) Les formations sédimentaires Tertiaires et Quaternaires de la cuvette tchadienne et les sols qui en dérivent. Mém No 43, ORSTOM, Paris, 407 pp, VIII pl, 1 carte géol 1/ 1 000 000, 2 feuilles
Picardi AC (1975) A system analysis of pastoralism in the West-African Sahel. Ph D Diss, Mass Inst of Technol, Cambridge, 337 pp
Piot JJ (1980) Management and utilization methods for ligneous forages: natural stands and artificial plantations. In: Le Houérou HN (ed) Browse in Africa. ILCA, Addis-Ababa, pp 339–350
Piot J, Diaité I (1983) Systèmes de production d'elevage au Sénégal. Etude du couvert ligneux. GRIZA/LAT, CTFT, Nogent/Marne, ISRA/LNERV, Dakar, 37 pp
Piot J, Millogo G (1980) Etude du ruissellement et de l'érosion dans l'Oudalan. CTFT/ORSTOM, Ouagadoudou, 33 pp
Piot J, Nebout JP, Nanot R, Toutain B (1980) Utilisation des ligneux sahéliens par les herbivores domestiques. CTFT, Nogent/Marne, 213 pp, 2 cartes 30 fig
Planchenault D (1981) Etude des systèmes de production d'élevage au Sénégal. Rapp 2e année. DGRST/GRIZA/LAT, IEMVT-ISRA/LNERV, Dakar, 29 pp
Planchenault D (1983) Etude des systèmes de production d'élevage au Sénégal. Etude zootechnique. C-R de fin d'étude. GRIZA/LAT/IEMVT-ISRA/LNERV, Dakar, 29 pp
Pollard JH (1977) A handbook of numerical and statistical techniques. Cambridge Univ. Press, London
Poncet Y (1986) Images spatiales et paysages sahéliens. Trav Doc, No 200, ORSTOM, Paris, 255 pp
Popov GB, Wood TG, Haggis MJ (1984) Insect pests of the Sahara. In: Cloudsley-Thompson JL (ed) Sahara Desert. Ch 12 Key environments, Pergamon Press, Oxford, UK, pp 145–174
Poulain JF (1976) Amélioration de la fertilité des sols agricoles du Mali. Bilan de 13 années de travaux (1962–1974). L'Agron Tropic, XXXI, 4:401–416
Poulain JF (1978) Synthèse des résultats agronomiques obtenus de 1971 à 1977 en Haute-Volta. Rappt IRAT/Haute-Volta, Minst Dev Rural, Ouagadougou, 175 pp
Poulet AR (1972) Recherches écologiques sur une savane sahélienne du Ferlo septentrional, Sénégal: les mammifères. Terre Vie, 26:440–472
Poulet AR (1974) Recherches écologiques sur une savane du Ferlo septentrional, Sénégal: quelques effets de la sécheresse sur le peuplement mammalien. Terre Vie, 28:124–130
Poulet AR (1978) Evolution of rodent population of a dry bush savanna in the Senegalese Sahel from 1969 to 1977. Bull Carnegie Mus Nat His 6:113–117
Poulet AR (1982) Pullulation de rongeurs dans le Sahel. Mécanismes et déterminismes du cycle d'abondance de *Taterillus pygargus* et d'*Arvicanthis niloticus* (Rongeurs Gerbillidés et Muridés) dans le Sahel du Sénégal de 1975 à 1977. ISBN 0638-4, ORSTOM, Paris, 368 pp, 46 fig
Poulet AR, Poupon H (1978) L'invasion d'*Arvicanthis niloticus* dans le Sahel Sénégalais en 1975–1976, et ses conséquences pour la strate ligneuse. Terre Vie, 31:161–195
Poupon H (1977a) Recherches écologiques sur une savane sahélienne du Ferlo septentrional, Sénégal: premières données sur *Commiphora africana* (Rich) Engl La Terre et la Vie 31:127–162

Poupon H (1977b) Evolution d'un peuplement *d'Acacia senegal* de 1972 à 1976. Cah ORSTOM, Sér Biol 124:283–291

Poupon H (1980) Structure et dynamique de la strate ligneuse d'une steppe sahélienne du Nord-Sénégal. Trav Doc, n° 115, ORSTOM, Paris, 351 pp, 7 cartes

Poupon H, Bille JC (1974) Recherches écologiques sur une savane sahélienne du Ferlo septentrional, Sénégal: Influence de la sécheresse de l'année 1972–1973 sur la strate ligneuse. Terre Vie, 28:49–75

Pratt DJ, Gwynne MD (1977) Rangeland management and ecology in East Africa. Hodder and Stoughton, London, 310 pp

Quézel P (1968a) Premiers résultats de l'exploitation botanique du Gourgeil (Rép du Soudan). CR Acad Sces, D 266:2061–2063

Quézel P (1968b) Flore et végétation des plateaux du Darfur Nord-Occidental et du Jebel Gourgeil. RCP n° 45, CNRS, Paris, 146 pp, 15 fig

Quézel P (1970) A preliminary description of the vegetation in the Sahel region in north Darfur. Sudan Notes Rec 51:119–125

Rabinowitch E (1959) La photosynthèse. Gauthiers-Villars, Paris, 172 pp

Radwanski SA, Wickens GE (1967) The ecology of *Acacia albida* on mantle soils in Zalingei, Jebel Marra, Sudan. J Appl Ecol 4:569–579

Rattray JM (1960) The grass cover of Africa. Agr Pap 49, FAO, Rome 170 pp, 1 map 1/10 000 000

Raynal J (1964) Etudes botaniques du Centre de Recherches Zootechniques de Dahra-Djoloff. ORSTOM, Dakar, 99 pp, 1 carte, 2 tabl

Receveur P (1960) Hydraulique pastorale: bases d'une politique de l'eau en zone sahélienne. Rapp Inédit, Serv de l'Elevage, Niamey (cité par E. Bernus, 1981 and B Peyre de Fabrègues, 1963/75), 27 pp

Receveur P (1965) Définition d'un programme d'aménagement hydropastoraux dans la zone sylvopastorale (Sénégal). Minist Coop, Paris, 69 pp

Riou C (1975) La détermination pratique de l'evaporation. Application à l'Afrique Centrale. Mém 80, ORSTOM, Paris, 236 pp

Rippstein G, Peyre de Fabrègues B (1972) Modernisation de la zone pastorale du Niger. Et Agrostol No. 33, IEMVT, Maisons-Alfort, 307 pp, 8 phot, 2 cartes 1/1 000 000

Roberty G (1940) Contribution à l'étude phytogéographique de l'Afrique Occidentale Française. *Candollea,* VIII: 83–137

Roberty G (1946) Les associations végétables du moyen-Niger. *Veröffentl Geobot Rübel,* Zürich, 22, Heft, 1946, 168 pp, 2 figs

Roberty G (1950) La végétation du Ferlo. Bull IFAN, 14, A (3):777–798, 1 carte

Roberty G (1954) Petite flore de l'Ouest-Africian. Larose, Paris, 441 pp

Roberty G (1955) Carte de la végétation de l'Afrique Occidentale Française: Diafarabé. 1/250 000, ORSTOM, Paris

Roberty G (1956) La végétation de l'Afrique tropicale occidentale (au S du 16ème parallèle, à l'W du méridien de Greenwich) IFAN/ORSTOM, 24 pp

Roberty G (1950/1960) Cartes de la végétation de l'AOF 1/1 000 000 Feuilles de Dakar, Thiès et Bouaké, ORSTOM, Paris

Roberty G (1960) Les régions naturelles de l'Afrique tropicale occidentale. Bull IFAN, Sér A, 22:95–136

Rodier JA (1964) Régimes hydrologiques de l'Afrique Noire à l'Ouest du Congo. Mém 6, ORSTOM, Paris. 138 pp, 24 pl

Rodier JA (1975) Evaluation de l'écoulement annuel dans le Sahel Tropical Africain, Trav Doc n° 46, ORSTOM, Paris, 122 pp

Rodier JA (1982) Evaluation of annual runoff in tropical African Sahel. Trav Doc n° 145, ORSTOM, Paris, 212 pp, 29 fig

Rognon P (1976) Essais d'interprètation des variations climatiques au Sahara depuis 40 000 ans. Rev Géogr Phys Géol Dyn 2–3:251–282

Roose EJ (1977) Erosion et ruissellement en Afrique de l'Ouest. Vingt années de mesures en petites parcelles expérimentales. Trav Doc n° 78, ORSTOM, Paris, 108 pp

Roose EJ (1978) Pédogénèse actuelle d'un sol ferrugineux issu de granite sous une savane arborée du Plateau Mossi (Haute-Volta), Gonsé, Campagnes 1968–1974. ORSTOM, Paris, 121 pp

Roose EJ (1981) Dynamique actuelle des sols ferrallitiques et ferrugineux tropicaux d'Afrique Occidentale. Trav Doc 130 ORSTOM, Paris, 569 pp

Roose EJ, Piot J (1984) Runoff, erosion and soil fertility restoration on the Mossi Plateau (Central Upper-Volta). Proc Harare Symp Challenges Afric Hydrol Water Resour, Publ; 144, IAHS, pp 485–498
Rose-Innes R (1966) The concept of woody pasture in low altitude tropical tree savanna environments. Proc IXth Int Grassland Congr: 1419–1423, Sao Paulo
Rose-Innes R, Mabey GL (1964–1966) Studies in browse plants in Ghana. Emp J Exp Agric:
1. 1964: Chemical composition: XXXII, 126, 114–120
2. 1964: Digestibility of *Griffonia simpliciflora* XXXII, 126:125–130
   1964: Digestibility of *Baphia nitida*, XXXII, 128: 274–278
   1966: Digestibility of *Antiaris africana*, (2):27–32
   1966: Digestibility of *Grewia carpinifolia* (2):113–117
3. 1964: Browse/grass ingestion, XXXII, 127:180–190

Rossetti C (1962) Prospection écologique, études en Afrique Occidentale. Observations sur la végétation. Conclusions sur les travaux entrepis de 1951 à 1959. UNSF/DL/ES/5, FAO, Rome, 71 pp
Rossetti C (1959) Prospections écologiques; études en Afrique Occidentale. Observations sur la végétation du Mali Oriental. UNSF/DL/ES/4, FAO, Rome, 68 pp, 24 phot
Santoir CH (1977) L'espace pastoral dans la région du Fleuve Sénégal. ORSTOM, Dakar, 60 pp
Santoir CH (1980) Sédentarisation des nomades et hydraulique pastorale dans le Djoloff. ORSTOM, Dakar, 72 pp
Santoir CH (1981) Contribution à l'étude de l'exploitation du cheptel du Ferlo, Sénégal. GRIZA/LAT, ORSTOM, Dakar
Santoir CH (1983) Raison pastorale et développement: les problèmes des Peuhl sénégalais face aux aménagements. 1 ere partie: Sédentarisation des nomades et hydraulique pastorale dans le Djoloff. Trav Doc 166, ORSTOM, Paris
Sarniguet J, Tyc J (1965) Exploitation du cheptel bovin au Mali. Secr d'Etat Coop, Paris, 296 pp, 2 cartes
Sarniguet J, Tyc J, Peyredieu du Charlat F, Lacrouts M (1975) Approvisionnement en viande de l'Afrique de l'Ouest. 4 vol, Secr d'Etat Coop; SEDES, Paris
Sayer JA (1977) Conservation of large mammals in the Republic of Mali. Biol Conserv 12:245–263
Schomber HW, Koch D (1961) Wildlife protection and hunting in Tunisia, Afr Wild Life 15:137–150
Seif el Din AG (1965) The natural regeneration of *Acacia sénégal* (L) Wild. M Sc; Diss, Univ of Khartoum
Seif el Din AG (1979) Rainfall distribution and vegetation in the Sahel. Proc Symp role for Sahel, IDRC, Otawa, pp 43–49
Seif el Din AG, Mubarak OM (1971) Ecological studies of the vegetation of the Sudan. II – Germination of seeds and establishment of seedlings of *Acacia senegal* (L) Wild under controlled conditions. J Appl Ecol, 8 (1):91–201. IV – The effect of simulated grazing on the growth of *Acacia senegal* (L) Wild seedlings. J Appl Ecol, 8 (1):211–216
Seifert WW, Kamrany NM (1974) A framework for evaluating long-term development strategies for Sahel-Sudan area. Interim report, part I. Center of Policy Alternatives, Mass Inst of Technol Cambridge, Massachussetts
Serres H (1977) Essai de bilan des politiques d'hydraulique pastorale. IEMVT, Maisons-Alfort, 135 pp
Serres H (1980) Politiques d'hydraulique pastorale. PUF, Paris, 121 pp
Setzer HW (1956) Mammals of the Anglo-Egyptian Sudan. Proc US Nat Mus, Vol 106, n° 3377: 447–588, Smithonian Institution, Washington, DC
Sharman MJ (1982) Rapport sur les vols systématiques de reconnaissance de la zone sylvopastorale du Nord Sénégal. EP/SEN/001; FAO, Rome; UNEP, Nairobi, 48 pp, 17 figs
Sharman MJ (1983a) Comparaison de quatre vols systématiques de reconnaissance au Ferlo. In: Vanpraet CL (ed) Méthodes d'inventaire et du surveillance continue des écosystemes pastoraux sahéliens, ISRA, Dakar, pp 127–148
Sharman MJ (1983b) Approche systémique du suivi continu d'une zone pastorale sahélienne. In: Vanpraet CL (ed) Méthodes d'inventaire et du surveillance continue des écosystemes pastoraux sahéliens, ISRA, Dakar, pp 149–160

Sharman MJ, Gning M (1983) Comportement du cheptel au Ferlo: résultats des suivis quotidiens. In: Vanpraet CL (ed) Méthodes d'inventaire et du surveillance continue des écosystemes pastoraux sahéliens, ISRA, Dakar, pp 209–222
Sharman MJ, Vanpraet CL (1983) Mesure de la production primaire par utilisation du spectroradiomètre. In: Vanpraet CL (ed) Méthodes d'inventaire et du surveillance continue des écosystemes pastoraux sahéliens, ISRA, Dakar, pp 321–332
Shepherd WO (1968) Range and pasture management. Report to the government of Sudan. TA Rept n° 2468, FAO, Rome, 49 pp, 2 maps
Shepherd WO, Baasher MM (1966) Forage resources of the Sudan savanna: potentials and problems. Proc UNDP meeting at Khartoum, on Savanna develop FAO, Rome, pp 136–145
Sicot M, Grouzis M (1981) Pluviométrie et production des pâturages naturels sahéliens. ORSTOM, Ouagadougou, 33 pp
Sidibé M (1978) Contribution à l'étude du phosphore dans le cadre de l'amélioration des pâturages naturels sahéliens. Thèse de 3ème cycle, Ec Norm Sup, Bamako 232 pp
Smith GEJ (1981) Some thoughts on sampling design. In: Low level aerial survey techniques. Monogr 4 ILCA, Addis-Ababa, pp 159–165
Smith J (1949) Distribution of tree species in the Sudan in relation to rainfall and soil texture. Bull 4, Ministr of Agric, Khartoum, 68 pp
Smith RI (1985) Inventory and monitoring of Sahelian pastoral systems in Dakar, Senegal. Inst Terrestr Ecol Nat Envir Counc mim rep, GEMS, UNEP, Nairobi
SODESP (1980) Présentation technique du projet de "Développement Intégré de l'Elevage dans la Zone sylvo-Pastorale". Note Tech n° 1; SODESP, Dakar, 24 pp
SODESP (1984) Bilan technique et économique au 15 avril 1984. Note Tech n° XVI, SODESP, Dakar, 35 pp
Sørensen T (1948) A method of establishing groups of equal amplitude in plant sociology based on similarity of species content. Det Kong Danske Vidensk Selsk Biol (Copenhagen) 5 (4):1–34
Squires V (1981) Livestock management in the arid zone. Inkata Press, London, 271 pp
Taiti SW (1981) Aerial survey methods: the experience in Kenya from 1968 to 1978. In: Low level aerial survey techniques, monogr 4, ILCA, Addis-Ababa, pp 27–32
Talbot MR (1980) Environmental response to climatic change in the West-African Sahel over the past 20 000 years. In: Williams MAJ, Faure H (eds) The Sahara and the Nile. Balkema, Rotterdam, pp 37–62
Terrible M (1978) Végétation de la haute-Volta au 1/1 000 000 Contribution à la connaissance de la Haute-Volta. Bobo-Dioulasso, 40 pp
Tetzlaff G, Adams LJ (1983) Present-day and early Holocene evaporation of Lake Chad. In: Street-Perrott A et al. (eds): Variations in the global water budget. Reidel Publ Comp, pp 347–360
Thevenin P, Risopoulos SA (1977) Les pays sahéliens: vol 1: Développement et vulgarisation dans le domaine pastoral. 134 pp; Vol II: Education et Formation dans le domaine pastoral, 211 pp EMASAR, Phase II, UNEP/FAO, Rome
Thiollay JM (1978) Les migrations de rapaces en Afrique Occidentale: Adaptations écologiques aux fluctuations saisonnières de production des écosystèmes. Terre Vie, 32:89–133
Tluczykont S, Gningue DI (1986) Un modèle de lutte intégrée contre la désertification pour la sauvegarde de l'environement et l'auto-suffisance alimentaire et énergétique. Projet Senegalo-Allemand de reboisement et d'aménagement sylvo-pastoral de la Zone Nord St Louis, 20 pp
Tothill JP (1948) (2nd edn 1954) – Agriculture in the Sudan. Oxford Univ Press, London, 974 pp, 408 figs, 1 map
Toupet C (1971/72) Les variations interannuelles des précipitations en Mauritanie Centrale. CR Somm Séances Soc Biogéogr, 48:39–47
Toupet C (1973) L'évolution du climat en Mauritanie du Moyen-Age jusqu'à nos jours C R Coll Nouakchott désertification Sud Sahara, Nelles Edit Afric, Dakar-Abidjan, pp 56–63
Toupet C (1977) La sédentarisation des nomades en mauritanie Centrale sahélienne. Champion, Paris, 490 pp
Toupet C, Michel P (1979) Sécheresse et Aridité; l'exemple de la Mauritanie et du Sénégal. Geo-Eco-Trop, 3 (2):137–157
Touré IA, Dia IP, Maldague M (1986) La Problématique et les stratégies sylvo-pastorales au Sahel. CIEM, Univ de Laval, Quebec, UNESCO, Paris, 353 pp

Toutain B (1974) Implantation d'un ranch d'embouche en Haute-Volta, Région de Léo. Et Agrostol 40, IEMVT, Maisons-Alfort, 195 pp
Toutain B (1976) Notice de la carte des ressources fourragères au 1/50 000, DGRST/ORSTOM/IEMVT, Maisons-Alfort, 60 pp, 1 carte
Toutain B (1977a) Essais de régénération mécanique de quelques parcours sahéliens dégradés. Rev Elev Médec Vétér Pays Trop 30 (2):191–198
Toutain B (1977b) Pâturages de l'ORD du Sahel et de la zone de délestage du Nord-Est de Fada N'Gourma. IEMVT, Maisons-Alfort, 119 pp
Toutain B (1980) The role of browse plants in animal production in the Sudanian Zone of West Africa. In: Le Houérou HN (ed) Browse in Africa. ILCA, Addis-Ababa, pp 103–108
Toutain B, De Wispelaere G (1978) Etude et cartographie des pâturages de l'ORD du Sahel et de la zone de délestage au Nord-Est de Fada N'Gourma (Haute-Volta). 3 vol Et Agropastorale 51, IEMVT, Maisons-Alfort
Toutain B, Bortoli L, Dulieu D, Forgiarini G, Menaut JC, Piot J (1983) Espèces ligneuses et herbacées dans les écosystèmes pâturés sahéliens de Haute-Volta. GERDAT, CTFT, ENS, CNRST, IRBET, Paris, Ougadougou, 194 pp
Traoré G (1978) Evolution de la disponibilité et de la qualité de fourrage au cours de la transhumance de Diafarabé. Thèse de 3ème cycle Ec Norm Sup, Bamako, 87 pp
Tricart J (1965) Reconnaissance géomorphologique de la vallée moyenne du Niger. Mém 72, IFAN, Dakar, 196 pp, 23 figs, 5 cartes, 7 pl
Tricart J (1969) Géomorphologie dynamique de la moyenne vallée du Niger. Ann Géogr Fr 368:333–443
Trochain JL (1940) Contribution à l'étude de la végétation du Sénegal. Larose, Paris, 433 pp, XXX pl
Trochain JL (1952) Les territoires phytogéographiques de l'Afrique Noire Française d'après leur pluviométrie. Rec Trav Lab Bot, Fac Sci, Montpellier, 5:113–124, 2 cartes, 2 tabl
Trochain JL (1957) Accord interafricain sur la définition des types de végétation de l'Afrique tropicale. Bull Inst Et Centrafr Nelle sér, 13–14:55–93, Brazzaville
Trochain JL (1969) Les territoires phytogéographiques de l'Afrique Noire d'après la trilogie: Climat, Flore, Végétation. CR Somm Soc Biogéogr, 402:139–157
Troupin G (1960) Etude phytocoenologique du Parc National de l'Akagera et du Ruanda Oriental. Recherches d'une méthode d'analyse appropriée de la végétation d'Afrique intertropicale. Publ n° 2 Inst Nat Rech Scient, Butaré, Ruanda, 293 pp
Trotignon J (1975) Le statut et la conservation de l'addax et de l'oryx et de la faune associée en Mauritanie. Rapport à l'UICN, Morges
Tubiana MJ (1985) Des troupeaux et des femmes. L'Harmattan, Paris, 390 pp
Tubiana MJ, Tubiana J (1977) The Zaghawa from an ecological perspective. Balkema, Rotterdam, 119 pp, 3 maps, phot
Tucker CJ (1979) Red and photographic infrared linear combinations for monitoring vegetation. Rem Sens Environ, 8:127–150
Tucker CJ (1980) A critical review of remote sensing and other methods for non-destructive estimation of standing crop biomass. Grass Forage Sci, 35:115–182
Tucker CJ, Holben BN, Elgin JH, McMurtey JE (1981) Remote sensing of total dry matter accumulation of winter wheat. Rem Sens Environ, 11:171–189
Tucker CJ, Vanpraet CL, Boerwinkel E, Gaston A (1983) Satellite remote sensing of total dry matter production in the Senegalese Sahel. Rem Sens Environ, 13:416–474
Tucker CJ, Townshend JRG, Goff TE (1985a) African land-cover classification using satellite data. Science, 227, 4685:369–375
Tucker CJ, Vanpraet C, Sharman MJ, Van Ittersum G (1985b) Satellite remote sensing of the total herbaceous biomass production in the Senegalese Sahel: 1980–1984. Remote Sens Environ, 17:233–249
Tucker CJ, Hielkema JU, Roffrey J (1985c) The potential of satellite remote sensing of ecological conditions for survey and forecasting desert locust activity. Int J Rem Sens 6, 1:127–138
Urvoy Y (1942) Les bassins du Niger. Etude de géographie physique et de paleogéographie. Mém 4, IFAN, Dakar, 144 pp, 6 pl, 4 cartes
Urvoy Y (1949) Histoire de l'Empire du Bornou. Mém 7, IFAN, Dakar, 164 pp

USDA (1975) Soil taxonomy. Agriculture handbook No. 436. Conservation service. US Dept of Agriculture, Washington, DC, 754 pp
Valentin G (1981) Systèmes de production d'élevage au Sénégal: evolution de la surface du sol: piétinement, érosion hydrique et éolienne. GRIZA/LAT/ORSTOM, 18 pp
Valentin G (1983) Effect of grazing and trampling around recently drilled water boreholes on soil deterioration in the Sahelian zone of northern Senegal. Int Conf Soil Erosion, Honolulu, 18 pp
Valentin G (1985) Organisations pelliculaires superficielles et quelques sols de régions subdésertiques, dynamique de formation et conséquences sur l'économie de l'eau. Thèse 3eme cycle, Paris VIII; Et and Thèses, ORSTOM, Paris, 259 pp, 22 tabl, 43 figs, 67 pl
Valenza J (1970) Etude dynamique de différents types de pâturages naturels en République du Sénégal. Proc XIth Grassland Congr 208; Surfers Paradise, Queensland, Australia
Valenza J (1975) The natural pasturelands of the sylvopastoral zone of the Senegal Sahel twenty years after their development. Proc Symp Evaluation Mapp Trop Afric rangelands ILCA, Addis-Ababa, pp 191–193
Valenza J (1981) Surveillance continue des pâturages naturels sahéliens. LNERV/ISRA, Rev Elev Med Vét Pays Trop, 34 (1):83–100
Valenza J (1984) Surveillance continue des pâturages naturels sahéliens sénégalais; résultats de 10 années d'observations. LNERV, ISRA, Dakar, 53 pp, 35 tabl, 50 figs
Valenza J, Diallo AK (1970) Etude des pâturages du runch de Doli (Rép du Sénégal). IEMVT, Dakar, Maisons-Alfort, 21 pp
Valenza J, Diallo AK (1972) Etude des pâturages naturels du Nord Sénégal. Et Agrostol n° 34, IEMVT, Maisons-Alfort, 311 pp, 1 carte 1/200 000, 3 feuilles
Valenza J, Fayolle F (1965) Note sur les essais de charge de pâturage en République du Sénégal. Rev. Elev Médec Vétér d Pays Trop 18 (3):321–327
Valverde JA (1957) Aves del Sahara Español. Estudio ecologico del desierto. Madrid, 487 pp
Valverde JA (1968) Ecological bases for fauna conservation in western Sahara. Pap n° 18, Int Symp, IBP, CT, sect, Hammamet Tunisia, London, 14 pp
Van Keulen H (1975) Simulation of water use and herbage growth in arid regions. PUDOC, Wageningen, 176 pp
Vanpraet CL (ed) (1983a) Méthodes d'inventaire et de surveillance continue des écosystèmes pastoraux sahéliens; application au développement. Min Rech: Sci Tech, Dakar, Sénégal, 439 pp
Vanpraet CL (1983b) De la surveillance continue des écosystèmes pastoraux à une gestion des pâturages. In: Vanpraet CL (ed) Méthodes d'inventaire et de surveillance continue des ecosystemes pastoraux saheliens, ISRA, Dakar, pp 31–40
Vanpraet CL (1985) Méthodes d'inventaire et de surveillance continue des écosystèmes pastoraux sahéliens. Rapport Technique Terminal. Diff Restr EP/SEN/001; FAO, Rome, UNEP, Nairobi, 595 pp, ISRA, Dakar
Vanpraet CL, Van Ittersum G (1983) Considérations sur les analyses dimensionnelles de quelques espèces ligneuses de la zone sylvopastorale au Sénégal. In: Vanpraet CL (ed) Méthodes d'inventaire. ISRA, Dakar, pp 243–250
Vanpraet CL, Sharman MJ, Tucker CJ (1983) Utilisation des images NOAA pour l'estimation de la production primaire en milieu sahélien. In: Vanpraet CL (ed) Méthodes d'inventaire, pp 299–320
Von Maydell HJ (1983) Arbres et arbustes du Sahel, leurs caractéristiques et leurs utilisations. Schriftenreihe 147, GTZ, Eschborn, Allemagne de l'Ouest, 531 pp, 100 pl
Vuillaume G (1968) Premiers résultats d'une étude analytique du ruissellement et de l'érosion en zone sahélienne. Bassin représentatif du Kountkouzout (Niger). Cah ORSTOM, sér Hydrol V, 2:33–56
Walker BH (1980) A review of browse and its role in livestock production in southern Africa. In: Le Houérou HN (ed) Browse in Africa. ILCA, Addis-Ababa, pp 7–24
Walter H, Lieth H (1967) Klimadiagram Weltatlas. Fisher, Jena, 200 pl
Watson RM, Tippett CI (1981a) Examples of low level aerial surveys conducted in Africa from 1968 to 1979: one firm's experience. In: Low level aerial survey techniques, monogr 4, pp 33–57, ILCA Addis-Ababa
Watson RM, Tippett CI (1981b) Aerial survey of range and livestock resources in central Somalia. Min Animal Prod Mogadishu, Resourc Mgt Res Ltd London

Watson RM, Tippet CI, Razk F, Jolly F, Beckett JJ, Scoles V, Cabson F (1977) Sudan National Livestock Survey and Resources Census. Mim Rept, Min Nat Res, Khartoum, Resour Mgt Res Ltd Nairobi
White F (1983) The vegetation of Africa. Nat Res XX, UNESCO, Paris, 356 pp, 1 map 1/5 000 000 (3 sheets)
White LP (1970) Brousse tigrée patterns in southern Niger. J Ecology 58:549–553
Wickens GE (1968) Savanna development, Sudan: Plant Ecology. UNDP/SF/SUD 25, FAO, Rome, 22 pp
Wickens GE (1969) A study of *Acacia Albida.* Kew Bull, 23:181–202
Wickens GE (1975) Changes in the climate and vegetation of the Sudan since 20 000 BP *Boissiera,* 24:43–65
Wickens GE (1976) The flora of Jebel Marra (Sudan Republic) and its geographical affinities. Kew Bull, add ser V, Her Maj's Stationery Office, London, 368 pp, 24 figs, 208 pl of distrib maps
Wickens GE (1980a) The uses of the Baobab (*Adansonia digitata*) in Africa. In: Le Houérou HN (ed) Browse in Africa. ILCA, Addis-Ababa, pp 151–154
Wickens GE (1980b) Alternative uses of browse species. In: Le Houérou HN (ed) Browse in Africa. ILCA, Addis-Ababa, pp 155–184
Wickens GE, Collier FW (1971) Some vegetation patterns in the Republic of Sudan. *Geoderma,* 6:43–59
Wiegand CL, Richardson AJ (1984) Leaf area, light interception and yield estimates from spectral components analysis. Agron J, 76:543–548
Wiegand CL, Richardson AJ, Kanesmasu ET (1979) Leaf area index estimates from wheat from Landsat and their implications for evapotranspiration and crop modeling. Agron J, 71:336–342
Williamson G, Payne WJA (1965) An introduction to animal husbandry in the tropics 2nd edn, Longmans, London, 435 pp
Williamson DT, Williamson JE (1981) An assessment of the impact of fences on large herbivore biomass in the Kalahari. Botswana Notes Rec 13:107–110
Wilson RT (1976) Studies on the livestock of southern Darfur, Sudan. a — III — Production traits in sheep. Trop Anim Hlth Prod 8:103–114 b — IV — Production traits in goats. Trop Anim Hlth Prod 8:221–232
Wilson RT (1977) Temporal changes in livestock numbers and patterns of transhumance in southern Darfur, Sudan. J Develop Areas, 11:493–508
Wilson RT (1978a) Studies on the livestock of southern Darfur, Sudan. VI — Notes on equines. Trop Anim Hlth Prod, 10:183–189
Wilson RT (1978b) Studies on the livestock of southern Darfur, Sudan V — Notes on camels. Trop Anim Hlth Prod, 10:19–25
Wilson RT (1978c) The "Gizu" winter grazing in the South Libyan Desert. J Arid Environ 1:325–342
Wilson RT (1979a) Recent resource surveys for rural development in southern Darfur, Sudan. Geogr J, 145, 3:452–460
Wilson RT (1979b) Wildlife in southern Darfur, Sudan: distribution and status at present and in recent past. Mammalia, 43, 3:323–338
Wilson RT (1979c) The primates in Darfur, Republic of Sudan. *Folia Primatol* 31:219–226
Wilson RT (1980a) Wildlife in northern Darfur, Sudan: a review of its distribution and status in the recent past and at present. Biol Conserv 17:85–101
Wilson RT (1980b) Fuel wood consumption in a central Malian town and its effects on browse availability. In: Le Houérou HN (ed) Browse in Africa. ILCA, Addis-Ababa, pp 473–476
Wilson RT (1980c) Livestock production in central Mali: structure of the herds and flocks and some related demographic parameters. Progr Doc n° AZ 45 d, ILCA, Addis-Ababa
Wilson RT (1981) Distribution and importance of the domestic donkey in circum-Saharan Africa. Singapore J Tropic Géorgr, 2:136–143
Wilson RT (1982a) The economic and social importance of goats and their products in the semi-arid arc of northern tropical Africa. Proc 3th Int Conf Goat Prod Disease Coll Agric Univ of Arizona, Tucson, pp 186–195
Wilson RT (1982b) Livestock production in central Mali. ILCA Bull 15, ILCA, Addis-Ababa, 23 pp
Wilson RT (1983) The camel Longman, London, 223 pp

Wilson RT (1984) Demography, vital statistics and productivity of domestic animals under traditional management in arid and semi-arid northern tropical Africa. In: Le Houérou HN (ed) Advances in desert and arid land technology and development: range and livestock in Africa. Harwood Press, New York, 90 pp (In preparation)
Wilson RT, Bourzat D (eds) (1985) Small ruminants in African agriculture. ILCA, Addis-Ababa, 261 pp
Wilson RT, Clarke SE (1975) Studies on the livestock of southern Darfur, Sudan. 1. The ecology and livestock resources of the area. Trop Anim Hlth Prod, 7:165–187
Wilson RT, Clarke SE (1976) Studies on the livestock of southern Darfur, Sudan. 2. Production traits in cattle. Trop Anim Hlth Prod, 8:47–51
Wilson RT, Wagenaar KT (1983) Enquête préliminaire sur la démographie des troupeaux et sur la reproduction des animaux domestiques dans la zone du project "Gestion des Pâturages et Elevage de la République du Niger". Progr Doc AZ 80, ILCA, Addis-Ababa, 95 pp
Wilson RT, Bailey L, Hales J, Moles D, Watkins AE (1980) The cultivation-cattle complex in western Darfur. Agric Systems, 5:119–135
Wilson RT, De Leeuw PN, De Haan C (eds) (1983) Recherches sur les systèmes des zones arides du Mali: résultats préliminaires. Rapp Recherches 5, CIPEA/ILCA, Addis-Ababa, 189 pp, 70 figs
Wischmeier WH (1959) A rainfall erosion index for a universal soil-loss equation. Soil SC Soc Am Proc 23:246–249
Wischmeier WH, Smith DD (1978) Predicting rainfall erosion losses. A guide to conservation planning. USDA Handbook 537, US Dept of Agr, Washington, DC
Zeuner FE (1963) A history of domesticated animals. Hutchinson, London
Zolotarevsky B, Murat M (1938) Divisions naturelles du Sahara et sa limite méridionale. In: La vie dans la région désertique Nord-Tropical de l'Ancien Monde. Mém VI, Soc Biogérogr, Paris, pp 335–350

Subject Index

Index of Scientific Names

Animals

Plants